T0237624

Lecture Notes in Mathematics

Volume 2251

More information about this series at http://www.springer.com/series/304

Jialin Hong · Xu Wang

Invariant Measures for Stochastic Nonlinear Schrödinger Equations

Numerical Approximations and Symplectic Structures

 Springer

Jialin Hong
LSEC, ICMSEC, Academy of Mathematics
and Systems Science
Chinese Academy of Sciences
Beijing, China

School of Mathematical Sciences
University of Chinese Academy of Sciences
Beijing, China

Xu Wang
LSEC, ICMSEC, Academy of Mathematics
and Systems Science
Chinese Academy of Sciences
Beijing, China

School of Mathematical Sciences
University of Chinese Academy of Sciences
Beijing, China

ISSN 0075-8434   ISSN 1617-9692   (electronic)
Lecture Notes in Mathematics
ISBN 978-981-32-9068-6   ISBN 978-981-32-9069-3   (eBook)
https://doi.org/10.1007/978-981-32-9069-3

This Springer imprint is published by the registered company Springer Nature Singapore Pte Ltd.
The registered company address is: 152 Beach Road, #21-01/04 Gateway East, Singapore 189721, Singapore

Preface

Stochastic differential equations, originated from the work of A. Einstein and put on solider mathematical footing by K. Itô and R. L. Stratonovich, are now widely used to model various phenomena caused by random medium or stochastic external sources, such as propagation of nonlinear dispersive waves in random medium and unstable stock prices in financial market. There are fruitful results on the study of geometric structures, dynamical behaviors, statistical properties, and some other important internal properties for stochastic differential equations.

In this monograph, we take both the geometric structure and dynamical behavior, more precisely, the (conformal) multi-symplecticity and ergodicity, of stochastic nonlinear Schrödinger equations into consideration when constructing numerical approximations. Thus, the problems considered in this monograph are related to several fascinating research hotspots: numerical analysis, stochastic analysis, ergodic theory, partial differential equation theory, and so on.

The main object of this monograph is the analysis and numerical approximations of invariant distributions for stochastic nonlinear Schrödinger equations. The considered model shows good performances over longtime due to its dynamical behavior and geometric structure. Among these properties, ergodicity and stochastic (conformal) multi-symplecticity play fundamental roles in the study of the dynamical behavior and geometric structure for stochastic nonlinear Schrödinger equations. The stochastic multi-symplecticity is an extension of the multi-symplectic conservation law in the deterministic case (see [34, 117] and references therein). It implies that the spatio-temporal geometric structure of the system is preserved over longtime, and is an essential part of symplectic geometric theory.

Ergodic theory, as an important branch of mathematics, studies the asymptotic behavior of measure-preserving transformations. It originates from the study of the "ergodic hypothesis" which is the basic hypothesis in statistical mechanics and information theory, and is now developed as a powerful amalgam of methods used for the analysis of statistical properties of dynamical systems. Thus, it is widely investigated by researchers in dynamical systems, probability theory, physics, biology, chemistry, etc. (see [39, 53, 85, 129, 170, 189] and references therein).

We first give a systematic summary of the definitions and sufficient conditions of the existence and uniqueness of invariant measures as well as the ergodicity for stochastic Markov processes in Chap. 1. Moreover, stochastic Kubo oscillator and stochastic Hamiltonian dissipative systems are introduced as two examples to study their invariant measures, geometric structures, and numerical approximations.

To have a better understanding of the longtime behavior of stochastic differential equations (SDEs) with ergodic invariant measures, we then investigate the following fundamental problems:

- What kind of systems are ergodic, and what is the convergence rate between the temporal average $\frac{1}{T}\int_0^T \varphi(P_s x)ds$ and the spatial average $\int_{\mathbb{M}} \varphi d\mu$?
- Are there any numerical approximations which could inherit the ergodicity of the original system, and how to estimate the error between the numerical invariant measure and the original one?
- Is the temporal average of a numerical approximation (possibly not ergodic) a proper approximation of $\int_{\mathbb{M}} \varphi d\mu$?

Chapter 2 answers these questions with respect to finite dimensional stochastic differential equations, and take the stochastic Langevin equation as an example to construct its ergodic numerical schemes. Some more specific conditions which ensure the ergodicity of the solution of stochastic differential equations under proper assumptions are also given in Chap. 2. The first two chapters will help the readers who are not familiar with invariant measures and ergodicity to get a better understanding on these concepts and their properties.

However, the answers to above questions with respect to stochastic partial differential equations (SPDEs) are far from complete. The main contribution of this monograph is the investigation of ergodic numerical approximations for stochastic nonlinear Schrödinger equations, which gives answers to the last two questions mentioned above for stochastic nonlinear Schrödinger equations.

In Chap. 3, the local and global existence and uniqueness of the solution to stochastic nonlinear Schrödinger equations with both additive and multiplicative noises are briefly introduced. The continuous dependence of the solution on initial data is also given to ensure that the semigroup generated by the solution is Feller, which is a fundamental condition when studying the existence of invariant measures, geometric structures, and the large deviation principle. We recall some results about invariant measures for both deterministic and stochastic Schrödinger equations with weak damping. For the deterministic case, its invariant measure is constructed based on a finite approximation of the considered equation in the energy space $\mathbf{H}^1(0, L)$ with $L > 0$. For the stochastic case with weak damping, the solution is shown to be Gaussian for the linear case. As a result, the distribution of the solution converges to a unique $\mathbb{C}$-valued Gaussian distribution. For the cubic nonlinear case with weak damping, the considered model is shown to posses the mixing property with a unique invariant measure, and the rate of convergence to equilibrium is at least polynomial of any power due to the damping term. The existence result for ergodic invariant measures is extended to the high dimension and unbounded domain case with certain conditions on the relation between the nonlinear term and the spatial dimension.

In Chap. 4, we concentrate on the geometric structures for both deterministic and stochastic nonlinear Schrödinger equations, as well as their numerical approximations. Several temporal semi-discretizations and full discretizations are reviewed for the deterministic nonlinear Schrödinger equation. The stochastic symplectic and multi-symplectic conservation laws are proved for stochastic nonlinear Schrödinger equations with a linear multiplicative noise in Stratonovich sense. The numerical approximations for the deterministic case are generalized to the stochastic case. The temporal semi-discretization based on the Runge–Kutta methods is studied for stochastic nonlinear Schrödinger equations. The symplectic condition is also established, under which the numerical solution preserves the discrete stochastic symplectic structure. A full discretization based on the midpoint scheme in both spatial and temporal directions is given afterward to inherit the stochastic multi-symplectic structure of the stochastic nonlinear Schrödinger equation. While for stochastic nonlinear Schrödinger equation with weak damping, its geometric structure is shown to satisfy a conformal stochastic multi-symplectic principle instead of the multi-symplectic conservation law.

Chapter 5 is devoted to studying the numerical approximations of the stochastic nonlinear Schrödinger equation with additional damped term and an infinite dimensional additive noise. This model is shown to be ergodic with a unique invariant measure [70], but is apparently not multi-symplectic anymore. We show that the system possesses a conformal multi-symplectic geometric structure instead of the multi-symplectic structure in conservative case. We are interested in the investigation of the geometric structure for this damped equation, and aim to give a computable approximation to the original invariant measure, as well as the approximate error defined in the standard way.

We consider in Chap. 6 the stochastic nonlinear Schrödinger equation with linear multiplicative noise, which also possesses the charge conservation law almost surely. For this conservative equation, it has been shown to possess the multi-symplectic structure [119], while its ergodicity turns to be an open problem. Our purposes are to find an ergodic finite dimensional approximation of the original system, and to approximate its ergodic limit through an ergodic, multi-symplectic full discretization. In addition, the temporal average of the fully discrete scheme is shown to converge to the spatial average of the finite dimensional approximation with a specific rate. These results may help to find a new way to show the ergodicity of conservative equations as an afterthought.

Last but not least, we find it an efficient way to give a visible description of the ergodic limit via numerical experiments, and several numerical experiments are given in Chaps. 5 and 6 to make it clearer to the reader what the longtime behaviors of ergodic processes are.

Beijing, China   Jialin Hong
May 2017   Xu Wang

Acknowledgements It is our pleasure to thank the authors of our references for their important contributions and motivating us to get into these wonderful mathematical fields. We would like to thank all the referees, whose comments and suggestions are of great help in improving this monograph. We would also like to thank Dr. Chuchu Chen, Yulan Lu, Jianbo Cui, Liying Sun, Chuying Huang, Diancong Jin and Derui Sheng for their careful reading and pointing out a lot of typos in the draft of this monograph. We also acknowledge the National Natural Science Foundation of China (No. 11021101, No. 11290142, No. 91130003, No. 91530118 and No. 91630312) for its financial support.

Contents

Notation and Symbols

$\mathbb{H}$	Hilbert space
$\mathbb{S}$	Unit sphere in the corresponding space
$\mathbb{N}$ (**resp.** $\mathbb{N}_+$)	Set of all nonnegative (resp. positive) integers
$\mathbb{R}$ (**resp.** $\mathbb{R}_+$)	Set of all real numbers (resp. set of all positive real numbers)
$\mathbb{R}^d$ (**resp.** $\mathbb{T}^d$)	d-dimensional real space (resp. d-dimensional torus)
$\mathscr{B}(\mathbb{H})$	Borel σ-algebra on $\mathbb{H}$
$\mathscr{P}(\mathbb{H})$	Space of all probability measures on $(\mathbb{H}, \mathscr{B}(\mathbb{H}))$
$\mathscr{L}(\mathbb{H}, \mathbb{K})$	Space of all linear bounded operators from space $\mathbb{H}$ to $\mathbb{K}$, denoted by $\mathscr{L}(\mathbb{H})$ if $\mathbb{K} = \mathbb{H}$
$\mathbf{L}^p(\mathbb{H}; \mathbb{K})$	Space of all $\mathbb{K}$-valued functions defined on $\mathbb{H}$ which are pth integrable
$\mathbf{L}^p(\mathbb{H}, \mu)$	Space of all functions defined on $\mathbb{H}$ which are pth integrable with respect to measure μ, also denoted by $\mathbf{L}^p(\mathbb{H})$ for simplicity
$\mathbf{B}_b(\mathbb{H})$	Banach space of all Borel bounded functions on $\mathbb{H}$
$\mathbf{C}_b(\mathbb{H})$	Banach space of all uniformly continuous and bounded functions on $\mathbb{H}$
$\mathbf{C}_b^\infty(\mathbb{H})$	Space of all smooth and bounded functions with bounded derivatives of any order
$\mathbf{C}_p^\infty(\mathbb{H})$	Space of all smooth functions with polynomial growth
$\mathbf{W}^{k,p}(\mathbb{H})$	Sobolev space consisting of all $\mathbb{R}$-valued functions on $\mathbb{H}$ whose first k weak derivatives are functions in $\mathbf{L}^p$
$\mathbf{W}^{k,\infty}(\mathbb{H})$	Sobolev spaces of all functions on $\mathbb{H}$ whose derivatives up to order k have finite $\mathbf{L}^\infty$ norm
$\mathbf{H}^p(\mathbb{R}^d)$	Sobolev spaces of all functions on $\mathbb{R}^d$ whose derivatives up to order p are square integrable
$(\Omega, \mathscr{F}, \mathbb{P})$	Probability space
$(\mathbb{M}, \mathscr{G}, \mu)$	Measure space
$\pi_t(x, G)$	Transition probability for a stochastic process starting from x to hit set G at time t

$N(0,1)$	Standard normal distribution
$\lvert \cdot \rvert$	Absolute value in $\mathbb{R}$ or $\mathbb{C}$
$\lVert \cdot \rVert$	Euclidean norm in the corresponding finite dimensional real space
$\lVert \cdot \rVert_F$	Frobenius norm of matrices in the corresponding space
$\langle M \rangle$	Quardratic variation process of stochastic process M
$\langle \cdot, \cdot \rangle_{\mathbb{H}}$	Inner product in some Hilbert space $\mathbb{H}$; in particular, if $\mathbb{H}$ is a finite dimensional real-valued space, denote $\langle \cdot, \cdot \rangle$ for simplicity
$(\cdot, \cdot)$	Inner product in a finite dimentional complex-valued space
E^c	Complementary set of set E
$\overline{M}$	Closure of M under the corresponding Euclidean norm
$\overline{U}$	Conjugate of complex number U
$s \wedge t$ (**resp.** $s \vee t$)	Minimum (resp. maximum) of $s, t \in \mathbb{R}$
$\mathfrak{R}(u)$ (**resp.** $\mathfrak{J}(u)$)	Real (resp. imaginary) part of u
$\mathrm{Dom}(A)$	Domain of operator A

Chapter 1
Invariant Measures and Ergodicity

1.1 Basic Definitions in Measure Spaces

In classical ergodic theory, the behaviors of automorphisms, endomorphisms, flows, and semiflows are studied in measure spaces.

Let $(\mathbb{M}, \mathscr{G}, \mu)$ be a measure space with a normalized measure μ. This section introduces some basic concepts concerning invariant measures and ergodicity for endomorphisms and semiflows. We refer to [53] for more details about the ergodic theory of dynamical systems in measure spaces.

Definition 1.1 An *endomorphism* of $\mathbb{M}$ is a surjection $P : \mathbb{M} \to \mathbb{M}$ such that for any $A \in \mathscr{G}$,

$$P^{-1}A \in \mathscr{G} \quad \text{and} \quad \mu(P^{-1}A) = \mu(A)$$

with $P^{-1}A$ denoting the inverse image of A. The measure μ is said to be *invariant* under the endomorphism P.

For continuous dynamical systems, a family of endomorphisms will be taken into consideration with measure μ still being invariant under these endomorphisms.

Definition 1.2 Let $\{P_t\}_{t \geq 0}$ be a one-parameter semigroup of endomorphisms on $(\mathbb{M}, \mathscr{G}, \mu)$ satisfying

$$P_{t+s}x = P_t \circ P_s x$$

for all $t, s \geq 0$ and $x \in \mathbb{M}$. Then $\{P_t\}_{t \geq 0}$ is called a *semiflow* if $\varphi(P_t x)$ is measurable on $\mathbb{M} \times \mathbb{R}_+$ for any measurable function φ on $\mathbb{M}$.

Remark 1.1 It shows immediately that μ is an invariant measure of $\{P_t\}_{t \geq 0}$, i.e.,

$$\int_{\mathbb{M}} \varphi(P_t x)\mu(dx) = \int_{\mathbb{M}} \varphi(x)\mu(dx)$$

© Springer Nature Singapore Pte Ltd. 2019
J. Hong and X. Wang, *Invariant Measures for Stochastic Nonlinear Schrödinger Equations*, Lecture Notes in Mathematics 2251,
https://doi.org/10.1007/978-981-32-9069-3_1

for any measurable and bounded function φ on $\mathbb{M}$, according to the approximation by simple functions. In fact, for indicator function $\mathbf{1}_A$ of measurable set $A \in \mathscr{G}$, we have

$$\int_{\mathbb{M}} \mathbf{1}_A(P_t x)\mu(dx) = \int_{\mathbb{M}} \mathbf{1}_{P_t^{-1}A}(x)\mu(dx) = \mu(P_t^{-1}A).$$

Note that $\mu(P_t^{-1}A) = \mu(A)$ since P_t is an endomorphism, which leads to

$$\int_{\mathbb{M}} \mathbf{1}_A(P_t x)\mu(dx) = \mu(A) = \int_{\mathbb{M}} \mathbf{1}_A(x)\mu(dx).$$

In some circumstances, we also use the notation $\int_{\mathbb{M}} \varphi d\mu := \int_{\mathbb{M}} \varphi(x)\mu(dx)$ for simplicity.

The expression "dynamical system" could stand for either endomorphisms or semiflows, or even for automorphisms and flows whose definitions are omitted to make the content simple and clear. In this case, the measure space $(\mathbb{M}, \mathscr{G}, \mu)$ is known as the phase space.

Definition 1.3 A set $G \in \mathscr{G}$ is said to be invariant with respect to $\{P_t\}_{t \geq 0}$ if $x \in G$ is equivalent to $P_t x \in G$ for any $t \geq 0$, i.e.,

$$\mathbf{1}_G(P_t x) = \mathbf{1}_G(x)$$

for any $t \geq 0$ and for all $x \in \mathbb{M}$.

In addition, the dynamical system $\{P_t\}_{t \geq 0}$ is said to be *ergodic* if all the invariant set G are *trivial*, i.e.,

$$\mu(G) = 0 \quad \text{or} \quad \mu(G) = 1.$$

The measure μ is called an ergodic invariant measure for $\{P_t\}_{t \geq 0}$.

When considering the asymptotic behavior of endomorphism P or semiflow $\{P_t\}_{t \geq 0}$, the Birkhoff–Khinchin ergodic theorem (see e.g. [22, 53]) shows that the limit

$$\lim_{N \to \infty} \frac{1}{N} \sum_{n=0}^{N-1} \varphi(P^n x) \quad \text{or} \quad \lim_{T \to \infty} \frac{1}{T} \int_0^T \varphi(P_s x)ds$$

always exists for μ-almost every $x \in \mathbb{M}$.

Theorem 1.1 (Birkhoff–Khinchin ergodic theorem) *For $\varphi \in \mathbf{L}^1(\mathbb{M}, \mu)$, the limits*

$$\lim_{N \to \infty} \frac{1}{N} \sum_{n=0}^{N-1} \varphi(P^n x)$$

and

$$\lim_{T \to \infty} \frac{1}{T} \int_0^T \varphi(P_s x) ds$$

exist for μ-almost every $x \in \mathbb{M}$ when P is an endomorphism and $\{P_t\}_{t \geq 0}$ is a semiflow.

The proof of above theorem can also be used to prove the Von Neumann theorem introduced in the next section and is given in Appendix B for the reader's convenience.

Denote by

$$\tilde{\varphi}(x) := \lim_{T \to \infty} \frac{1}{T} \int_0^T \varphi(P_s x) ds. \tag{1.1}$$

Based on the results above, we claim that $A_a := \{\tilde{\varphi}(x) < a\}$ is an invariant set for any $a \in \mathbb{R}$, that is, for any $x \in A_a$, $P_t x \in A_a$ for any $t \geq 0$, and vice versa. In fact,

$$\tilde{\varphi}(P_t x) = \lim_{T \to \infty} \frac{1}{T} \int_0^T \varphi(P_s \circ P_t x) ds = \lim_{T \to \infty} \frac{1}{T} \int_0^T \varphi(P_{s+t} x) ds$$

$$= \lim_{T \to \infty} \frac{1}{T} \int_0^T \varphi(P_s x) ds = \tilde{\varphi}(x),$$

which verifies the claim. It then leads to the following corollary when the semiflow $\{P_t\}_{t \geq 0}$ is ergodic.

Corollary 1.1 *If $\{P_t\}_{t \geq 0}$ is ergodic, then $\tilde{\varphi}(x)$ equals a constant μ-almost surely, and the constant is exactly the spatial average of φ, i.e.,*

$$\tilde{\varphi}(x) = \int_{\mathbb{M}} \tilde{\varphi} d\mu = \int_{\mathbb{M}} \varphi d\mu.$$

Proof Since $\tilde{\varphi}(x)$ exists for μ-almost every $x \in \mathbb{M}$, we derive according to above claim and the ergodicity of semiflow $\{P_t\}_{t \geq 0}$ that there must exist $a_0 \neq b_0 \in \mathbb{R}$ such that $\mu(A_{a_0}) = 0$ and $\mu(A_{b_0}) = 1$. As a result,

$$a_0 \leq \tilde{\varphi}(x) \leq b_0, \quad \mu\text{-a.s.}$$

Choose any $c \in (a_0, b_0)$, then $\mu(A_c) = 0$ or 1 since the ergodicity of the semiflow. If $\mu(A_c) = 0$, we denote $a_1 = c$ and $b_1 = b_0$, otherwise we denote $a_1 = a_0$ and $b_1 = c$. It then ensures that

$$a_1 \leq \tilde{\varphi}(x) \leq b_1, \quad \mu\text{-a.s.}$$

Following this procedure, we get a family of closed intervals $\{[a_n, b_n]\}_{n \in \mathbb{N}}$. According to the principle of nested intervals, one derives

$$\lim_{n \to \infty} a_n = \lim_{n \to \infty} b_n,$$

which ensures that $\tilde{\varphi}$ is a constant μ-almost surely.

Moreover, integrating (1.1) with respect to μ, we obtain

$$\tilde{\varphi}(x) = \int_{\mathbb{M}} \tilde{\varphi} d\mu = \lim_{T \to \infty} \frac{1}{T} \int_0^T \int_{\mathbb{M}} \varphi(P_s x)\mu(dx)ds$$

$$= \lim_{T \to \infty} \frac{1}{T} \int_0^T \int_{\mathbb{M}} \varphi(x)\mu(dx)ds = \int_{\mathbb{M}} \varphi d\mu,$$

where we have used Remark 1.1 in the third step. □

Remark 1.2 The converse of above corollary is also true: assume that the following limit is a constant μ-almost surely

$$\lim_{T \to \infty} \frac{1}{T} \int_0^T \varphi(P_t x)dt = \int_{\mathbb{M}} \varphi d\mu.$$

Then for any invariant set G, we choose $\varphi = \mathbf{1}_G$ in above equation and get

$$\mu(G) = \int_{\mathbb{M}} \mathbf{1}_G d\mu = \lim_{T \to \infty} \frac{1}{T} \int_0^T \mathbf{1}_G(P_t x)dt = \mathbf{1}_G(x),$$

which indicates that G is trivial and thus $\{P_t\}_{t \geq 0}$ is ergodic.

It then gives an equivalent definition of ergodicity:

$$\lim_{T \to \infty} \frac{1}{T} \int_0^T \varphi(P_t x)dt = \int_{\mathbb{M}} \varphi d\mu$$

for $\varphi \in \mathbf{L}^1(\mathbb{M}, \mu)$.

For the initial value problem of a specific differential equation defined on $\mathbb{M} = \mathbb{R}^d$

$$\frac{dX(t)}{dt} = f(X(t)), \quad X(0) = x, \quad t \geq 0$$

one can easily define a Markov semigroup $P_t x := X_x(t)$, where $X_x(\cdot)$ denotes the solution starting from the initial value x. In the following, for some test function φ, we use the notation $P_t \varphi(x)$ instead of $\varphi(P_t x)$ to emphasize the linearity of the operator P_t for any $t \geq 0$.

For Markov semigroup $\{P_t\}_{t \geq 0}$ generated by the solution of differential equations or stochastic differential equations, its invariant measures and ergodicity are also widely studied to investigate the longtime behavior of its ensemble average. Different from the results given above for endomorphisms, the existence of invariant probability measures for Markov semigroup $\{P_t\}_{t \geq 0}$ is not trivial.

In the following sections, we give definitions of invariant measures and ergodicity in view of stochastic differential equations by introducing the Markov semigroup generated by the solution of a stochastic differential equation. Some sufficient conditions on existence and uniqueness of the invariant measure and ergodicity for stochastic processes or Markov chains are also incorporated. We refer to [58, 60, 120, 143, 182] and references therein for more details.

1.2 Invariant Measures for Stochastic Processes

Let $(\Omega, \mathscr{F}, \mathbb{P})$ be a complete probability space with filtration $\{\mathscr{F}_t\}_{t \geq 0}$ and $\mathbb{H}$ be a separable Hilbert space. For an initial value problem of the following stochastic differential equation

$$
\begin{cases}
dX(t) = b(X(t))dt + \sigma(X(t))dW(t), \\
X(0) = x \in \mathbb{H}
\end{cases}
\tag{1.2}
$$

with W being the standard Wiener process associated to filtration $\{\mathscr{F}_t\}_{t \geq 0}$, there exists a unique solution $X : \Omega \times [0, T] \to \mathbb{H}$ which is an $\mathbb{H}$-valued stochastic Markov process under certain conditions.

We denote by $X_x(t)$ the solution at time t starting from x when it is necessary to point out the dependence of the initial value x, and eliminate the variable $\omega \in \Omega$. Assumptions on the drift coefficient b and the diffusion coefficient σ will be given in the next subsection to ensure the well-posedness of (1.2).

For $\varphi \in \mathbf{B}_b(\mathbb{H})$ with $\mathbf{B}_b(\mathbb{H})$ denoting the space of all measurable and bounded functions defined on $\mathbb{H}$, we define the transition semigroup $\{P_t\}_{t \geq 0}$ as

$$
P_t \varphi(x) = \mathbb{E}[\varphi(X_x(t))],
\tag{1.3}
$$

which is a Markov semigroup on $\mathbf{B}_b(\mathbb{H})$. Here $\mathbb{E}[\cdot]$ denotes the expectation of a random variable. In addition, we denote by

$$
\pi_t(x, G) = \mathbb{P}(X_x(t) \in G)
$$

the transition probability that $X_x(t)$ hits G for any $G \in \mathscr{B}(\mathbb{H})$, where $\mathscr{B}(\mathbb{H})$ is the Borel σ-algebra. Then $\pi_t(x, \cdot)$ is a probability measure (also called a *probability kernel*, see [58, 116]) and satisfies

$$
P_t \varphi(x) = \int_\Omega \varphi(X_x(t))d\mathbb{P} = \int_\mathbb{H} \varphi(y)\pi_t(x, dy).
\tag{1.4}
$$

In addition, choosing $\varphi = \mathbf{1}_G$ for some $G \in \mathscr{B}(\mathbb{H})$, we have

$$
\pi_t(x, G) = P_t \mathbf{1}_G(x) = \mathbb{E}[\mathbf{1}_G(X_x(t))] = \mathbb{P}(X_x(t) \in G).
$$

We define the discrete version of π_t similarly: for a discrete $\mathbb{H}$-valued Markov chain $\{X_n\}_{n \in \mathbb{N}}$, for example, the solution of some numerical scheme for (1.2), we denote by

$$\pi_n(x, G) = \mathbb{P}(X_n \in G | X_0 = x)$$

the discrete transition kernel for $x \in \mathbb{H}$ and $G \in \mathscr{B}(\mathbb{H})$.

Under certain assumptions which ensure that the solution $X_x(t)$ to (1.2) is uniformly continuous with respect to the initial value x, the Markov semigroup $\{P_t\}_{t \geq 0}$ defined in (1.3) is *Feller*: for any $\varphi \in \mathbf{C}_b(\mathbb{H})$, $P_t\varphi \in \mathbf{C}_b(\mathbb{H})$ (see e.g. Chap. 3) with $\mathbf{C}_b(\mathbb{H})$ being the space of all continuous and bounded functions. We also call X a Feller process if the Markov semigroup $\{P_t\}_{t \geq 0}$ associated to X is Feller.

Next, we introduce some useful conditions which ensure the existence of invariant measures for Markov Feller processes. We denote by $\mathbb{H}$ a separate Hilbert space and $\{P_t\}_{t \geq 0}$ the Markov transition semigroup associated to the solution X.

Definition 1.4 ([58]) A probability measure $\mu \in \mathscr{P}(\mathbb{H})$ is said to be *invariant* for X if

$$\int_{\mathbb{H}} P_t\varphi(x)\mu(dx) = \int_{\mathbb{H}} \varphi d\mu \tag{1.5}$$

for all $\varphi \in \mathbf{B}_b(\mathbb{H})$ and $t \geq 0$, where $P_t\varphi(x) := \mathbb{E}[\varphi(X_x(t))]$ and $\mathscr{P}(\mathbb{H})$ denotes the space of all the probability measures on $\mathbb{H}$.

In the sequel, by saying "invariant measure", we always mean "invariant probability measure".

Remark 1.3 If Markov semigroup $\{P_t\}_{t \geq 0}$ possesses an invariant measure, it can be uniquely extended to a family of linear bounded operators on $\mathbf{L}^p(\mathbb{H}, \mu)$, $p \geq 1$, such that

$$\|P_t\|_{\mathscr{L}(\mathbf{L}^p(\mathbb{H},\mu))} \leq 1, \quad t \geq 0.$$

One could find immediately that if μ is an invariant measure of X, then

$$\mu(G) = \int_{\mathbb{H}} \pi_t(x, G)\mu(dx), \quad \forall t \geq 0 \tag{1.6}$$

by choosing $\varphi = \mathbf{1}_G$, and in addition, the integration of $\mathbb{E}[\varphi(X_x(t))]$ with respect to μ keeps a constant for any $t \geq 0$.

An intuitive description of invariant measures is that if the probability distribution $\pi_t(x, \cdot)$ of $X_x(t)$ converges to a probability measure μ, then μ is an invariant measure of $X_x(t)$. The following theorem gives a sufficient condition for the existence of invariant measures, which is less restrictive than the description above.

Definition 1.5 ([58]) If a subset $\Lambda \subset \mathscr{P}(\mathbb{H})$ satisfies that, for any $\varepsilon > 0$, there exists a compact set K_ε such that $\mu(K_\varepsilon) \geq 1 - \varepsilon$ for all $\mu \in \Lambda$, then Λ is said to be *tight*.

The following theorem indicates the existence of invariant measures under the tightness assumption taking advantage of the Prokhorov theorem (see Theorem 6.7, [58]).

Theorem 1.2 (Krylov–Bogoliubov theorem) *If X is a Markov Feller process and for some initial value $x_0 \in \mathbb{H}$, the set $\{\mu_T\}_{T>0}$ with definition*

$$\mu_T(G) := \frac{1}{T}\int_0^T \mathbb{E}[\mathbf{1}_G(X_{x_0}(t))]dt = \frac{1}{T}\int_0^T \pi_t(x_0, G)dt, \quad \forall\, G \in \mathscr{B}(\mathbb{H})$$

is tight, then there exists an invariant measure $\mu \in \mathscr{P}(\mathbb{H})$ for X.

Proof Since $\{\mu_T\}_{T>0} \subset \mathscr{P}(\mathbb{H})$ is tight, according to the Prokhorov theorem, there exist $\{\mu_{T_k}\}_{k\in\mathbb{N}} \subset \{\mu_T\}_{T>0}$ and $\mu \in \mathscr{P}(\mathbb{H})$ such that $\{\mu_{T_k}\}_{k\in\mathbb{N}}$ is weakly convergent to μ. That is, for $\psi \in \mathbf{C}_b(\mathbb{H})$,

$$\lim_{k\to\infty}\int_{\mathbb{H}}\psi d\mu_{T_k} = \int_{\mathbb{H}}\psi d\mu.$$

On the other hand,

$$\lim_{k\to\infty}\int_{\mathbb{H}}\psi d\mu_{T_k} = \lim_{k\to\infty}\frac{1}{T_k}\int_0^{T_k}\int_{\mathbb{H}}\psi(x)\pi_t(x_0, dx)dt = \lim_{k\to\infty}\frac{1}{T_k}\int_0^{T_k}P_t\psi(x_0)dt.$$

For any $\varphi \in \mathbf{C}_b(\mathbb{H})$, utilizing the Feller property of $\{P_t\}_{t\geq 0}$, we choose $\psi = P_s\varphi \in \mathbf{C}_b(\mathbb{H})$ in above equations for $s \geq 0$ and derive from above equations that

$$\int_{\mathbb{H}}P_s\varphi d\mu = \lim_{k\to\infty}\frac{1}{T_k}\int_0^{T_k}P_{t+s}\varphi(x_0)dt$$

$$= \lim_{k\to\infty}\frac{1}{T_k}\left[\int_0^{T_k}P_t\varphi(x_0)dt + \int_{T_k}^{T_k+s}P_t\varphi(x_0)dt - \int_0^s P_t\varphi(x_0)dt\right]$$

$$= \int_{\mathbb{H}}\varphi d\mu,$$

which indicates that μ is an invariant measure for X. $\qquad\square$

Remark 1.4 Note that the tightness of transition probabilities $\{\pi_t(x_0, \cdot)\}_{t>0}$ ensures the tightness of $\{\mu_T\}_{T>0}$ defined in the theorem above, which can be obtained through the Definition 1.5. Thus, it also ensures the existence of invariant measures for X.

As a consequence of the theorem above, another frequently used sufficient condition for the existence of invariant measures, or even the existence of ergodic invariant measures, is stated below utilizing Lyapunov functionals.

Theorem 1.3 (Proposition 7.10, [58]) *If there exist some $x_0 \in \mathbb{H}$ and a constant $C = C(x_0) > 0$ such that*

$$\mathbb{E}[V(X_{x_0}(t))] \leq C(x_0), \quad \forall\, t \geq 0,$$

then there exists an invariant measure for X, where $V : \mathbb{H} \to [1, +\infty]$ *is a Borel function (Lyapunov functional) whose level sets*

$$K_a := \{x \in \mathbb{H} : V(x) \leq a\}$$

are compact for any a > 0.

Proof For any $\varepsilon > 0$, let $a(\varepsilon) = \frac{C(x_0)}{\varepsilon}$. We obtain for any $t > 0$ that the complementary set $K^{\mathsf{c}}_{a(\varepsilon)}$ of $K_{a(\varepsilon)}$ satisfies

$$
\begin{aligned}
\pi_t(x_0, K^{\mathsf{c}}_{a(\varepsilon)}) &= \int_{\{V(x) > a(\varepsilon)\}} \pi_t(x_0, dy) \\
&\leq \int_{\mathbb{H}} \left[\frac{V(y)}{a(\varepsilon)} \right] \pi_t(x_0, dy) \\
&= \frac{1}{a(\varepsilon)} \mathbb{E}[V(X_{x_0}(t))] \leq \varepsilon,
\end{aligned}
$$

which yields apparently

$$\pi_t(x_0, K_{a(\varepsilon)}) > 1 - \varepsilon.$$

Hence $\{\pi_t(x_0, \cdot)\}_{t>0}$ is tight, which ensures the existence of invariant measures according to Remark 1.4. $\qquad\square$

The assumption in Theorem 1.3 is usually called the Lyapunov condition, which is often characterized by some other sufficient conditions. For example, a dissipative condition $x \cdot b(x) \leq -\beta |x|^2 + C$ for some $\beta > 0$ and $C > 0$ is used in [179]. The details of other conditions which ensure the existence of invariant measures will be discussed in Chap. 2. Furthermore, the theorem above could also be applied to a discrete Markov chain $\{X_n\}_{n \in \mathbb{N}}$ to gain the existence of invariant measures if the condition

$$\mathbb{E}[V(X_n)] \leq C(X_0)$$

is satisfied for some constant $C(X_0)$ depending on initial data X_0.

Example 1.1 Consider the stochastic differential equation

$$dX(t) = -X(t)dt + \sqrt{2}\,dB(t), \quad X(0) = \xi \in \mathbb{R}, \quad t \geq 0, \qquad (1.7)$$

where B is a one-dimensional standard Brownian motion. Applying Itô's formula to $|X(t)|^2$ with $|\cdot|$ denoting the absolute value, we have

$$\mathbb{E}|X(t)|^2 = e^{-2t}\mathbb{E}|\xi|^2 + 1 - e^{-2t} \leq \mathbb{E}|\xi|^2 + 1,$$

which implies the existence of invariant measures based on Theorem 1.3 by choosing the Lyapunov functional $V(\cdot) = |\cdot|^2 + 1$.

It is well known that (1.7) can be solved explicitly, whose solution is given by the Ornstein–Uhlenbeck process

$$X(t) = e^{-t}\xi + \int_0^t \sqrt{2}e^{-(t-s)}dB(s).$$

If the initial value ξ is a constant, then the probability kernel of $X(t)$ is a Gaussian distribution satisfying

$$\pi_t(\xi, \cdot) = N(e^{-t}\xi, 1 - e^{-2t}) \to N(0, 1) \quad \text{as } t \to \infty.$$

If the initial value ξ is an $N(0, 1)$-distributed random variable, then $X(t)$ is also an $N(0, 1)$-distributed random variable for any $t \geq 0$, which implies the invariance of the measure $\mu = N(0, 1)$ (see also [58]).

More precisely, we claim that μ satisfies the identity in Definition 1.4. In fact, by choosing φ in Definition 1.4 as $\varphi(x) = e^{\mathrm{i}hx}$ for any $h \in \mathbb{R}$ which can be chosen as proper approximations of functions in $\mathbf{C}_b(\mathbb{R})$, we have

$$\int_{\mathbb{R}} P_t\varphi(x)\mu(dx) = \int_{\mathbb{R}} \mathbb{E}\left[e^{\mathrm{i}h(e^{-t}x + \int_0^t \sqrt{2}e^{-(t-s)}dB(s))}\right]\frac{1}{\sqrt{2\pi}}e^{-\frac{x^2}{2}}dx$$
$$= \frac{1}{\sqrt{2\pi}}\int_{\mathbb{R}} e^{\frac{1}{2}(e^{-t}h + \mathrm{i}x)^2 - \frac{h^2}{2}}dx =: f(t).$$

Since $f'(t) \equiv 0$, we have

$$\int_{\mathbb{R}} P_t\varphi(x)\mu(dx) = f(t) = f(0) = \frac{1}{\sqrt{2\pi}}\int_{\mathbb{R}} e^{\mathrm{i}hx - \frac{x^2}{2}}dx = \int_{\mathbb{R}} \varphi(x)\mu(dx),$$

which completes the proof of the claim.

Next, we give a brief introduction of ergodicity for both stochastic processes and invariant measures, based on which we will show that the invariant measure of (1.7) is also ergodic, strongly mixing and unique.

1.3 Ergodicity

As an important asymptotic behavior, ergodicity is studied in a number of different research areas, while we concentrate on the ergodicity of the stochastic flows generated by stochastic differential equations. In this section, we introduce definitions of ergodicity and mixing property for the solution X to (1.2) (see also [58,

74, 143]), which characterize different convergence rates of the temporal average of $P_t\varphi(x) = \mathbb{E}[\varphi(X_x(t))]$.

For Markov semigroup $\{P_t\}_{t\geq 0}$, similar to the results concerned on semiflows shown in the Birkhoff–Khinchin ergodic theorem, the limit

$$\lim_{T\to\infty}\frac{1}{T}\int_0^T P_t\varphi(x)dt =: \tilde{\varphi}(x)$$

exists with $\tilde{\varphi}(x)$ satisfying

$$\int_{\mathbb{H}} \tilde{\varphi}d\mu = \int_{\mathbb{H}} \varphi d\mu$$

based on the Von Neumann theorem (see e.g. [58, 60]). Following are some definitions related to this limit.

Definition 1.6 *(see e.g. [60])* Let μ be an invariant measure of X.

(i) X is said to be *ergodic* on $\mathbb{H}$ if

$$\lim_{T\to\infty}\frac{1}{T}\int_0^T \mathbb{E}[\varphi(X_x(t))]dt = \int_{\mathbb{H}} \varphi d\mu \quad \text{in } \mathbf{L}^2(\mathbb{H}, \mu) \qquad (1.8)$$

for all $\varphi \in \mathbf{L}^2(\mathbb{H}, \mu)$.

(ii) X is said to be *strongly mixing* on $\mathbb{H}$ if

$$\lim_{t\to\infty} \mathbb{E}[\varphi(X_x(t))] = \int_{\mathbb{H}} \varphi d\mu \quad \text{in } \mathbf{L}^2(\mathbb{H}, \mu)$$

for all $\varphi \in \mathbf{L}^2(\mathbb{H}, \mu)$.

(iii) X is said to be *exponentially mixing* on $\mathbb{H}$ if there exist $\rho > 0$ and a positive function $C(\cdot)$ such that for any bounded Lipschitz continuous function φ on $\mathbb{H}$, all $t > 0$ and all $x \in \mathbb{H}$,

$$\left|\mathbb{E}[\varphi(X_x(t))] - \int_{\mathbb{H}} \varphi d\mu\right| \leq C(x)L_\varphi e^{-\rho t},$$

where L_φ denotes the Lipschitz constant of φ.

The convergence in $\mathbf{L}^2(\mathbb{H}, \mu)$ here and below is interpreted with respect to the initial value $x \in \mathbb{H}$. The ergodicity of a stochastic process is usually described as the case its temporal average equals its spatial average. The spatial average is also known as the ergodic limit, which is approximated in Chap. 6. The spatial average is also

A stochastic differential equation is usually called ergodic (resp. strongly mixing, exponentially mixing) if its solution is ergodic (resp. strongly mixing, exponentially mixing). Different definitions above show that $P_t\varphi(x)$ converges to the spatial average $\int_{\mathbb{H}} \varphi d\mu$ in different senses or with different rates.

Example 1.2 We still consider the equation in Example 1.1 with a deterministic initial value ξ. The probability kernel of the solution $X_\xi(t)$ is

$$\pi_t(\xi, dx) = \frac{1}{\sqrt{2\pi(1 - e^{-2t})}} e^{-\frac{(x - e^{-t}\xi)^2}{2(1 - e^{-2t})}} dx.$$

Hence, we can calculate

$$\mathbb{E}[\varphi(X_\xi(t))] = \int_{\mathbb{R}} \varphi(x)\pi_t(\xi, dx) = \frac{1}{\sqrt{2\pi(1 - e^{-2t})}} \int_{\mathbb{R}} \varphi(x) e^{-\frac{(x - e^{-t}\xi)^2}{2(1 - e^{-2t})}} dx$$

$$\xrightarrow{t \to \infty} \frac{1}{\sqrt{2\pi}} \int_{\mathbb{R}} \varphi(x) e^{-\frac{x^2}{2}} dx = \int_{\mathbb{R}} \varphi d\mu$$

with μ denoting the Gaussian distribution. As a result, the solution $X_\xi(t)$ to (1.7) is strongly mixing, and thus is ergodic.

For a discrete $\mathbb{H}$-valued Markov chain $\{X_n\}_{n \in \mathbb{N}}$, we define its ergodicity if it possesses an invariant measure μ and in addition it satisfies that

$$\lim_{N \to \infty} \frac{1}{N} \sum_{n=1}^{N} \mathbb{E}[\varphi(X_n)|X_0 = x] = \int_{\mathbb{H}} \varphi d\mu \quad \text{in } \mathbf{L}^2(\mathbb{H}, \mu).$$

We would like to mention that the following is an equivalent definition about ergodicity with respect to invariant measure μ according to the Birkhoff–Khinchin ergodic theorem introduced in Sect. 1.1.

Definition 1.7 Let μ be an invariant measure of X. Then μ is said to be *ergodic* if any invariant set is trivial. More precisely, if $G \in \mathscr{B}(\mathbb{H})$ satisfies

$$\mathbb{P}(X_x(t) \in G) = \mathbf{1}_G(x), \quad \mu\text{-a.s.}$$

for any $t \geq 0$, then $\mu(G) = 0$ or $\mu(G) = 1$.

It is worth noticing that the definition above coincides with those in Definition 1.6 (i) according to Corollary 1.1 and Remark 1.2. The set G is invariant since its characteristic function $\mathbf{1}_G$ is invariant under $\{P_t\}_{t \geq 0}$: recall that

$$\mathbb{P}(X_x(t) \in G) = \mathbb{E}[\mathbf{1}_G(X_x(t))] = P_t \mathbf{1}_G(x).$$

Thus, the condition in Definition 1.7 turns to be

$$P_t \mathbf{1}_G(x) = \mathbf{1}_G(x), \quad \mu\text{-a.s.}$$

for any $t \geq 0$.

The relationship between invariant measures and ergodic measures is stated in the following proposition, which can also be regard as a sufficient condition of the ergodicity of an invariant measure.

Proposition 1.1 (Theorem 5.16, [58]) *Assume that the invariant measure μ for X is unique, then μ is ergodic.*

Proof Assume by contradiction that μ is not ergodic. Then there exists a nontrivial invariant set G such that $0 < \mu(G) < 1$ based on Definition 1.7. Define a measure μ_G based on the invariant set G

$$\mu_G(A) := \frac{\mu(G \cap A)}{\mu(G)}, \quad A \in \mathscr{B}(\mathbb{H}).$$

We claim that $\mu_G \neq \mu$ is also an invariant measure for X, which gives rise to a contradiction that μ is the unique invariant measure.

To prove the claim, we only need to verify that (1.5) holds when $\varphi = \mathbf{1}_A$ for any $A \in \mathscr{B}(\mathbb{H})$, i.e.,

$$\int_{\mathbb{H}} P_t \mathbf{1}_A d\mu_G = \int_{\mathbb{H}} \mathbf{1}_A d\mu_G.$$

Note that the left hand side of above equation shows

$$\int_{\mathbb{H}} P_t \mathbf{1}_A d\mu_G = \frac{1}{\mu(G)} \int_G P_t \mathbf{1}_A d\mu = \frac{1}{\mu(G)} \left[\int_G P_t \mathbf{1}_{A \cap G} d\mu + \int_G P_t \mathbf{1}_{A \cap G^c} d\mu \right]$$

$$= \frac{1}{\mu(G)} \int_{\mathbb{H}} P_t \mathbf{1}_{A \cap G} d\mu = \frac{1}{\mu(G)} \int_{\mathbb{H}} \mathbf{1}_{A \cap G} d\mu = \frac{\mu(A \cap G)}{\mu(G)},$$

while the right hand side shows

$$\int_{\mathbb{H}} \mathbf{1}_A d\mu_G = \mu_G(A) = \frac{\mu(A \cap G)}{\mu(G)}.$$

Here, we have used the fact that

$$0 \leq P_t \mathbf{1}_{A \cap G^c} \leq P_t \mathbf{1}_{G^c} = \mathbf{1}_{G^c} \quad \mu\text{-a.s.}$$

since G is an invariant set. Thus, $P_t \mathbf{1}_{A \cap G^c} = 0$ μ-almost surely on G. $\qquad\square$

It follows directly from Example 1.2 that the invariant measure obtained in Example 1.1 is unique.

Taking advantage of above proposition, we derive a stronger result than that in Theorem 1.3 if the Lyapunov condition holds uniformly.

Theorem 1.4 *If there exists a positive constant C_0 such that*

$$\mathbb{E}[V(X_x(t))] \leq C_0, \quad \forall x \in \mathbb{H}, \; t \geq 0,$$

then the set of all the invariant measures $\Lambda \subset \mathscr{P}(\mathbb{H})$ is tight and there exists an ergodic invariant measure for X.

Proof Based on Theorem 1.3, Λ is not empty. Note that there is a natural embedding of $\mathscr{P}(\mathbb{H})$ into $\mathbf{C}_b^*(\mathbb{H})$ by setting

$$F_\mu(\varphi) := \int_{\mathbb{H}} \varphi \, d\mu, \quad \forall \, \varphi \in \mathbf{C}_b(\mathbb{H}),$$

which satisfies

$$\|F_\mu\| := \sup_{\|\varphi\|=1} \left| \int_{\mathbb{H}} \varphi \, d\mu \right| \leq 1.$$

Then Λ is a convex subset of $\mathbf{C}_b^*(\mathbb{H})$.

Step 1. We first show that Λ is tight. For any $\delta > 0$, $t > 0$ and $\mu \in \Lambda$, by choosing the test function $\varphi = \frac{V}{1+\delta V}$ and utilizing the fact that μ is an invariant measure, we have

$$\int_{\mathbb{H}} \frac{V(x)}{1+\delta V(x)} \mu(dx) = \int_{\mathbb{H}} P_t \left(\frac{V(x)}{1+\delta V(x)} \right) \mu(dx)$$

$$= \int_{\mathbb{H}} \mathbb{E}\left[\frac{V(X_x(t))}{1+\delta V(X_x(t))} \right] \mu(dx)$$

$$\leq \int_{\mathbb{H}} \mathbb{E}[V(X_x(t))] \mu(dx) \leq C_0.$$

Let $\delta \to 0$, we derive

$$\sup_{\mu \in \Lambda} \int_{\mathbb{H}} V \, d\mu \leq C_0.$$

For any $\varepsilon > 0$, denoting $a(\varepsilon) = \frac{C_0}{\varepsilon}$ and $K_{a(\varepsilon)}$ as the definition in Theorem 1.3, we have

$$\mu(K_{a(\varepsilon)}^c) = \int_{\{V(x)>a(\varepsilon)\}} \mu(dx) \leq \int_{\mathbb{H}} \frac{V(x)}{a(\varepsilon)} \mu(dx) \leq \varepsilon,$$

which indicates

$$\mu(K_{a(\varepsilon)}) > 1 - \varepsilon, \quad \forall \, \mu \in \Lambda.$$

Step 2. Now we show the existence of ergodic invariant measures for X. The Krein–Milman theorem (see e.g. [190]) says that a non-empty compact convex subset of a locally convex linear topological space has at least one extremal point. It then suffices to show that the extremal points of Λ coincide with the ergodic invariant measures for X.

To show that all the extremal points of Λ are ergodic invariant measures for X, we denote by μ an extremal point of Λ, that is, if there exist $\mu_1, \mu_2 \in \Lambda$ and $\alpha \in (0, 1)$ such that

$$\mu = \alpha \mu_1 + (1 - \alpha)\mu_2,$$

then $\mu_1 = \mu_2 = \mu$. Assume by contradiction that μ is not ergodic. Then following the same procedure in the proof of Proposition 1.1, there exists a nontrivial invariant set G such that $\mu_G, \mu_{G^c} \in \Lambda$ with the definitions

$$\mu_G(A) = \frac{\mu(G \cap A)}{\mu(G)}, \quad \mu_{G^c}(A) = \frac{\mu(G^c \cap A)}{\mu(G^c)}, \quad \forall A \in \mathscr{B}(\mathbb{H})$$

and $\alpha := \mu(G) \in (0, 1)$. It then give rise to a contradiction since

$$\mu = \alpha \mu_G + (1 - \alpha)\mu_{G^c}$$

and $\mu_G \neq \mu_{G^c}$.

We finally show that if μ is ergodic, then it is an extremal point in Λ. Let μ be an ergodic invariant measure for X. Assume by contradiction that it is not an extremal point in Λ, i.e., there exist $\mu_1, \mu_2 \in \Lambda$ with $\mu_1 \neq \mu_2$ and $\alpha \in (0, 1)$ such that

$$\mu = \alpha \mu_1 + (1 - \alpha)\mu_2.$$

It is clear that μ_1 and μ_2 are absolutely continuous with respect to μ, that is, if $A \in \mathscr{B}(\mathbb{H})$ satisfies $\mu(A) = 0$, then $\mu_1(A) = \mu_2(A) = 0$, denoted by $\mu_1 \ll \mu$ and $\mu_2 \ll \mu$. Then for any $A \in \mathscr{B}(\mathbb{H})$, according to the Birkhoff–Khinchin ergodic theorem in Sect. 1.1 and that μ is ergodic, we have

$$\lim_{T \to \infty} \frac{1}{T} \int_0^T P_t \mathbf{1}_A(x)dt = \int_{\mathbb{H}} \mathbf{1}_A d\mu = \mu(A), \quad \mu\text{-a.s.,}$$

which also holds μ_1-almost surely since $\mu_1 \ll \mu$. Hence integrating above equation with respect to μ_1 yields

$$\mu(A) = \lim_{T \to \infty} \frac{1}{T} \int_0^T \int_{\mathbb{H}} P_t \mathbf{1}_A(x)d\mu_1 dt$$
$$= \lim_{T \to \infty} \frac{1}{T} \int_0^T \int_{\mathbb{H}} \mathbf{1}_A(x)d\mu_1 dt = \mu_1(A),$$

where we used the fact that $\mu_1 \in \Lambda$ in the second step. It then indicates $\mu_1 = \mu$, and also $\mu_2 = \mu$ in the same procedure, which is a contradiction to $\mu_1 \neq \mu_2$. $\qquad \square$

1.4 Strong Feller and Irreducibility Properties

Strong Feller and irreducibility properties are frequently utilized to ensure the uniqueness of the invariant measure, as well as ergodicity. More details about the sufficient conditions of strong Feller and irreducibility properties will be given in the following chapter.

Definition 1.8 *(see [58, 60])* Let $X(t)$ be the solution to (1.2) with the probability kernel $\pi_t(\cdot, \cdot)$.

(i) $X(t)$ is *strong Feller* if $\int_{\mathbb{H}} \varphi(y)\pi_t(\cdot, dy) \in \mathbf{C}_b(\mathbb{H})$ for any $\varphi \in \mathbf{B}_b(\mathbb{H})$.
(ii) $X(t)$ is *irreducible* if $\pi_t(x, B(x_0, r)) > 0$ for all $x_0, x \in \mathbb{H}, r > 0$ and any $t > 0$, where $B(x_0, r)$ denotes the ball centred at x_0 with radius r.

In other words, the strong Feller property implies that $\pi_t(x, \cdot)$ is continuous in $x \in \mathbb{H}$, while the irreducibility property means that any open sets are reachable with positive probability by the solution $X_x(t)$ started at any $x \in \mathbb{H}$. These two properties yield the following properties of μ, whose proof is based on [58, 60].

Theorem 1.5 *If X is strong Feller and irreducible, and possesses an invariant measure μ, then*

(i) *μ is equivalent to the probability kernel $\pi_t(x, \cdot)$ for any $t > 0$ and all $x \in \mathbb{H}$.*
(ii) *μ is an ergodic measure of $X(t)$.*
(iii) *μ is the unique invariant measure of $X(t)$.*

Here, two measures μ and ν are called equivalent if each is absolutely continuous with respect to the other, i.e., $\mu \ll \nu$ and $\nu \ll \mu$. More precisely, for every $G \in \mathscr{B}(\mathbb{H})$, $\mu(G) = (>)0 \Leftrightarrow \nu(G) = (>)0$.

Proof (i) **Step 1**. We first prove that for $t > 0$, $\pi_t(x, \cdot)$ are equivalent for all $x \in \mathbb{H}$. To this end, it suffices to show that for any fixed $x \in \mathbb{H}$, if $G \in \mathscr{B}(\mathbb{H})$ such that $\pi_t(x, G) > 0$, then $\pi_t(y, G) > 0$ for all $y \in \mathbb{H}$. In fact, as

$$\pi_t(x, G) = \int_{\mathbb{H}} \pi_s(y, G)\pi_{t-s}(x, dy) > 0$$

for any $s \in (0, t)$ based on the Chapman–Kolmogorov equation of Markov processes, there exists some $y_0 \in \mathbb{H}$ such that $\pi_s(y_0, G) > 0$. It follows from the strong Feller property that $\pi_s(y, \cdot)$ is continuous in y. Thus, there exists $r > 0$ such that $\pi_s(w, G) > 0$ for all $w \in B(y_0, r)$, and also, $\pi_t(y, B(y_0, r)) > 0$ for any $t > 0$ and $y \in \mathbb{H}$ based on the irreducibility property. Hence, it shows

$$\pi_t(y, G) = \int_{\mathbb{H}} \pi_s(w, G)\pi_{t-s}(y, dw) \geq \int_{B(y_0, r)} \pi_s(w, G)\pi_{t-s}(y, dw) > 0.$$

Step 2. μ is equivalent to $\pi_t(x, \cdot)$ for any $t > 0$ and $x \in \mathbb{H}$. Firstly, if $G \in \mathscr{B}(\mathbb{H})$ such that $\pi_t(x, G) = 0$ for some $x \in \mathbb{H}$, then we have $\pi_t(y, G) = 0$ for all $y \in \mathbb{H}$ based on Step 1. It yields $\mu \ll \pi_t(x, \cdot)$ as

$$\mu(G) = \int_{\mathbb{H}} \pi_t(y, G)\mu(dy) = 0$$

based on (1.6). On the other hand, if $\mu(G) = 0$, then there must exist some $v_0 \in \mathbb{H}$ such that $\pi_t(v_0, G) = 0$. We hence get $\pi_t(x, \cdot) \ll \mu$ for all $x \in \mathbb{H}$ based on Step 1.

(ii) μ is ergodic. Let $G \in \mathscr{B}(\mathbb{H})$ be an invariant set of $X(t)$ with $\mu(G) > 0$, whose definition is given in Definition 1.7. Thus,

$$\mathbb{P}(X_x(t) \in G) = \pi_t(x, G) > 0, \quad \forall x \in \mathbb{H}$$

according to the fact $\pi_t(x, G)$ is equivalent to μ for all $x \in \mathbb{H}$ proved in (i). Since G is an invariant set, we derive

$$\mathbf{1}_G(x) = \mathbb{P}(X_x(t) \in G) > 0, \quad \mu\text{-a.e.},$$

which implies $\mu(G) = 1$. It indicates that any invariant set is trivial, so μ is ergodic.

(iii) μ is unique. Assume by contradiction that there is another invariant measure v for $X(t)$ which also satisfies the assumptions in this theorem, then v is equivalent to μ and is also ergodic. Let $G \in \mathscr{B}(\mathbb{H})$ be such that $\mu(G) \neq v(G)$. We set

$$E := \left\{ x \in \mathbb{H} \,\middle|\, \lim_{T \to \infty} \frac{1}{T} \int_0^T \mathbb{E}[\mathbf{1}_G(X_x(t))]dt = \mu(G) \right\},$$

$$F := \left\{ x \in \mathbb{H} \,\middle|\, \lim_{T \to \infty} \frac{1}{T} \int_0^T \mathbb{E}[\mathbf{1}_G(X_x(t))]dt = v(G) \right\}.$$

Then based on the ergodicity of μ and v with $\varphi = \mathbf{1}_G$ in Definition 1.6, we have $\mu(E) = 1$, $v(F) = 1$ and $E \cap F = \varnothing$. It then yields that $F \subset E^c$, which leads to $\mu(F) = \mu(E^c) = 0$, which is in contradiction to the fact that μ and v are equivalent. $\qquad\square$

Above conditions and properties can also be defined for a discrete Markov chain $\{X_n\}_{n \in \mathbb{N}}$:

(i) $\{X_n\}_{n \in \mathbb{N}}$ is *strong Feller* if $\int_{\mathbb{H}} \varphi(y)\pi_n(\cdot, dy) \in \mathbf{C}_b(\mathbb{H})$ for any $\varphi \in \mathbf{B}_b(\mathbb{H})$.
(ii) $\{X_n\}_{n \in \mathbb{N}}$ is *irreducible* if $\pi_n(x, B(x_0, r)) > 0$ for all $x, x_0 \in \mathbb{H}$, $r > 0$ and any $n \in \mathbb{N}_+$.

In particular, if $\mathbb{H}$ is a finite dimensional Banach space with norm $\| \cdot \|_{\mathbb{H}}$, another kind of sufficient condition for ergodicity is stated as follows.

Corollary 1.2 (Theorem 5.5, [182]) *Assume that the homogeneous Markov chain $\{X_n\}_{n \in \mathbb{N}}$ is strong Feller and irreducible. In addition, there exists a compact set $G_0 \in \mathscr{B}(\mathbb{H})$ with positive Legesgue measure $m(G_0) > 0$ such that*

$$\mathbb{E}\big[\|X_{n+1}\|_{\mathbb{H}} - \|X_n\|_{\mathbb{H}} \mid X_n = x\big] \leq \begin{cases} -C_1, & \forall\, x \notin G_0, \\ C_2 < \infty, & \forall\, x \in G_0 \end{cases} \tag{1.9}$$

for some positive constants C_1 and C_2. Then $\{X_n\}_{n\in\mathbb{N}}$ is ergodic.

It is easy to check that condition (1.9) is a sufficient condition for the Lyapunov condition in Theorem 1.3, and thus this corollary also leads to the results given in Theorem 1.5. In fact, for $X_n = x \notin G_0$, we have $\mathbb{E}[\|X_{n+1}\|_{\mathbb{H}}] \leq \mathbb{E}[\|X_n\|_{\mathbb{H}}]$, while for $X_n = x \in G_0$, we get

$$\mathbb{E}[\|X_{n+1}\|_{\mathbb{H}}] \leq \alpha \mathbb{E}[\|X_n\|_{\mathbb{H}}] + (1-\alpha)\mathbb{E}[\|X_n\|_{\mathbb{H}}] + C_2 \leq \alpha \mathbb{E}[\|X_n\|_{\mathbb{H}}] + (1-\alpha)C + C_2$$

for some $\alpha \in (0,1)$, since G_0 is bounded. The desired result is then obtained by recurrence.

1.5 Invariant Measures for Hamiltonian Systems

A class of specific stochastic dynamical systems—stochastic Hamiltonian systems—plays an important role in stochastic dynamics, whose geometric structure has been investigated by many authors. For instance, stochastic Hamiltonian systems with multiplicative noises and additive noises are considered in [145] and [146], respectively. The symplectic structure of the systems, as well as numerical methods with the same property, are studied in these papers. The author in [149] studies symplectic methods obtained by composition of stochastic flows of simpler Hamiltonian systems. The author in [134] shows that an averaging principle holds for a completely integrable stochastic Hamiltonian system. Authors in [119] propose the stochastic multi-symplectic conservation law for stochastic Hamiltonian partial differential equations, and develop a stochastic multi-symplectic method for the stochastic nonlinear Schrödinger equation driven by a linear multiplicative noise. The ergodic theory of Hamiltonian systems could be tracked back to the study of statistical ensembles, i.e., a collection of stationary distributions on the phase space for a given deterministic Hamiltonian system, which is also called 'monode' by Boltzmann (see e.g. [23, 85]).

For classical mechanical systems, the Lagrangian is denoted by

$$L(q, \dot{q}) = T(q, \dot{q}) + U(q)$$

with $q = (q_1, \cdots, q_d)^\top \in \mathbb{R}^d$ being the generalized coordinate, T and U being the kinetic energy and potential energy, respectively. According to the Hamilton's principle, one seeks the extremal q such that the action functional

$$\int_0^T L(q, \dot{q})\, dt$$

is stationary under variation with $\delta q(0) = \delta q(T) = 0$. Hence, L and q obey the Lagrange equation

$$\frac{d}{dt}\left(\frac{\partial L}{\partial \dot{q}}\right) = \frac{\partial L}{\partial q}.$$

By introducing the generalized momentum, which is also known as the Poisson's variables, given by the Legendre transform

$$p = \frac{\partial L}{\partial \dot{q}}(q, \dot{q}) \tag{1.10}$$

and the Hamiltonian

$$H(p, q) := p^{\mathsf{T}}\dot{q} - L(q, \dot{q}),$$

one get the following equations

$$\dot{p} = -\frac{\partial H}{\partial q}(p, q), \quad \dot{q} = \frac{\partial H}{\partial p}(p, q),$$

which constitute the Hamiltonian systems. Equivalently, we have

$$\dot{X} = J^{-1}\nabla H(X)$$

by denoting $X = (p^{\mathsf{T}}, q^{\mathsf{T}})^{\mathsf{T}} \in \mathbb{R}^{2d}$ and the $2d$-dimensional standard symplectic matrix

$$J = \begin{pmatrix} 0 & I_d \\ -I_d & 0 \end{pmatrix}_{2d \times 2d}.$$

Here, I_d denotes the d-dimensional identity matrix, and the following condition is required that (1.10) defines a continuously differentiable bijection $p \leftrightarrow \dot{q}$ for any q, which is called the Legendre transform.

For an autonomous Hamiltonian system with $\mathbb{M}$ being the phase space, its phase flow $\phi_t : x \to X_x(t)$ preserves phase volume

$$\int_{\mathbb{M}} f(\phi_t(x))dS = \int_{\mathbb{M}} f\,dS,$$

where dS denotes the phase volume element of $\mathbb{M}$. This identity is also referred to as Liouville's theorem and ϕ_t is called an equi-measure transformation (see e.g. [190]). Then the ergodic hypothesis of Boltzmann that the temporal average of a physical quantity should be equal to the spatial average of this physical quantity can be expressed by

$$\lim_{T \to \infty} \frac{1}{T} \int_0^T f(\phi_t(x))dt = \frac{\int_{\mathbb{M}} f\,dS}{\int_{\mathbb{M}} dS}$$

with the assumption that the total volume of the system is finite $\int_M dS < \infty$. A generalization of above argument to the Markov process is the condition of the existence of invariant measures for the considered systems.

Hence, the existence and ergodicity of invariant measures for stochastic Hamiltonian systems (see e.g. [179]), as well as its numerical approximations, are also of vital importance and remain to be further investigated. It motivates us to consider both geometric structure and dynamical behavior of stochastic nonlinear Schrödinger equations over longtime numerically introduced in Chaps. 5 and 6.

In this section, we investigate the invariant measures, geometric structures and symplectic integrators for two simple Hamiltonian systems–stochastic Kubo oscillator and stochastic dissipative Hamiltonian systems.

1.5.1 Stochastic Kubo Oscillator

In this subsection, we denote $\mathbb{H} = \mathbb{R}^{2d}$, and investigate the geometric structure as well as existence of invariant measures for the Kubo oscillator. Consider the following $\mathbb{R}^{2d}$-valued stochastic differential equation in the Stratonovich sense

$$\begin{cases} dp = - q dt - q \circ dB(t), & p(0) = p_0, \\ dq = p dt + p \circ dB(t), & q(0) = q_0 \end{cases} \tag{1.11}$$

with deterministic initial value $(p_0^\top, q_0^\top)^\top \in \mathbb{R}^{2d}$ and B being a one-dimensional $\mathbb{R}$-valued standard Brownian motion.

Denoting $X = (p^\top, q^\top)^\top = (p^1, \cdots, p^d, q^1, \cdots, q^d)^\top \in \mathbb{R}^{2d}$ and

$$H(X) = H(p, q) := \frac{1}{2}(\|p\|^2 + \|q\|^2)$$

with $\|\cdot\|$ denoting the Euclidean norm in the corresponding Euclidean spaces, we can rewrite (1.11) into a compact form

$$dX = J^{-1}\nabla H(X)dt + J^{-1}\nabla H(X) \circ dB(t) \tag{1.12}$$

with $x := X(0) = (p_0^\top, q_0^\top)^\top$. Here J denotes the $2d$-dimensional standard symplectic matrix.

Energy conservation law. Equations of the form (1.12) are called stochastic Hamiltonian equations. We can find out that the Hamiltonian $H(p, q)$ of this system is a conserved quantity (or invariant or first integral). Actually, the solution of (1.11) can be expressed explicitly

$$\begin{cases} p(t) = - \sin(t + B(t))q_0 + \cos(t + B(t))p_0, \\ q(t) = \cos(t + B(t))q_0 + \sin(t + B(t))p_0. \end{cases}$$

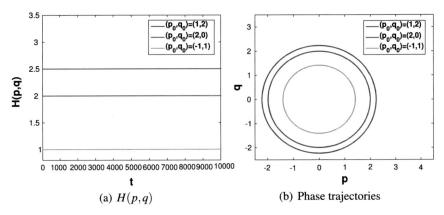

Fig. 1.1 a Hamiltonian $H(p,q)$ and **b** phase trajectories starting from different initial values $(d = 1, T = 10^4)$

Its Hamiltonian, as well as its phase trajectory, can be simulated directly as in Fig. 1.1.

Moreover, if the drift and diffusion coefficients involve the same Hamiltonian function, it always holds that the Hamiltonian function is a first integral. More precisely, for any $t \geq 0$, the solution $X_x(t)$ of (1.12) with $H(x) = c_0$ lies on the isoenergetic surface

$$\Sigma_{c_0} := \{X \in \mathbb{H} \mid H(X) = c_0\}$$

almost surely, i.e.,

$$H(p(t), q(t)) = H(p_0, q_0) \quad \mathbb{P}\text{-a.s.}$$

In fact, based on the Stratonovich chain rule, we get

$$
\begin{aligned}
dH(X(t)) &= \nabla H(X(t))^\top \circ dX(t) \\
&= \nabla H(X(t))^\top J^{-1} \nabla H(X(t)) dt + \nabla H(X(t))^\top J^{-1} \nabla H(X(t)) \circ dB(t) \\
&= 0
\end{aligned}
$$

since J^{-1} is skew symmetric.

In the sequel, we omit the notation '$\mathbb{P}$-a.s.' or 'almost surely' unless it is necessary.

Symplectic conservation law. The phase flow $\phi_t : x \to X$ of Eq. (1.12) is linear, and possesses stochastic symplectic structure

$$\left(\frac{\partial X}{\partial x}\right)^\top J \left(\frac{\partial X}{\partial x}\right) = J,$$

or equivalently,

$$\mathrm{d}p(t) \wedge \mathrm{d}q(t) := \sum_{i=1}^{d} \mathrm{d}p^i(t) \wedge \mathrm{d}q^i(t) = \mathrm{d}p_0 \wedge \mathrm{d}q_0$$

with 'd' denoting the exterior derivative and $\mathrm{d}p \wedge \mathrm{d}q$ being a differential 2-form.

We introduce here the geometric description of symplectic transformations for Hamiltonian systems. The reader is referred to [10, 82, 95] and references therein for more details.

The 2-form $\mathrm{d}p(t) \wedge \mathrm{d}q(t)$ on $\mathbb{R}^{2d}$ are expressed as the sum of d 2-forms $\mathrm{d}p^i(t) \wedge \mathrm{d}q^i(t)$, $i = 1, \cdots, d$, on $\mathbb{R}^2$. For the two dimensional case, i.e. $d = 1$, the preservation of the symplectic structure is equivalent to the preservation of the phase area.

In fact, we set two vectors $\xi = (\xi_1, \xi_2)^\top, \eta = (\eta_1, \eta_2)^\top \in \mathbb{R}^2$, which span a parallelogram with oriented area

$$or.area(\xi, \eta) := \det \begin{pmatrix} \xi_1 & \eta_1 \\ \xi_2 & \eta_2 \end{pmatrix} = \xi_1 \eta_2 - \xi_2 \eta_1 = \xi^\top J \eta.$$

The area of the parallelogram under the projection of the symplectic mapping A satisfying $A^\top J A = J$ turns to be

$$or.area(Aa, Ab) = (Aa)^\top J (Ab) = a^\top (A^\top J A) b = or.area(a, b),$$

which indicates the area preservation property of A. One can find from Fig. 1.2 that the area keeps invariant under the linear flow ϕ_t of (1.12).

Symplectic integrators. A natural question in application is to construct numerical integrators which could also preserve the stochastic symplectic structure. We next introduce two equivalent ways of proving the symplecticity of numerical integrators.

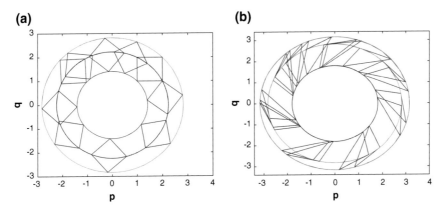

Fig. 1.2 Images under the phase flow with initial image being (**a**) a square determined by $(1, 1), (1, 2), (2, 2)$ and $(2, 1)$ or (**b**) a triangle determined by $(1, 1.5), (2, 2)$ and $(3, 1)$ $(d = 1,$ $T = 10)$

Definition 1.9 ([95]) A numerical one-step method is called symplectic if the one-step map $Y_1 = \phi_n(Y_0)$ is symplectic whenever the method is applied to a smooth Hamiltonian system.

As an example, we consider the midpoint scheme applied to (1.12)

$$X_{n+1} = X_n + J^{-1}\nabla H\left(\frac{X_{n+1} + X_n}{2}\right)(\tau + \triangle_{n+1}B),$$

where τ denotes the uniform time step-size and $\triangle_{n+1}B = B(t_{n+1}) - B(t_n)$ is the increment of the Brownian motion B with $t_n = n\tau$ and $n \in \mathbb{N}$.

Denote $f = J^{-1}\nabla H$, and

$$Y_{n+1} := \frac{X_{n+1} + X_n}{2} = X_n + \frac{1}{2}f(Y_{n+1})(\tau + \triangle_{n+1}B).$$

By noting that

$$\frac{\partial X_{n+1}}{\partial X_n} = I_{2d} + \frac{\partial f(Y_{n+1})}{\partial X_n}(\tau + \triangle_{n+1}B)$$

and

$$\frac{\partial Y_{n+1}}{\partial X_n} = I_{2d} + \frac{1}{2}\frac{\partial f(Y_{n+1})}{\partial X_n}(\tau + \triangle_{n+1}B),$$

we conclude the symplecticity of the midpoint scheme:

$$\left(\frac{\partial X_{n+1}}{\partial X_n}\right)^{\top} J\left(\frac{\partial X_{n+1}}{\partial X_n}\right)$$

$$= J + J\left(\frac{\partial f(Y_{n+1})}{\partial X_n}\right)(\tau + \triangle_{n+1}B) + \left(\frac{\partial f(Y_{n+1})}{\partial X_n}\right)^{\top} J(\tau + \triangle_{n+1}B)$$

$$+ \left(\frac{\partial f(Y_{n+1})}{\partial X_n}\right)^{\top} J\left(\frac{\partial f(Y_{n+1})}{\partial X_n}\right)(\tau + \triangle_{n+1}B)^2$$

$$= J + \left[\left(\frac{\partial Y_{n+1}}{\partial X_n} - \frac{1}{2}\frac{\partial f(Y_{n+1})}{\partial X_n}(\tau + \triangle_{n+1}B)\right)^{\top} J\left(\frac{\partial f(Y_{n+1})}{\partial X_n}\right)\right](\tau + \triangle_{n+1}B)$$

$$+ \left[\left(\frac{\partial f(Y_{n+1})}{\partial X_n}\right)^{\top} J\left(\frac{\partial Y_{n+1}}{\partial X_n} - \frac{1}{2}\frac{\partial f(Y_{n+1})}{\partial X_n}(\tau + \triangle_{n+1}B)\right)\right](\tau + \triangle_{n+1}B)$$

$$+ \left(\frac{\partial f(Y_{n+1})}{\partial X_n}\right)^{\top} J\left(\frac{\partial f(Y_{n+1})}{\partial X_n}\right)(\tau + \triangle_{n+1}B)^2$$

$$= J + \left[\left(\frac{\partial Y_{n+1}}{\partial X_n} \right)^\top J \left(\frac{\partial f(Y_{n+1})}{\partial X_n} \right) + \left(\frac{\partial f(Y_{n+1})}{\partial X_n} \right)^\top J \left(\frac{\partial Y_{n+1}}{\partial X_n} \right) \right] (\tau + \Delta_{n+1} B)$$

$$= J + \left[\left(\frac{\partial Y_{n+1}}{\partial X_n} \right)^\top \left(\nabla^2 H(Y_{n+1}) + \nabla^2 H(Y_{n+1}) J^{-\top} J \right) \left(\frac{\partial Y_{n+1}}{\partial X_n} \right) \right] (\tau + \Delta_{n+1} B)$$

$$= J,$$

where in the last step we used the fact $J^{-\top} J = -I_{2d}$.

Equivalently, the symplecticity of integrators can also be proved utilizing differential 2-forms by rewriting the midpoint scheme above as

$$\begin{cases} p_{n+1} = p_n - H_q \tilde{\Delta}_{n+1}, \\ q_{n+1} = q_n + H_p \tilde{\Delta}_{n+1}, \end{cases}$$

where $\tilde{\Delta}_{n+1} := \tau + \Delta_{n+1} B$, and $H_p := \partial_p H$ and $H_q := \partial_q H$ are evaluated at $(\frac{p_{n+1}+p_n}{2}, \frac{q_{n+1}+q_n}{2})$. Then the total differentials of p_{n+1} and q_{n+1} read

$$\begin{cases} dp_{n+1} = dp_n - \left(H_{qp} \dfrac{dp_{n+1} + dp_n}{2} + H_{qq} \dfrac{dq_{n+1} + dq_n}{2} \right) \tilde{\Delta}_{n+1}, \\ dq_{n+1} = dq_n + \left(H_{pp} \dfrac{dp_{n+1} + dp_n}{2} + H_{pq} \dfrac{dq_{n+1} + dq_n}{2} \right) \tilde{\Delta}_{n+1}. \end{cases}$$

Note that $H_{pp} = H_{qq} = 1$ and $H_{pq} = H_{qp} = 0$ for the stochastic Kubo oscillator. It then shows immediately

$$\left(-\frac{\tilde{\Delta}_{n+1}}{2} dp_{n+1} + dq_{n+1} \right) \wedge \left(dp_{n+1} + \frac{\tilde{\Delta}_{n+1}}{2} dq_{n+1} \right)$$

$$= \left(\frac{\tilde{\Delta}_{n+1}}{2} dp_n + dq_n \right) \wedge \left(dp_n - \frac{\tilde{\Delta}_{n+1}}{2} dq_n \right),$$

which indicates that

$$dp_{n+1} \wedge dq_{n+1} = dp_n \wedge dq_n, \quad \forall n \in \mathbb{N}$$

since $1 + \frac{\tilde{\Delta}_{n+1}^2}{4} > 0$ for any $n \in \mathbb{N}$.

Invariant measures. The existence of invariant measures for (1.11) can be proved based on the Krylov–Bogoliubov Theorem in Sect. 1.2. The invariant measure for (1.11) is not unique in the whole space $\mathbb{R}^{2d}$ since solutions starting from initial values with different norms lie in different spheres.

Theorem 1.6 *The Kubo oscillator (1.11) possesses invariant measures.*

Proof Fix some initial value $x \in \mathbb{R}^{2d}$ which satisfies that $\|x\| = c_0$. We consider the transition probability $\pi_t(x, \cdot) \in \mathscr{P}(\mathbb{R}^{2d})$. We claim that the subset $\{\mu_T\}_{T>0} \subset \mathscr{P}(\mathbb{R}^{2d})$ defined through

$$\mu_T(G) = \frac{1}{T} \int_0^T \pi_t(x, G)dt$$

is tight if the set $\{\pi_t(x, \cdot)\}_{t>0}$ is tight. In fact, if $\{\pi_t(x, \cdot)\}_{t>0}$ is tight, then for any $\varepsilon > 0$, there exists a compact set K_ε such that

$$\mu_T(K_\varepsilon) = \frac{1}{T} \int_0^T \pi_t(x, K_\varepsilon)dt \geq \frac{1}{T} \int_0^T (1 - \varepsilon)dt = 1 - \varepsilon,$$

which verifies the claim. The tightness of $\{\mu_T\}_{T>0}$ ensures that (1.11) possesses an invariant measure according to Theorem 1.2.

It then suffices to show the tightness of $\{\pi_t(x, \cdot)\}_{t>0}$. Note that the isoenergetic surface $\Sigma_{c_0} \subset \mathbb{R}^{2d}$ is tight, then for any $\varepsilon > 0$,

$$\pi_t(x, \Sigma_{c_0}) = \mathbb{P}(X_x(t) \in \Sigma_{c_0}) = 1 > 1 - \varepsilon,$$

which completes the proof.   □

1.5.2 Stochastic Dissipative Hamiltonian Systems

In this section, we aim to introduce an fundamental procedure of showing the existence and uniqueness of invariant measures for stochastic differential equations with non-global Lipschitz drift coefficients and degenerate noises. Hence, we turn to consider stochastic Hamiltonian systems with dissipative terms of the type

$$\begin{cases} dp = - H_q(p, q)dt - F(p, q)H_p(p, q)dt + dW(t), & p(0) = p_0, \\ dq = H_p(p, q)dt, & q(0) = q_0 \end{cases}$$

with $p, q \in \mathbb{R}^d$ and W being a d-dimensional standard Brownian motion.

Here we only focus on the one-dimensional case with $F(p, q) \equiv \alpha > 0$ and $H(p, q) = \frac{1}{2}p^2 + \frac{1}{4}q^4$, i.e.,

$$\begin{cases} dp = (-q^3 - \alpha p)dt + dW(t), & p(0) = p_0 \in \mathbb{R}, \\ dq = pdt, & q(0) = q_0 \in \mathbb{R} \end{cases} \tag{1.13}$$

to make everything clear. We refer to [179] for the general case with certain assumptions on F and H.

Geometric structure. For the system (1.13) with dissipative term, the symplectic conservation law is not satisfied anymore. Instead, its differential 2-form $dp(t) \wedge dq(t)$ decays exponentially.

Theorem 1.7 *The phase flow of (1.13) satisfies the conformal symplectic conservation law, that is,*

$$dp(t) \wedge dq(t) = e^{-\alpha t} dp_0 \wedge dq_0, \quad \forall\, t \geq 0.$$

Proof By calculating

$$
\begin{aligned}
d\,(dp(t) \wedge dq(t)) &= d[dp(t)] \wedge dq(t) + dp(t) \wedge d[dq(t)] \\
&= d[(-H_q(p,q) - \alpha p)dt + dW(t)] \wedge dq(t) + dp(t) \wedge d[p(t)dt] \\
&= [(-H_{qq}(p,q)dq(t) - \alpha dp(t)) \wedge dq(t) + dp(t) \wedge dp(t)]dt \\
&= -\alpha[dp(t) \wedge dq(t)]dt,
\end{aligned}
$$

we get the result immediately, where we have used the fact that

$$A\,dp \wedge dp = 0$$

for any symmetric matrices $A \in \mathbb{R}^{d \times d}$ and $p \in \mathbb{R}^d$. $\square$

Invariant measures. To show the existence of invariant measures for a Markov process utilizing the Krylov–Bogoliubov theorem, the process need to be Feller in addition. That is, the solution of the considered model continuously depends on the initial data. This property is ensured by the local Lipschitz coefficients and the uniform boundedness of the solution stated below following the idea of Talay [179].

Lemma 1.1 *If the initial value $(p_0, q_0) \in \mathbb{R}^2$ has finite moments of all order. Then the moments of all order for the solution of (1.13) are bounded uniformly in time, i.e., for any $k \in \mathbb{N}$, there exists a constant $C = C(k, \alpha, p_0, q_0)$ such that*

$$\sup_{t \geq 0} \mathbb{E}\left[\left(|p(t)|^2 + |q(t)|^2\right)^k\right] < C.$$

Proof We first set the Lyapunov functional V as

$$
\begin{aligned}
V(p,q) &:= H(p,q) + \frac{\alpha}{2} pq + \left(\frac{\alpha^2 + 1}{4}\right)^2 \\
&= \frac{1}{4}\left(p^2 + q^2\right) + \left(\frac{1}{2}p + \frac{\alpha}{2}q\right)^2 + \left(\frac{1}{2}q^2 - \frac{\alpha^2 + 1}{4}\right)^2 \\
&\geq \frac{1}{4}\left(p^2 + q^2\right).
\end{aligned}
$$

Then Itô's formula applied to $V(p(t), q(t))$ yields that

$$dV(p(t), q(t)) = V_p(p(t), q(t))dp(t) + V_q(p(t), q(t))dq(t) + \frac{1}{2}V_{pp}dt$$

$$= \left(-\frac{\alpha}{2}p(t)^2 - \frac{\alpha}{2}q(t)^4 - \frac{\alpha^2}{2}p(t)q(t) + \frac{1}{2} \right) dt$$

$$+ \left(p(t) + \frac{\alpha}{2}q(t) \right) dW(t).$$

Taking expectation to both sides of above equation, we obtain

$$d\mathbb{E}[V(p(t), q(t))] = \left[-\alpha\mathbb{E}[V(p(t), q(t))] - \frac{\alpha}{4}q(t)^4 + \alpha\left(\frac{\alpha^2+1}{4} \right)^2 + \frac{1}{2} \right] dt$$

$$\leq -\alpha\mathbb{E}[V(p(t), q(t))]dt + C(\alpha)dt,$$

which indicates that

$$\mathbb{E}[V(p(t), q(t))] \leq e^{-\alpha t}\mathbb{E}[V(p_0, q_0)] + C(\alpha) \int_0^t e^{-\alpha(t-s)}ds$$

$$\leq \mathbb{E}[V(p_0, q_0)] + C(\alpha).$$

This give the result for $k = 1$. For higher moments, assume that

$$\sup_{t \geq 0} \mathbb{E}\left[V^k(p(t), q(t)) \right] \leq C.$$

Then for $k + 1$, we have

$$d\mathbb{E}\left[V^{k+1}(p(t), q(t)) \right]$$

$$= (k+1)\mathbb{E}\left[V^k(p(t), q(t)) \left(-\frac{\alpha}{2}p(t)^2 - \frac{\alpha}{2}q(t)^4 - \frac{\alpha^2}{2}p(t)q(t) + \frac{1}{2} \right) \right] dt$$

$$+ \frac{1}{2}(k+1)k\mathbb{E}\left[V^{k-1}(p(t), q(t)) \left(p(t) + \frac{\alpha}{2}q(t) \right)^2 \right] dt$$

$$\leq -\alpha(k+1)\mathbb{E}\left[V^{k+1}(p(t), q(t)) \right] dt + Cdt,$$

where we have used the fact

$$\left(p + \frac{\alpha}{2}q \right)^2 \leq 2p^2 + \frac{\alpha^2}{2}q^2 \leq 4\left(2 \vee \frac{\alpha^2}{2} \right) V(p, q)$$

in the last step. The proof is complete by induction.   □

Moreover, the lemma above also ensures the existence of invariant measures for (1.13) according to Theorem 1.3. To gain the uniqueness of the invariant measure, we next show that all the invariant measures for the solution of (1.13) admit strictly positive densities with respect to the Lebesgue measure.

Lemma 1.2 (Lemma 2.2, [179]) *For any stochastic initial value $X_0 := (p_0, q_0) \in \mathbb{R}^2$ and $t > 0$, the distribution of the solution $X(t) := (p(t), q(t))$ has a strictly positive density ρ_t with respect to the Lebesgue measure.*

Proof For any stochastic initial value X_0 with initial probability distribution ν_0, the distribution of the solution $X(t)$ reads

$$\pi_t^{\nu_0}(A) = \int_{\mathbb{R}^2} \pi_t(x, A)\nu_0(dx), \quad \forall\, A \in \mathscr{B}(\mathbb{R}^2).$$

As a result, we only need to show that for any deterministic initial value $x := (p_0, q_0)$, the transition probability $\pi_t(x, \cdot)$ admits a positive density $\tilde{\rho}_t(x, \cdot) \in \mathbf{L}^1(\mathbb{R}^2)$ for any $t > 0$ which ensures that $\pi_t^{\nu_0}(\cdot)$ also has a positive density

$$\rho_t(y) = \int_{\mathbb{R}^2} \tilde{\rho}_t(x, y)\nu_0(dx)$$

since

$$\begin{aligned}
\pi_t^{\nu_0}(A) &= \int_{\mathbb{R}^2} \int_A \tilde{\rho}_t(x, y)dy\nu_0(dx) \\
&= \int_A \left(\int_{\mathbb{R}^2} \tilde{\rho}_t(x, y)\nu_0(dx) \right) dy, \quad \forall\, A \in \mathscr{B}(\mathbb{R}^2).
\end{aligned}$$

One can check that the span of the vector fields and their Lie brackets of system (1.13) is equal to the whole space, which is known as the Hörmander condition. It ensures the existence of the jointly continuous density $\tilde{\rho}_t(x, \cdot)$ for the law $\pi_t(x, \cdot)$ according to Theorem 2.2. This result will be discussed in detail in Chap. 2, so we omit it here. We only show the positivity of $\tilde{\rho}_t(x, \cdot)$.

Following the argument used in [139], we consider the associated control problem

$$\begin{cases} d\tilde{p} = (-\tilde{q}^3 - \alpha\tilde{p})dt + du(t), \\ d\tilde{q} = \tilde{p}dt \end{cases}$$

with a smooth control function $u \in C^1(0, T)$, which has the following equivalent form

$$\frac{d^2\tilde{q}}{dt^2} + \alpha\frac{d\tilde{q}}{dt} + \tilde{q}^3 = \frac{du}{dt}.$$

For any fixed $T > 0$ and any vectors $x_0 = (p_0, q_0)$, $x^+ = (p^+, q^+) \in \mathbb{R}^2$, we construct $\tilde{q} \in \mathbf{C}^\infty(0, T)$ such that

$$\left(\tilde{q}(0), \frac{d\tilde{q}}{dt}(0) \right) = (q^-, p^-) \quad \text{and} \quad \left(\tilde{q}(T), \frac{d\tilde{q}}{dt}(T) \right) = (q^+, p^+)$$

using polynomial interpolation. Hence we can also get the control function $u \in$ $C^\infty(0, T)$ with $u(0) = 0$.

Denoting $X := (p, q)$, $\tilde{X} := (\tilde{p}, \tilde{q})$,

$$F(X) := \begin{pmatrix} p \\ -q^3 - \alpha p \end{pmatrix} \quad \text{and} \quad \Gamma := \begin{pmatrix} 0 & 0 \\ 0 & 1 \end{pmatrix}. \tag{1.14}$$

We achieve that

$$X(t) - \tilde{X}(t) = \int_0^t F(X(s)) - F(\tilde{X}(s))ds + \Gamma(W(t) - u(t)), \quad \forall\, t \in [0, T].$$

Note that the following property holds for Brownian motions

$$\mathbb{P}\left(\sup_{0 \le t \le T} \left| W(t) - u(t) \right| < \varepsilon \right) > 0, \quad \forall\, \varepsilon > 0.$$

If the event $\{\omega \in \Omega : \sup_{0 \le t \le T} \left| W(t) - u(t) \right| < \varepsilon\}$ happens, then the Gronwall inequality in a small time interval and its continuation yield that for some $\delta(\varepsilon)$,

$$\sup_{0 \le t \le T} \left| X(t) - \tilde{X}(t) \right| < \delta(\varepsilon)$$

according to the facts that F is continuously differentiable and thus locally Lipschitz, and that the ranges of $X(t)$ and $\tilde{X}(t)$ ($t \in [0, T]$) are both compact sets.

As a result, for any $\delta > 0$, by choosing $\varepsilon > 0$ small enough, we finally obtain

$$\mathbb{P}\left(\left| X(T) - x^+ \right| < \delta \right) = \pi_T(x_0, B(x^+, \delta)) > 0,$$

where $\tilde{X}(T) = x^+$ and $B(x^+, \delta)$ denotes the open ball centered at x^+ with radius δ. $\square$

The proof of Lemma 1.2 also shows that X is strong Feller and irreducible, based on which one can get the uniqueness of the invariant measure.

Theorem 1.8 *There exists a unique invariant measure for the solution of (1.13), which admits a strictly positive density with respect to the Lebesgue measure.*

Proof For an invariant measure μ, choosing initial values which satisfy the distribution μ, then the solution X at any time $t \ge 0$ admits the same law since

$$\mu(A) = \int_{\mathbb{R}^2} \mathbf{1}_A(x)\mu(dx) = \int_{\mathbb{R}^2} P_t \mathbf{1}_A(x)\mu(dx)$$

$$= \int_{\mathbb{R}^2} \mathbf{1}_A(X_x(t))\mu(dx) = \mathbb{P}(X(t) \in A), \quad \forall\, A \in \mathscr{B}(\mathbb{R}^2),\ t \ge 0.$$

It is then deduced from the lemma above that μ has a strictly positive density with respect to the Lebesgue measure.

Based on the proof of Lemma 1.2, X is strong Feller and irreducible which ensures that μ is the unique invariant measure and is ergodic according to Theorem 1.5. □

Summary

This chapter gives the definitions of invariant measures and ergodicity for stochastic processes, as well as several sufficient conditions for the existence and uniqueness of invariant measures, which provide fundamental tools of studying the ergodicity of differential equations and their numerical approximations.

For a deterministic Hamiltonian system, it is well known that its Hamiltonian and symplectic structure are conserved by the flow of the system. As a result, the solution lies in the isoenergetic surface, which indicates that the invariant measure of the system will not be unique. It is also the case for the stochastic Kubo oscillator driven by standard Brownian motions introduced in Sect. 1.5.2. Moreover, even though symplectic schemes possess the discrete symplectic conservation law and perform better than non-symplectic schemes over long time, they could not preserve the Hamiltonian in general. Hence, the ergodicity for Hamiltonian systems as well as their numerical approximations remains unclear.

By bringing in a dissipative term, [179] considers a kind of stochastic dissipative Hamiltonian systems driven by degenerate noises, whose unique global solution is an ergodic process. The implicit Euler scheme applied to the considered model is shown to converge to the equilibrium exponentially. As a specific example of stochastic dissipative Hamiltonian systems, the stochastic Langevin equation will be studied in the following chapter, and high order schemes possessing ergodicity and the conformal symplectic structure will be constructed.

If the stochastic differential equations are driven by rough paths instead of standard Brownian motions, the solution is not a Markov process any more. The invariant measures and ergodicity for stochastic differential equations driven by rough paths are studied in [42, 97, 98] and references therein. It is still unclear and worth considering whether it is possible to approximate the equilibrium of rough differential equations through a proper numerical approximation.

Chapter 2
Invariant Measures for Stochastic Differential Equations

For ergodic SDEs, when solving them numerically, it is extremely important to choose proper schemes which could possess the properties under consideration and be applicable to practice. To the best of our knowledge, the numerical analysis of ergodic SDEs usually follows two directions. One is to construct numerical schemes which could inherit the ergodicity of the original system, and then to give the approximate error between the numerical invariant measure and the original one. We refer to [139] for the study of ergodic numerical schemes of SDEs, to [178] and [179] for the approximations of invariant measures of general SDEs and stochastic Hamiltonian systems, respectively, to [2] for high order approximations of invariant measures of ergodic SDEs, to [31] for approximation of the invariant measure for parabolic SPDEs, and to [50] for stochastic nonlinear Schrödinger equation case. The other one is to investigate the convergence rate of the temporal average for the numerical solution to the ergodic limit for the underlying SDEs, see [148] for the research about the Langevin equation, [140] for general SDEs, and [32] for parabolic SPDEs.

This chapter encompasses a brief introduction on above results concerning ergodicity of both SDEs and numerical schemes as well as approximate error of invariant measures and the ergodic limit, mainly based on [139, 140, 178] and references therein.

In Sect. 2.1, several sufficient conditions related to the drift and diffusion terms are introduced to ensure the existence and uniqueness of invariant measures for SDEs. More specific SDEs, i.e., non-degenerate SDEs and stochastic Langevin equations, as well as their numerical approximations, are studied in Sects. 2.2 and 2.3 to illustrate the analysis in Sect. 2.1. For an ergodic numerical approximation, the error between its invariant measure and the one of the original equation is studied in Sect. 2.4. For general (probably not ergodic) numerical approximations, the temporal average of the numerical solution may also be a proper approximation of the invariant measure of the original equation, which is stated in Sect. 2.5.

© Springer Nature Singapore Pte Ltd. 2019
J. Hong and X. Wang, *Invariant Measures for Stochastic Nonlinear Schrödinger Equations*, Lecture Notes in Mathematics 2251,
https://doi.org/10.1007/978-981-32-9069-3_2

Let $\mathbb{H} = \mathbb{R}^d$ for some $d \in \mathbb{N}$ throughout this chapter. We still denote by $X_x(t)$ the solution of the following initial value problem

$$\begin{cases} dX(t) = b(X(t))dt + \sigma(X(t))dW(t), & t \geq 0 \\ X(0) = x \in \mathbb{H} \end{cases} \tag{2.1}$$

with $b : \mathbb{H} \to \mathbb{H}$, $\sigma : \mathbb{H} \to \mathbb{R}^{d \times m}$ and W an m-dimensional Wiener process. In the following sections, precise assumptions on coefficients b and σ will be given to get certain properties (see e.g. Theorems 2.2 and 2.3).

We denote by $\{X_n\}_{n \in \mathbb{N}}$ a homogeneous Markov chain with deterministic initial value $X_0 = x$, which may be the solution of a numerical approximation for (2.1).

2.1 Ergodicity of Solutions to General Stochastic Differential Equations

It has been shown in Chap. 1 that the Lyapunov condition ensures the existence of invariant measures for (2.1) while the uniqueness of the invariant measure is obtained if the strong Feller and irreducibility properties are satisfied. Now we introduce some conditions which concentrate on a compact set instead of the whole state space.

2.1.1 Existence of Invariant Measures

We denote by $\mathscr{L}$ the infinitesimal generator of the process X in (2.1):

$$\mathscr{L} := \sum_{i=1}^{d} b^i \frac{\partial}{\partial x_i} + \frac{1}{2} \sum_{i,j=1}^{d} (\sigma \sigma^\top)_{ij} \frac{\partial^2}{\partial x_i \partial x_j}. \tag{2.2}$$

The following assumption on b and σ is a sufficient condition of the Lyapunov condition given in Theorem 1.3. We denote by $\| \cdot \|$ the Euclidean norm of vectors in $\mathbb{H}$.

Assumption 2.1 ([139]) There exist real numbers $a_1, a_2 \in (0, \infty)$ and a function $V : \mathbb{H} \to [1, \infty)$ with

$$\lim_{\|y\| \to \infty} V(y) = \infty$$

such that

$$\mathscr{L}V(y) \leq -a_1 V(y) + a_2, \quad \forall\, y \in \mathbb{H}.$$

Proposition 2.1 ([139]) *If the infinitesimal generator $\mathscr{L}$ of X in (2.1) satisfies Assumption 2.1, we have*

$$\mathbb{E}[V(X_n)] \leq \alpha\mathbb{E}[V(X_{n-1})] + \beta \tag{2.3}$$

with $X_n := X(nT)$ for some real numbers $T \in (0, +\infty)$, $\alpha \in (0, 1)$ and $\beta \in [0, +\infty)$, which leads to the Lyapunov condition in Theorem 1.3.

Proof Itô's formula applied to $V(X(t))$ yields that

$$dV(X(t)) = \mathscr{L}V(X(t))dt + \nabla V(X(t))^\top \sigma(X(t))dW(t)$$
$$\leq -a_1 V(X(t))dt + a_2 dt + \nabla V(X(t))^\top \sigma(X(t))dW(t).$$

Then taking expectation to both sides of above formula and using the Gronwall inequality, we obtain

$$\mathbb{E}[V(X(t))] \leq e^{-a_1(t-s)}\mathbb{E}[V(X(s))] + \frac{a_2}{a_1}\left(1 - e^{-a_1(t-s)}\right),$$

which implies (2.3) with $t = nT$, $s = (n-1)T$, $\alpha = e^{-a_1 T}$ and $\beta = \frac{a_2}{a_1}$. Based on this recursive formula, it indicates

$$\mathbb{E}[V(X_n)] \leq \alpha\mathbb{E}[V(X_{n-1})] + \beta \leq \alpha^n\mathbb{E}[V(X_0)] + \beta(1 + \alpha + \cdots + \alpha^{n-1})$$
$$= \alpha^n\mathbb{E}V(X_0) + \beta\frac{1 - \alpha^n}{1 - \alpha} \xrightarrow{n \to \infty} \frac{\beta}{1 - \alpha},$$

which implies the uniform boundedness of $\mathbb{E}[V(X_n)]$ and coincides with the Lyapunov condition given in Theorem 1.3. $\qquad\square$

In particular, if the Markov chain $\{X_n\}_{n \in \mathbb{N}}$ is generated through a numerical scheme $X_{n+1} = F(X_n, \tau, \triangle_{n+1}W)$ of (2.1) with time step-size τ, increment $\triangle_{n+1}W = W(t_{n+1}) - W(t_n)$ and X_n being an approximation of $X(t_n)$ at $t_n = n\tau$, then the following proposition is frequently used to test whether the scheme could inherit the Lyapunov structure. The idea of its proof comes from that of Theorem 7.2 in [139].

Proposition 2.2 *If $F \in \mathbf{C}^\infty(\mathbb{R}^d \times \mathbb{R}^m, \mathbb{H})$ and Assumption 2.1 holds for an essentially quadratic function V:*

$$C_1(1 + \|y\|^2) \leq V(y) \leq C_2(1 + \|y\|^2), \quad \|\nabla V(y)\| \leq C_3(1 + \|y\|), \quad \forall\, y \in \mathbb{H}$$

for some constants $C_1, C_2, C_3 > 0$. Then (2.3) holds for the proposed scheme when applied to Eq. (2.1) if

(i) there exist $c_1, \varepsilon > 0$, independent of $\tau > 0$, such that $\mathbb{E}\|X(t_1) - X_1\|^2 \leq c_1(1 + \|x\|^2)\tau^{\varepsilon+2}$ for any initial value $x \in \mathbb{H}$,

(ii) there exists $c_2 = c_2(r) > 0$, independent of $\tau > 0$, such that $\mathbb{E}\|X_1\|^r \le c_2(1 + \|x\|^r)$ for all $r \ge 1$ and $x \in \mathbb{H}$.

Proof Noting that

$$C_1(1 + \mathbb{E}\|X(t)\|^2) \le \mathbb{E}[V(X(t))] \le e^{-a_1 t} V(x) + \frac{a_2}{a_1}\left(1 - e^{-a_1 t}\right),$$

we have $\mathbb{E}\|X(t)\|^2 \le C(1 + \|x\|^2)$, where C denotes generic constants which may be different from one to another. For some fixed numerical solution X_{n-1}, we denote by $X_{X_{n-1}}(\tau)$ the exact solution of (2.1) at time τ starting from $X(0) = X_{n-1}$. Thus, based on conditions above, we conclude

$$\mathbb{E}[V(X_n)] \le \mathbb{E}[V(X_{X_{n-1}}(\tau))] + \mathbb{E}|V(X_{X_{n-1}}(\tau)) - V(X_n)|$$

$$\le e^{-a_1 \tau}\mathbb{E}[V(X_{n-1})] + C + \mathbb{E}\left|\int_0^1 \langle \nabla V(\theta X_n + (1-\theta)X_{X_{n-1}}(\tau)), X_{X_{n-1}}(\tau) - X_n\rangle d\theta\right|$$

$$\le e^{-a_1 \tau}\mathbb{E}[V(X_{n-1})] + C\left(\mathbb{E}[1 + \|X_n\|^2 + \|X_{X_{n-1}}(\tau)\|^2]\mathbb{E}[\|X_{X_{n-1}}(\tau) - X_n\|^2]\right)^{\frac{1}{2}} + C$$

$$\le e^{-a_1 \tau}\mathbb{E}[V(X_{n-1})] + C\left(1 + \mathbb{E}\|X_{n-1}\|^2\right)\tau^{\frac{\varepsilon}{2}+1}$$

$$\le e^{-\tilde{a}_1 \tau}\mathbb{E}[V(X_{n-1})] + C,$$

where we have used the fact that $e^{-a_1\tau} + C\tau^{\frac{\varepsilon}{2}+1} \le e^{-\tilde{a}_1\tau}$ for some $\tilde{a}_1 > 0$ and sufficiently small τ in the last step.   □

For examples which satisfy conditions in Proposition 2.2, we refer to Corollaries 7.4 and 7.5 in [139].

2.1.2 Uniqueness of the Invariant Measure

Based on the probability kernel $\pi_t(x, \cdot)$ (resp. $\pi_n(x, \cdot)$) for Markov process $\{X_x(t)\}_{t\ge 0}$ (resp. Markov chain $\{X_n\}_{n\in\mathbb{N}}$) defined in Sect. 1.2, the following assumption given in a compact set is used as a necessary condition for strong Feller and irreducibility properties.

Assumption 2.2 ([139]) Suppose for some fixed compact set $G_0 \in \mathscr{B}(\mathbb{H})$ that

(i) for some $x_0 \in \text{int}(G_0)$ and any $r > 0$, there exists $t_* = t_*(r)$ such that

$$\pi_{t_*}(x, B(x_0, r)) > 0, \quad \forall x \in G_0;$$

(ii) $\pi_t(x, \cdot)$ possesses a density $p_t(x, y)$, i.e.,

$$\pi_t(x, G) = \int_G p_t(x, y)dy, \quad \forall x \in G_0, \ \forall G \in \mathscr{B}(\mathbb{H}) \cap \mathscr{B}(G_0)$$

with $p_t(x, y)$ jointly continuous in $(x, y) \in G_0 \times G_0$.

Assumption 2.2, together with Assumption 2.1, gives an alternative way to Theorems 1.3 and 1.5 of showing the existence and uniqueness of the invariant measure, which is stated in the following theorem.

Theorem 2.1 (Theorem 2.5, [139]) *If X satisfies Assumptions 2.1 and 2.2, or alternatively, $\{X_n\}_{n \in \mathbb{N}}$ defined in Proposition 2.1 satisfies (2.3) and Assumption 2.2 by replacing t with n, then $\{X(t)\}_{t \geq 0}$ (resp. $\{X_n\}_{n \in \mathbb{N}}$) possesses a unique invariant measure with*

$$G_0 = \left\{ y \,\middle|\, V(y) \leq \frac{2\beta}{\rho - \alpha} \right\}$$

for some $\rho \in (\alpha^{\frac{1}{2}}, 1)$. Here, α and β are the same as those in Proposition 2.1.

In general, Assumption 2.2 is not easy to verify. Thus, a sufficient condition stated in the following proposition is frequently used to verify Assumption 2.2 (ii). That is, a Lie bracket condition, which is also known as the Hörmander condition (see [111, 168] and references therein), implies the hypoelliptic setting of the generator $\mathscr{L}$ and that π_t has a continuous density (Theorem 38.16, [168]).

Theorem 2.2 (Hörmander's theorem) *Assumption 2.2 (ii) is satisfied by the transition probability of (2.1) if*

(i) *$b, \sigma_i \in C^\infty(\mathbb{H})$, $i = 1, \cdots, m$ with $\sigma = (\sigma_1, \cdots, \sigma_m)$;*
(ii) *the Hörmander condition is satisfied, more precisely, $\Lambda_n(y) = \mathbb{H}$ for some $n \in \mathbb{N}$ and for all $y \in \mathbb{H}$, where*

$$\Lambda_0(y) = \mathrm{span}\{b(y), \sigma_i(y), i = 1, \cdots, m\},$$
$$\Lambda_{n+1}(y) = \mathrm{span}\{f(y), [g, f](y) : f \in \Lambda_n, g \in \Lambda_0\}$$

and $[f, g](y) = (\nabla g(y))f(y) - (\nabla f(y))g(y)$ denotes the Lie bracket.

Remark 2.1 Operator $\mathscr{L}$ is called hypoelliptic if $\mathscr{L}f \in C^\infty(\mathbb{H})$ implies that $f \in C^\infty(\mathbb{H})$, but this condition is not easy to verify. Thus, one usually uses the Hörmander condition, which actually is a sufficient condition of hypoelliptic operators (Theorem 1.1, [111]), as the hypoelliptic setting.

As applications of the above theory, the following two sections concern two kinds of SDEs: non-degenerate SDEs with dissipative conditions and the stochastic Langevin equation with degenerate noises. We give their ergodicity results under specific conditions.

2.2 Non-degenerate Stochastic Differential Equations and Ergodic Schemes

For general SDEs (2.1) or Markov chains on $\mathbb{H}$, we give some other sufficient conditions (see also [2, 140, 178–180] and references therein), which are easy to verify for a specific equation, to show its ergodicity.

Theorem 2.3 *Assume that*

(i) *functions b and σ are smooth and have bounded derivatives to any order, and furthermore, σ is also bounded itself;*

(ii) *there exist a constant $\zeta > 0$ and a compact set G_0 such that*

$$\langle y, b(y) \rangle \le -\zeta \|y\|^2, \quad \forall\, y \in \mathbb{H} - G_0;$$

(iii) *the associated operator $\mathcal{L}$ is uniformly elliptic, i.e., for some $\eta > 0$,*

$$y^\top (\sigma(z)\sigma(z)^\top) y \ge \eta \|y\|^2, \quad \forall\, y, z \in \mathbb{H}.$$

Then (2.1) is ergodic with a unique invariant measure μ, and furthermore, μ possesses a smooth density.

Proof To verify the Lyapunov condition, we denote $V(y) = \|y\|^2$. Then $V(y)$ is bounded for $y \in G_0$. For $y \in \mathbb{H} - G_0$, we have

$$\mathcal{L}V(y) = 2\langle y, b(y) \rangle + \sum_{i,j=1}^{d} (\sigma(y)\sigma(y)^\top)_{ij}$$
$$\le -2\zeta V(y) + \|\sigma(y)\|_F^2,$$

where $\|\cdot\|_F$ denotes the Frobenius norm of matrices. It immediately implies that Assumption 2.1 holds as σ is uniformly bounded, and ensures the existence of invariant measures. In addition, the elliptic setting ensures that the Dirichlet problem for following equation $\mathcal{L}v = 0$ admits a unique solution (see e.g. Lemma 3.4 and Remark 3.10, [120]). This fact, together with the Existence Theorem for weak solutions (see e.g. Theorem 4, Chap. 6.2, [77]) in classical theory of differential equations, indicates that

$$\mathcal{L}^* \rho = 0$$

possesses a unique nontrivial solution $\rho_\infty \in \mathbf{C}^\infty(\mathbb{H})$ satisfying $\int_\mathbb{H} \rho_\infty dx = 1$, where $\mathcal{L}^*$ denotes the $\mathbf{L}^2(\mathbb{H})$-adjoint operator of $\mathcal{L}$.

We then construct a probability measure μ satisfying $d\mu(x) = \rho_\infty(x)dx$. Denote $u(x, t) = \mathbb{E}[\varphi(X_x(t))]$, then Itô's formula yields that

$$u(x, t) = u(x, 0) + \int_0^t \mathcal{L}u(x, s)ds.$$

Taking the $\mathbf{L}^2(\mathbb{H})$-inner product between both sides of above equation and ρ_∞, and denoting $u(t) = u(x, t)$ for short, we have

$$
\begin{aligned}
\langle u(t), \rho_\infty \rangle_{\mathbf{L}^2(\mathbb{H})} &= \langle u(0), \rho_\infty \rangle_{\mathbf{L}^2(\mathbb{H})} + \int_0^t \langle \mathcal{L}u(s), \rho_\infty \rangle_{\mathbf{L}^2(\mathbb{H})}ds \\
&= \langle u(0), \rho_\infty \rangle_{\mathbf{L}^2(\mathbb{H})} + \int_0^t \langle u(s), \mathcal{L}^*\rho_\infty \rangle_{\mathbf{L}^2(\mathbb{H})}ds = \langle u(0), \rho_\infty \rangle_{\mathbf{L}^2(\mathbb{H})},
\end{aligned}
$$

which implies that

$$\int_\mathbb{H} \mathbb{E}[\varphi(X_x(t))]d\mu(x) = \int_\mathbb{H} \varphi d\mu(x)$$

and thus μ is the unique invariant measure of (2.1) as a result of the uniqueness of ρ_∞. □

Furthermore, it holds that if (2.1) possesses an invariant measure with a smooth density ρ_∞ with respect to the Lebesgue measure, then ρ_∞ is a solution of the steady state Fokker–Planck equation $\mathcal{L}^*\rho_\infty = 0$ (see e.g. Theorem 4, Chap. IX, [174]).

The assumptions in Theorem 2.3 are extremely strong for a stochastic process to be ergodic. Actually, it can be weakened onto some compact set.

Theorem 2.4 *Assume that there exists a bounded domain $U \subset \mathbb{H}$ with regular boundary such that*

(i) in U and some neighborhood thereof, denoted by $\tilde{U}$,

$$y^\top (\sigma(z)\sigma(z)^\top)y \geq \eta\|y\|^2, \quad \forall y \in \mathbb{H}, \ z \in \tilde{U};$$

(ii) defining a stopping time $\tau_x = \inf\{t : X_x(t) \in U\}$, one has $\mathbb{E}[\tau_x] < \infty$ for any $x \in \mathbb{H} - U$, and $\sup_{x \in K} \mathbb{E}[\tau_x] < \infty$ for any compact set $K \subset \mathbb{H}$.

Then (2.1) is also ergodic with a unique invariant measure μ.

We refer to [120] for the proof of this theorem: the existence of invariant measures is shown in Theorem 4.1, the ergodicity of the solution is proved in Theorem 4.2 and Corollary 4.3, while the uniqueness of the invariant measure is given in Corollary 4.4.

For ergodic SDEs (2.1), a natural question is whether one can construct numerical schemes which could inherit this dynamical property. We can find through Theorem 2.1 that the Lyapunov structure and the strong Feller and irreducibility properties are crucial and easy to verify when proving the ergodicity of (2.1). Numerical schemes are also constructed in order to inherit these properties and gain a unique numerical invariant measure.

Let $\sigma = (\sigma_1, \cdots, \sigma_m) \in \mathbf{C}_b^{\infty}(\mathbb{H}, \mathbb{R}^{d \times m})$ and $W = (W_1, \cdots, W_m)^{\top} \in \mathbb{R}^m$. We now show the ergodicity of the Euler–Maruyama scheme with sufficiently small time step-size τ:

$$X_{n+1} = X_n + b(X_n)\tau + \sigma(X_n)\triangle_{n+1}W, \tag{2.4}$$

where $\triangle_{n+1}W = W(t_{n+1}) - W(t_n)$. It is apparent that the numerical solution $\{X_n\}_{n \in \mathbb{N}}$ of (2.4) exists and is an $\mathscr{F}_{t_n}$-adapted homogenous Markov chain.

Theorem 2.5 *The numerical solution of (2.4) applied to (2.1), under the assumptions in Theorem 2.3, is ergodic for sufficiently small time step-size τ.*

Proof We can find out that the matrix $\sigma(x)$ is of full rank for any initial value $x \in \mathbb{R}^d$ according to the assumption (iii) in Theorem 2.3. It also implies that, for any open ball $B^o \subset G_0$, the transition probability

$$\pi_1(x, B^o) = \mathbb{P}(x + b(x)\tau + \sigma(x)\triangle_1 W \in B^o) > 0$$

since $\triangle_1 W$ is Gaussian. Furthermore, since the distribution of $\triangle_{n+1}W$ has a smooth density with respect to the Lebesgue measure for any $n \in \mathbb{N}$, the transition kernel $\pi_n(x, \cdot) = [\pi_1(x, \cdot)]^n$ has a smooth density. We then conclude that (2.4) satisfies Assumption 2.2.

Based on the assumptions (i) and (ii) in Theorem 2.3, we have that (2.3) holds: for $X_n \notin G_0$

$$
\begin{aligned}
\mathbb{E}\|X_{n+1}\|^2 &= \mathbb{E}\|X_n\|^2 + 2\tau\mathbb{E}\langle X_n, b(X_n)\rangle + \tau^2\mathbb{E}\|b(X_n)\|^2 + \mathbb{E}\|\sigma(X_n)\triangle_{n+1}W\|^2 \\
&\leq (1 - 2\zeta\tau)\mathbb{E}\|X_n\|^2 + \tau^2\mathbb{E}\|b(0) + \nabla b(\theta X_n)X_n\|^2 + C\tau \\
&\leq (1 - C\tau)\mathbb{E}\|X_n\|^2 + C\tau
\end{aligned}
$$

for sufficiently small τ independent of n, and $\theta \in [0, 1]$. If $X_n \in G_0$, it can be deduced that

$$\mathbb{E}\|X_{n+1}\|^2 \leq (1 - C\tau)\mathbb{E}\|X_n\|^2 + C\tau\mathbb{E}\|X_n\|^2 + C\tau,$$

where $\mathbb{E}\|X_n\|^2$ is bounded as G_0 is compact.

We complete the proof taking advantage of Theorem 2.1.   □

Example 2.1 Recall the one-dimensional Ornstein–Uhlenbeck process given through (1.7) in Example 1.1. It is apparent that this equation satisfies the non-degenerate assumptions in Theorem 2.3 and thus possesses a unique invariant measure $\mu = N(0, 1)$. Scheme (2.4) applied to (1.7) reads

$$X_{n+1} = X_n - X_n \tau + \sqrt{2} \triangle_{n+1} B \tag{2.5}$$

with time step-size $\tau \in (0, 1)$, deterministic initial value $X_0 = \xi \in \mathbb{R}$ and the increments $\triangle_{n+1} B = B(t_{n+1}) - B(t_n)$. Thus, the numerical solution $\{X_n\}_{n \in \mathbb{N}}$ obeys the following distribution

$$X_1 \sim N\left((1 - \tau)\xi, 2\tau\right) =: \pi_1(\xi, \cdot),$$
$$X_2 \sim N\left((1 - \tau)^2 \xi, 2\tau(1 - \tau)^2 + 2\tau\right) =: \pi_2(\xi, \cdot),$$
$$X_3 \sim N\left((1 - \tau)^3 \xi, 2\tau\left((1 - \tau)^4 + (1 - \tau)^2 + 1\right)\right) =: \pi_3(\xi, \cdot),$$
$$\cdots$$
$$X_n \sim N\left((1 - \tau)^n \xi, \frac{2}{2 - \tau}\left(1 - (1 - \tau)^{2n}\right)\right) =: \pi_n(\xi, \cdot),$$
$$\cdots$$

We finally obtain that the distributions of X_n converge to $\mu^\tau := N\left(0, \frac{2}{2-\tau}\right)$ as $n \to \infty$. We claim that μ^τ is an invariant measure for $\{X_n\}_{n \in \mathbb{N}}$. In fact, if $X_n \sim N\left(0, \frac{2}{2-\tau}\right)$, then one can calculate through (2.5) that $X_{n+1} \sim N\left(0, \frac{2}{2-\tau}\right)$. In addition, $\{X_n\}_{n \in \mathbb{N}}$ can also be shown to be strong Feller and irreducible, taking advantage of Gaussian distributions π_n, which then yields that μ^τ is the unique invariant measure of (2.5).

It is worth noticing that the conditions given in Theorems 2.3 and 2.4 are still very strong, and are not satisfied by a wide class of models. In some cases where the conditions in Theorem 2.3 are not satisfied, the Lyapunov structure may be easily destroyed by numerical schemes, especially explicit ones, since it is a global structure instead of concentrating on some compact set. We refer to the one-dimensional stochastic differential equation $dX = -X^3 dt + dW$ as a counterexample given in [139], whose drift coefficient has polynomial growth. The Euler–Maruyama scheme applied to this equation fails to be ergodic (Lemma 6.3, [139]), and even not converge in strong sense ([113, 114]).

As a result, numerical schemes need to be specially constructed when conditions in Theorems 2.3 and 2.4 are not satisfied. In the following, we show the ergodicity, as well as its geometric structure, for a type of SDEs with degenerate noise.

2.3 Stochastic Langevin Equation and Its Discretizations

In this section, we study a stochastic dissipative Hamiltonian system similar to that studied in Sect. 1.5.2 utilizing the argument introduced in the former two sections.

We still denote $\mathbb{H} = \mathbb{R}^d$. Consider a stochastic Langevin equation with the hypoel-liptic setting (see Theorem 2.2 and Remark 2.1):

$$
\begin{cases}
dp = -\nabla F(q)dt - \gamma p dt - \displaystyle\sum_{i=1}^m \sigma_i dW_i, \\[2ex]
dq = p dt
\end{cases}
\tag{2.6}
$$

with $p, q \in \mathbb{H}$ denoting the position and momentum of a particle, respectively. It is a damped Hamiltonian system with degenerate additive noises and satisfies the stochas-tic conformal symplectic structure, which will be introduced in the following. Here $F \in \mathbf{C}^\infty(\mathbb{H}, \mathbb{R}_+), \gamma > 0, \sigma = (\sigma_1, \cdots, \sigma_m) \in \mathbb{R}^{d \times m}$ and $W = (W_1, \cdots, W_m)^\top$ is a standard m-dimensional Brownian motion. Equation (2.6) admits a separable Hamil-tonian function $H(p, q) = \frac{1}{2}\|p\|^2 + F(q)$. In particular, we assume that $m \geq d$ and $\text{rank}\{\sigma_i \in \mathbb{H}, i = 1, \cdots, m\} = d$.

2.3.1 Ergodicity for Exact and Numerical Solutions

The ergodicity of (2.6) has been considered by several authors (see e.g. [43, 120, 139, 167]). We introduce one of these proofs through the following three lemmas according to Theorem 1.5.

Lemma 2.1 *There exists a Lyapunov functional* $V : \mathbb{R}^{2d} \to [1, +\infty]$ *which satisfies Assumption 2.1 with infinitesimal generator* $\mathcal{L}$ *associated to (2.6) under condition*

$$
F(q) + \frac{3\gamma^2}{8}\|q\|^2 - \langle \nabla F(q), q \rangle \leq C
\tag{2.7}
$$

for any $q \in \mathbb{H}$ *and some constant* $C > 0$, *and thus ensures the existence of invariant measures for (2.6).*

Proof Define $V(p, q) = 2F(q) + c_1\|q\|^2 + c_2\|p\|^2 + c_3\langle p, q \rangle + 1$ with coeffici-ents $c_i, i = 1, 2, 3$, chosen such that Assumption 2.1 is satisfied. Then we have

$$
\mathcal{L}V(p, q) = \langle p, 2\nabla F(q) + 2c_1 q + c_3 p \rangle - \langle \nabla F(q) + \gamma p, 2c_2 p + c_3 q \rangle + c_2\|\sigma\|_F^2
$$
$$
= (2c_1 - \gamma c_3)\langle p, q \rangle + (c_3 - 2\gamma)\|p\|^2 - c_3\langle \nabla F(q), q \rangle + \|\sigma\|_F^2
$$

by choosing $c_2 = 1$ in the last step. We then choose $c_3 = \gamma$ and $2c_1 = \gamma c_3$ for simplicity, and obtain

$$\mathscr{L}V(p,q) = -\gamma\|p\|^2 - \gamma\langle\nabla F(q),q\rangle + \|\sigma\|_F^2$$

$$= -\frac{\gamma}{2}V(p,q) - \left\|\left(\frac{\gamma}{2}\right)^{\frac{3}{2}}q - \left(\frac{\gamma}{2}\right)^{\frac{1}{2}}p\right\|^2 + \gamma F(q) + \frac{3\gamma^3}{8}\|q\|^2$$

$$- \gamma\langle\nabla F(q),q\rangle + \frac{\gamma}{2} + \|\sigma\|_F^2$$

$$\le -\frac{\gamma}{2}V(p,q) + \gamma\left(F(q) + \frac{3\gamma^2}{8}\|q\|^2 - \langle\nabla F(q),q\rangle\right) + \frac{\gamma}{2} + \|\sigma\|_F^2.$$

According to condition (2.7), Assumption 2.1 holds for $V(p,q) = 2F(q) + \frac{\gamma^2}{2}\|q\|^2 + \|p\|^2 + \gamma\langle p,q\rangle + 1$. $\qquad\square$

The condition (2.7) can of course be modified by choosing other constants c_1, c_2 and c_3 in $V(p,q)$. We refer to [139] for a general form of condition (2.7). In the following lemma, we show the strong Feller property via the Hörmander condition under the assumption $m \ge d$. We can also find out through its proof that if $m < d$, there are not enough noises to drive the solution to any points in the state space.

Lemma 2.2 *Equation (2.6) satisfies the hypoelliptic setting and thus its solution is a strong Feller process.*

Proof This proof is a part of that of Theorem 3.2, [139]. Rewrite (2.6) into

$$d\begin{pmatrix}p\\q\end{pmatrix} = \begin{pmatrix}-\nabla F(q) - \gamma p\\p\end{pmatrix}dt - \sum_{i=1}^{m}\begin{pmatrix}\sigma_i\\0\end{pmatrix}dW_i$$

$$=: Y_0(p,q)dt + \sum_{i=1}^{m}Y_i dW_i. \qquad (2.8)$$

For $m \ge d$ and $\text{rank}\{Y_i,\ i = 1,\cdots,m\} = d$, we can deduce that

$$[Y_i, Y_0] = (\nabla Y_0)Y_i = \begin{pmatrix}-\gamma I_d & -\nabla^2 F(q)\\I_d & 0\end{pmatrix}\begin{pmatrix}-\sigma_i\\0\end{pmatrix} = \begin{pmatrix}\gamma\sigma_i\\-\sigma_i\end{pmatrix}$$

are independent of $Y_i, i = 1,\cdots,m$, and hence

$$\text{rank}\{Y_i, [Y_i, Y_0],\ i = 1,\cdots,m\} = 2d,$$

which completes the proof. $\qquad\square$

Lemma 2.3 (Lemma 3.4, [139]) *The solution to (2.6) is irreducible.*

In addition, one can also show that the invariant measure μ of (2.6) possesses a smooth Boltzmann–Gibbs density (see e.g. [139])

$$\rho(p,q) \propto \exp\left\{-\gamma\left[\frac{p^2}{2} + F(q)\right]\right\},$$

taking advantage of the Fokker–Planck equation. One can then calculate the ergodic limit based on the density ρ, precisely,

$$\int_{\mathbb{R}^{2d}} \varphi(p,q)d\mu(p,q) = \int_{\mathbb{R}^d \times \mathbb{R}^d} \varphi(p,q)\rho(p,q)dpdq,$$

which can be used as a reference value to see whether a numerical scheme could approximate the ergodic limit properly (see Sect. 2.5). We refer to [2, 25, 167] and references therein for more details.

For this kind of SDEs with degenerate additive noise, one can also construct ergodic numerical schemes benefiting especially from the additive noise. As is shown in (2.8), the Langevin equation (2.6) can be rewritten as

$$dX = Y_0(X)dt + YdW$$

by denoting $X = (p^\top, q^\top)^\top$, $Y_0(X) := Y_0(p,q)$, $Y = (Y_1, \cdots, Y_m) \in \mathbb{R}^{2d \times m}$ and $W = (W_1, \cdots, W_m)^\top$. The authors in [139] study three schemes: the Euler–Maruyama scheme

$$X_{n+1} = X_n + Y_0(X_n)h + Y\triangle_{n+1}W, \tag{2.9}$$

the backward Euler method

$$X_{n+1} = X_n + Y_0(X_{n+1})h + Y\triangle_{n+1}W, \tag{2.10}$$

the split-step backward Euler method

$$\begin{cases} X_* = X_n + Y_0(X_*)h, \\ X_{n+1} = X_* + Y\triangle_{n+1}W, \end{cases} \tag{2.11}$$

and obtain the ergodicity for the numerical solution $\{X_n\}_{n\in\mathbb{N}}$ as stated in the following theorem.

Theorem 2.6 (Corollary 7.4, [139]) *Consider the Langevin equation (2.6) with hypoelliptic setting. Assume that essentially quadratic function $F \in \mathbf{C}^\infty(\mathbb{R}^d, \mathbb{R}_+)$ satisfies condition (2.7) and that ∇F is globally Lipschitz. Solutions $\{X_n\}_{n\in\mathbb{N}}$ of schemes (2.9), (2.10) and (2.11) are all ergodic, and each of them possesses a unique invariant measure.*

2.3.2 Geometric Structure: Conformal Symplecticity

It is well known that Hamiltonian systems in both deterministic and stochastic cases involve symplectic structures (see [10, 47, 51, 56, 82, 95, 100, 103, 110, 119, 169,

187, 188, 195] and references therein). For the stochastic case, as an example, in the following form

$$dX = J^{-1}\nabla H_0(X)dt + J^{-1}\nabla H_1(X) \circ dW(t) \tag{2.12}$$

with initial value $X(0) = (p_0^\top, q_0^\top)^\top$, $p_0, q_0 \in \mathbb{H}$. Here, $W(t)$ is an m-dimensional standard Brownian motion and

$$J = \begin{pmatrix} 0 & I_d \\ -I_d & 0 \end{pmatrix}$$

is the standard symplectic matrix. If we decompose X as $X = (p^\top, q^\top)^\top$ with $p, q \in \mathbb{H}$ and $p(0) = p_0, q(0) = q_0$, then the phase flow $(p_0, q_0) \mapsto (p, q)$ of (2.12) preserves the symplectic structure

$$dp \wedge dq = dp_0 \wedge dq_0$$

almost surely with 'd' denoting the exterior derivative and $dp \wedge dq$ being a differential 2-form.

For the stochastic Langevin equation (2.6), it can be described as a damped Hamiltonian system (see e.g. [179]), and degenerates to a Hamiltonian system when the absorption coefficient $\gamma = 0$. Thus, it possesses another important geometric structure, that is, conformal symplectic structure. Actually, rewrite (2.6) as

$$dX = \left[J^{-1}\nabla H_0(X) + \Lambda X \right] dt + J^{-1}\nabla H_1(X) \circ dW \tag{2.13}$$

with $X = (p^\top, q^\top)^\top$, $x_0 := (p_0^\top, q_0^\top)^\top$,

$$H_0 = \frac{1}{2}\|p\|^2 + F(q), \quad H_1 = \sigma q \quad \text{and} \quad \Lambda = \begin{pmatrix} -\gamma I_d & 0 \\ 0 & 0 \end{pmatrix}.$$

Hence similar to [95] and based on (2.13), the phase flow $\phi_t : x_0 \mapsto X(t)$ of (2.13) satisfies

$$d\left[\left(\frac{\partial\phi_t}{\partial x_0}\right)^\top J \left(\frac{\partial\phi_t}{\partial x_0}\right) \right] = \left(d\frac{\partial\phi_t}{\partial x_0} \right)^\top J \left(\frac{\partial\phi_t}{\partial x_0}\right) + \left(\frac{\partial\phi_t}{\partial x_0}\right)^\top J \left(d\frac{\partial\phi_t}{\partial x_0} \right)$$

$$= \left[J^{-1}\nabla^2 H_0(\phi_t)\frac{\partial\phi_t}{\partial x_0} + \Lambda\frac{\partial\phi_t}{\partial x_0} \right]^\top J \left(\frac{\partial\phi_t}{\partial x_0}\right) dt + \left[J^{-1}\nabla^2 H_1(\phi_t)\frac{\partial\phi_t}{\partial x_0} \circ dW \right]^\top J \left(\frac{\partial\phi_t}{\partial x_0}\right)$$

$$+ \left(\frac{\partial\phi_t}{\partial x_0}\right)^\top J \left[J^{-1}\nabla^2 H_0(\phi_t)\frac{\partial\phi_t}{\partial x_0} + \Lambda\frac{\partial\phi_t}{\partial x_0} \right] dt + \left(\frac{\partial\phi_t}{\partial x_0}\right)^\top J \left[J^{-1}\nabla^2 H_1(\phi_t)\frac{\partial\phi_t}{\partial x_0} \circ dW \right]$$

$$= -\gamma \left[\left(\frac{\partial\phi_t}{\partial x_0}\right)^\top J \left(\frac{\partial\phi_t}{\partial x_0}\right) \right] dt,$$

where in the last step we have used the fact that $J^{-\top}J = -I_{2d}$ and $\Lambda J + J\Lambda = -\gamma J$. Thus, we obtain

$$\left(\frac{\partial \phi_t}{\partial x_0}\right)^\top J \left(\frac{\partial \phi_t}{\partial x_0}\right) = e^{-\gamma t} J,$$

which implies immediately the following stochastic conformal symplectic structure

$$dp(t) \wedge dq(t) = e^{-\gamma t} dp_0 \wedge dq_0.$$

It indicates that the phase flow of (2.6) is not area preserving any more, but possesses the exponentially decay area instead. Namely, denote by $\text{Vol}(t)$ the phase volume on the domain $D_t \subset \mathbb{R}^{2d}$ at time t, then

$$\text{Vol}(t) = \int_{D_t} dp(t) dq(t) = \int_{D_0} \det\left(\frac{\partial(p(t), q(t))}{\partial(p_0, q_0)}\right) dp_0 dq_0,$$

where the determinant of the Jacobian matrix $\frac{\partial(p(t), q(t))}{\partial(p_0, q_0)}$ satisfies

$$\det\left(\frac{\partial(p(t), q(t))}{\partial(p_0, q_0)}\right) = e^{-\gamma t d}.$$

Recall that schemes (2.9), (2.10) and (2.11) for (2.6) are all ergodic, but they fail to preserve the conformal symplectic structure of the original system. To construct numerical schemes which could inherit both the ergodicity and the conformal symplecticity, we introduce the following systems based on the splitting technique (see [147] and references therein)

$$\begin{aligned}
d\tilde{p} &= -\nabla F(\tilde{q}) dt - \sigma dW, \quad \tilde{p}(0) = \tilde{p}_0, \\
d\tilde{q} &= \tilde{p} dt, \quad \tilde{q}(0) = \tilde{q}_0;
\end{aligned} \tag{2.14}$$

$$\begin{aligned}
d\hat{p} &= -\gamma \hat{p} dt, \quad \hat{p}(0) = \hat{p}_0, \\
d\hat{q} &= 0, \quad \hat{q}(0) = \hat{q}_0.
\end{aligned} \tag{2.15}$$

We would like to mention that other kinds of splitting, for example, Lie–Trotter splitting method (see e.g. [3, 25]), may be also available to construct conformal symplectic schemes.

Note that the system (2.15) can be solved exactly and explicitly by $\hat{p}(t) = e^{-\gamma t}\hat{p}_0$ and $\hat{q}(t) = \hat{q}_0$. Furthermore, we apply the midpoint scheme to the system (2.14), which together with the exact solution of (2.15) yields the scheme

$$\begin{cases} \tilde{p}_{n+1} = p_n - \tau \nabla F \left(\dfrac{\tilde{q}_{n+1} + q_n}{2} \right) - \sigma \triangle_{n+1} W, \\[2mm] \tilde{q}_{n+1} = q_n + \dfrac{\tau}{2}(\tilde{p}_{n+1} + p_n), \\[2mm] p_{n+1} = e^{-\gamma \tau} \tilde{p}_{n+1}, \\[2mm] q_{n+1} = \tilde{q}_{n+1} \end{cases}$$

with time step-size τ, numerical solutions $\{p_n, q_n\}_{n \in \mathbb{N}}$ and $\triangle_{n+1} W = W(t_{n+1}) - W(t_n)$. The above scheme can be rewritten into a compact form by eliminating intermediate variables $\tilde{p}_{n+1}$ and $\tilde{q}_{n+1}$:

$$\begin{cases} p_{n+1} = e^{-\gamma \tau} p_n - \tau e^{-\gamma \tau} \nabla F \left(\dfrac{q_{n+1} + q_n}{2} \right) - e^{-\gamma \tau} \sigma \triangle_{n+1} W, \\[2mm] q_{n+1} = q_n + \dfrac{\tau}{2}(e^{\gamma \tau} p_{n+1} + p_n). \end{cases} \tag{2.16}$$

Theorem 2.7 *Let F be essentially quadratic. The solution $\{p_n, q_n\}_{n \in \mathbb{N}}$ of the proposed scheme (2.16) for (2.6) is ergodic with a unique invariant measure when sampled at even n, i.e., $\{p_{2n}, q_{2n}\}_{n \in \mathbb{N}}$ is ergodic. In addition, it also preserves the discrete stochastic conformal symplectic structure*

$$dp_{n+1} \wedge dq_{n+1} = e^{-\gamma \tau} dp_n \wedge dq_n.$$

Proof We show that the proposed scheme could inherit the Lyapunov structure according to Proposition 2.2, which is sufficient to show the existence of the invariant measure for scheme (2.16). Actually, it is not hard to verify that the condition (ii) in Proposition 2.2 holds for the proposed scheme. Based on Lemma 2.1 and the above fact, we can also get condition (i) in Proposition 2.2 with $\varepsilon = 1$.

To get the uniqueness of the invariant measure, we only need to verify that Assumption 2.2 holds for the Markov chain $\{p_{2n}, q_{2n}\}_{n \in \mathbb{N}}$. We first show that the transition probability $\pi_2(x_0, \cdot)$ has a smooth density with $x_0 = (p_0^\top, q_0^\top)^\top$. We define a function $K : \mathbb{R}^{2d} \times \mathbb{R}^{2d} \to \mathbb{R}^{2d}$ as

$$K(u^1, v^1, u^2, v^2) = \begin{pmatrix} u^2 - e^{-\gamma \tau} u^1 + \tau e^{-\gamma \tau} \nabla F \left(\frac{v^2 + v^1}{2} \right) + e^{-\gamma \tau} \sigma \triangle_{n+1} W \\[2mm] v^2 - v^1 - \frac{\tau}{2}(e^{\gamma \tau} u^2 + u^1) \end{pmatrix},$$

whose Jacobian matrix

$$\frac{\partial K(u^1, v^1, u^2, v^2)}{\partial (u^2, v^2)} = \begin{pmatrix} I_d & \frac{\tau}{2} e^{-\gamma \tau} \nabla^2 F(\frac{v^2 + v^1}{2}) \\[2mm] -\frac{\tau}{2} e^{\gamma \tau} & I_d \end{pmatrix}$$

is positive definite for any (u^1, v^1) and sufficiently small τ. Hence, based on the implicit function theorem, there exists a continuous differentiable function k such that $(p_{n+1}, q_{n+1}) = k(p_n, q_n)$. This fact ensures that the transition probability of

(p_{n+1}, q_{n+1}) possesses a continuous density, i.e., Assumption 2.2 (ii) holds, as $\triangle_{n+1} W$ possesses a smooth density.

Next we show that Assumption 2.2 (i) holds, i.e, $\pi_2(x_0, B(y, r)) > 0$ for any $x_0, y \in \mathbb{R}^{2d}$ and $r > 0$. For any fixed x_0 and $x_2 \in B(y, r)$, we consider the first two steps in the Markov chain $\{p_n, q_n\}_{n \in \mathbb{N}}$

$$p_2 = e^{-\gamma\tau} p_1 - \tau e^{-\gamma\tau} \nabla F \left(\frac{q_2 + q_1}{2} \right) - e^{-\gamma\tau} \sigma \triangle_2 W; \tag{2.17}$$

$$q_2 = q_1 + \frac{\tau}{2} (e^{\gamma\tau} p_2 + p_1), \tag{2.18}$$

$$p_1 = e^{-\gamma\tau} p_0 - \tau e^{-\gamma\tau} \nabla F \left(\frac{q_1 + q_0}{2} \right) - e^{-\gamma\tau}, \tag{2.19}$$

$$q_1 = q_0 + \frac{\tau}{2} (e^{\gamma\tau} p_1 + p_0) \sigma \triangle_1 W \tag{2.20}$$

with $(p_0^\top, q_0^\top)^\top = x_0$. From (2.18) and (2.20), p_1 and q_1 can be uniquely determined to ensure that $(p_2^\top, q_2^\top)^\top = x_2$. Thus, $\triangle_2 W$ and $\triangle_1 W$ can also be determined according to (2.17) and (2.19), respectively. As the Brownian motion hits any open ball with positive probability, we finally get that $\pi_2(x_0, B(y, r)) > 0$. We now conclude from all the above that the numerical solution possesses a unique invariant measure based on Theorem 2.1.

On the other hand, it is well known that the midpoint scheme preserves the discrete symplectic structure for stochastic Hamiltonian systems. Thus, we have

$$\mathrm{d} p_{n+1} \wedge \mathrm{d} q_{n+1} = e^{-\gamma\tau} \mathrm{d}\tilde{p}_{n+1} \wedge \mathrm{d}\tilde{q}_{n+1} = e^{-\gamma\tau} \mathrm{d} p_n \wedge \mathrm{d} q_n,$$

which shows the conformal symplecticity of scheme (2.16).   □

This approach is also available if we apply the symplectic Euler method to system (2.14), and an explicit scheme will be obtained in that circumstance. The midpoint scheme and symplectic Euler scheme applied to (2.6) are both of order one in weak convergence sense. Schemes of higher weak convergence orders are constructed based on generating functions and modified equations in the following subsection.

2.3.3 Schemes of High Weak Convergence Order

In this subsection, an approach of designing conformal symplectic schemes with a higher weak convergence order is introduced by transforming (2.6) to a homogenous Hamiltonian system. The generating function is then available to design symplectic schemes for Hamiltonian systems (see e.g. [7, 8, 185, 186, 188]).

Denote $\hat{X}(t) = e^{\gamma t} p(t)$ and $\hat{Y}(t) = q(t)$ with components X_l and $Y_l, l = 1, \cdots, d$, respectively. Then Itô's formula applied to $\hat{X}$ and $\hat{Y}$ yields

$$\begin{cases} d\hat{X} = -e^{\gamma t}\nabla F(\hat{Y})dt - e^{\gamma t}\sum_{i=1}^{m}\sigma_i dW_i(t), \\ d\hat{Y} = e^{-\gamma t}\hat{X}dt, \end{cases}$$

which forms a non-autonomous stochastic Hamiltonian system

$$\begin{cases} d\hat{X} = -\dfrac{\partial \hat{H}_0}{\partial \hat{Y}}dt - \sum_{i=1}^{m}\dfrac{\partial \hat{H}_i}{\partial \hat{Y}}\circ dW_i, \\ d\hat{Y} = \dfrac{\partial \hat{H}_0}{\partial \hat{X}}dt + \sum_{i=1}^{m}\dfrac{\partial \hat{H}_i}{\partial \hat{X}}\circ dW_i \end{cases}$$

with Hamiltonians

$$\hat{H}_0 = e^{\gamma t}F(\hat{Y}) + \frac{1}{2}e^{-\gamma t}|\hat{X}|^2, \quad \hat{H}_i = e^{\gamma t}\sigma_i \cdot \hat{Y}$$

and '$\circ$' indicating that equations above hold in Stratonovich integral sense.

Bringing in two new variables X_{d+1} and Y_{d+1} such that

$$dX_{d+1} = -\frac{\partial \hat{H}_0}{\partial t}dt - \sum_{i=1}^{m}\frac{\partial \hat{H}_i}{\partial t}\circ dW_i(t), \quad X_{d+1}(0) = F(q_0) + \frac{1}{2}|p_0|^2 + \sum_{i=1}^{m}\sigma_i \cdot q_0,$$

$$dY_{d+1} = dt, \quad Y_{d+1}(0) = 0,$$

and denoting $X = (\hat{X}^\top, X_{d+1})^\top = (X_1, \cdots, X_d, X_{d+1})^\top$ and $Y = (\hat{Y}^\top, Y_{d+1})^\top = (Y_1, \cdots, Y_d, Y_{d+1})^\top$, we get a $(2d+2)$-dimensional autonomous stochastic Hamiltonian system

$$\begin{cases} dX = -\dfrac{\partial H_0}{\partial Y}dt - \sum_{i=1}^{m}\dfrac{\partial H_i}{\partial Y}\circ dW_i, \\ dY = \dfrac{\partial H_0}{\partial X}dt + \sum_{i=1}^{m}\dfrac{\partial H_i}{\partial X}\circ dW_i, \end{cases} \tag{2.21}$$

where

$$H_0(X, Y) = e^{\gamma Y_{d+1}}F(\hat{Y}) + \frac{1}{2}e^{-\gamma Y_{d+1}}|\hat{X}|^2 + X_{d+1},$$

$$H_i(X, Y) = e^{\gamma Y_{d+1}}\sigma_i \cdot \hat{Y}.$$

Denote $X(0) = x$ and $Y(0) = y$ for convenience. It is revealed in [185] that the generating function $S(X, y, t)$ related to (2.21) is the solution of the following stochastic Hamilton–Jacobi partial differential equation

$$d_t S(X, y, t) = H_0 \left(X, y + \frac{\partial S}{\partial X} \right) dt + \sum_{i=1}^{m} H_i \left(X, y + \frac{\partial S}{\partial X} \right) \circ dW_i. \qquad (2.22)$$

Moreover, the mapping $(x, y) \mapsto (X(t), Y(t))$ defined by

$$X(t) = x - \frac{\partial S(X(t), y, t)}{\partial y}, \quad Y(t) = y + \frac{\partial S(X(t), y, t)}{\partial X} \qquad (2.23)$$

is the stochastic flow of (2.21). Based on the Itô representation theorem and stochastic Taylor-Stratonovich expansion, $S(X, y, t)$ has a series expansion (see e.g. [7])

$$S(X, y, t) = \sum_{\alpha} G_\alpha(X, y) J_\alpha^t, \qquad (2.24)$$

where

$$J_\alpha^t = \int_0^t \int_0^{s_l} \cdots \int_0^{s_2} \circ dW_{j_1}(s_1) \circ dW_{j_2}(s_2) \circ \cdots \circ dW_{j_l}(s_l)$$

with multi-index $\alpha = (j_1, j_2, \cdots, j_l) \in \{0, 1, \cdots, m\}^{\otimes l}, l \geq 1$ and $dW_0(s) := ds$. Before calculating coefficients $G_\alpha(X, y)$ in (2.24), we first specify some notations. Let $l(\alpha)$ denote the length of α, and $\alpha-$ be the multi-index resulting from discarding the last index of α. Define $\alpha * \alpha' = (j_1, \cdots, j_l, j_1', \cdots, j_{l'}')$ where $\alpha = (j_1, \cdots, j_l)$ and $\alpha' = (j_1', \cdots, j_{l'}')$. The concatenation '$*$' between a set of multi-indices Λ and α is $\Lambda * \alpha = \{\beta * \alpha | \beta \in \Lambda\}$. Furthermore, define

$$\Lambda_{\alpha, \alpha'} = \begin{cases} \{(j_1, j_1'), (j_1', j_1)\}, & \text{if } l = l' = 1, \\ \{\Lambda_{(j_1), \alpha'-} * (j_{l'}'), \alpha' * (j_1)\}, & \text{if } l = 1, l' \neq 1, \\ \{\Lambda_{\alpha-, (j_1')} * (j_l), \alpha * (j_1')\}, & \text{if } l \neq 1, l' = 1, \\ \{\Lambda_{\alpha-, \alpha'} * (j_l), \Lambda_{\alpha, \alpha'-} * (j_{l'}')\}, & \text{if } l \neq 1, l' \neq 1. \end{cases}$$

For $k > 2$, let $\Lambda_{\alpha_1, \cdots, \alpha_k} = \{\Lambda_{\beta, \alpha_k} : \beta \in \Lambda_{\alpha_1, \cdots, \alpha_{k-1}}\}$. Substituting (2.24) into (2.22) and taking Taylor expansions to H_i $(i = 0, 1, \cdots, m)$ at (X, y), we obtain $G_\alpha = H_i$ with $\alpha = (i)$ being a single index, and

$$G_\alpha(X, y) = \sum_{r=1}^{l(\alpha)-1} \frac{1}{r!} \sum_{k_1, \cdots, k_r=1}^{d+1} \frac{\partial^r H_{j_l}(X, y)}{\partial y_{k_1} \cdots \partial y_{k_r}} \sum_{\substack{l(\alpha_1) + \cdots + l(\alpha_r) \\ = l(\alpha) - 1 \\ \alpha- \in \Lambda_{\alpha_1, \cdots, \alpha_r}}} \frac{\partial G_{\alpha_1}}{\partial X_{k_1}} \cdots \frac{\partial G_{\alpha_r}}{\partial X_{k_r}}$$

for multi-indices $\alpha = (j_1, j_2, \cdots, j_l)$ with $l \geq 2$.

We refer to [7, 146, 185] for more details about generating functions and weakly convergent symplectic numerical schemes obtained by truncating the generating function. In this case, simulation of multiple integrals is involved to obtain schemes with high weak convergence orders.

To reduce the simulation of multiple integrals, it is more convenient to consider the modified stochastic Hamiltonian system

$$
\begin{cases}
dX^M = -\dfrac{\partial H_0^M(X^M, Y^M)}{\partial Y^M}dt - \displaystyle\sum_{i=1}^{m} \dfrac{\partial H_i^M(X^M, Y^M)}{\partial Y^M} \circ dW_i, \quad X^M(0) = x, \\[4mm]
dY^M = \dfrac{\partial H_0^M(X^M, Y^M)}{\partial X^M}dt + \displaystyle\sum_{i=1}^{m} \dfrac{\partial H_i^M(X^M, Y^M)}{\partial X^M} \circ dW_i, \quad Y^M(0) = y.
\end{cases}
$$

(2.25)

Here,

$$
H_i^M(X^M, Y^M) = H_i(X^M, Y^M) + H_i^{(1)}(X^M, Y^M)\tau + \cdots + H_i^{(\ell)}(X^M, Y^M)\tau^\ell
$$

(2.26)

with some fixed step-size $\tau \in (0, 1)$ and undetermined functions $H_i^{(j)}$ for $j = 1, \cdots, \ell$ and $i = 0, \cdots, m$. Rewrite the multiple Stratonovich integrals J_α^t as

$$
J_\alpha^t =
\begin{cases}
\displaystyle\sum_\beta C_\alpha^\beta I_\beta^t, & l(\alpha) \geq 2, \\[4mm]
I_\alpha^t, & l(\alpha) = 1,
\end{cases}
$$

where $\beta = (i_1, i_2, \cdots, i_l) \in \{0, 1, \cdots, m\}^{\otimes l}$ are indices of length $l \geq 1$, C_α^β are certain constants given in [122], and

$$
I_\beta^t := \int_0^t \int_0^{s_l} \cdots \int_0^{s_2} dW_{i_1}(s_1)dW_{i_2}(s_2) \cdots dW_{i_l}(s_l)
$$

are multiply integrals in the Itô sense. Denote by

$$
\hat{S}(X, y, t) = \sum_\alpha \hat{G}_\alpha(X, y) \sum_{l(\beta) \leq k} C_\alpha^\beta I_\beta^t,
$$

(2.27)

the truncated modified generating function (see e.g. [7, 122]), where

$$
\hat{G}_\alpha(X, y) = \sum_{r=1}^{l(\alpha)-1} \frac{1}{r!} \sum_{k_1, \cdots, k_r=1}^{d+1} \frac{\partial^r H_{jl}^M(X, y)}{\partial y_{k_1} \cdots \partial y_{k_r}} \sum_{\substack{l(\alpha_1) + \cdots + l(\alpha_r) \\ = l(\alpha) - 1 \\ \alpha- \in \Lambda_{\alpha_1, \cdots, \alpha_r}}} \frac{\partial \hat{G}_{\alpha_1}}{\partial X_{k_1}} \cdots \frac{\partial \hat{G}_{\alpha_r}}{\partial X_{k_r}}
$$

for $l(\alpha) \geq 2$, and $\hat{G}_{(i)} = H_i^M$ for $i = 0, 1, \cdots, m$. Then (2.27) determines a one-step approximation

$$X_1 = x - \frac{\partial \hat{S}(X_1, y, \tau)}{\partial y}, \quad Y_1 = y + \frac{\partial \hat{S}(X_1, y, \tau)}{\partial X_1}. \tag{2.28}$$

By choosing proper functions $H_i^{(j)}$ in (2.26), we will be able to construct symplectic schemes approximating (2.21) with weak order $k + k'$, that is, the local weak error is of order $k + k' + 1$

$$|\mathbb{E}\phi(X(\tau), Y(\tau)) - \mathbb{E}\phi(X_1, Y_1)| = O(\tau^{k+k'+1}), \tag{2.29}$$

even though the scheme is only of weak order k when approximating the modified Eq. (2.25).

We take the case $k = k' = 1$ as an example in which Hamiltonians H_i, $i = 0, \cdots, m$, should be modified as in (2.26) with $\ell = 1$. Let $\phi \in \mathbf{C}_p^6(\mathbb{R}^{2d+2}, \mathbb{R})$. It then suffices to determine functions $H_i^{(1)}$, $i = 0, \cdots, m$. Utilizing the Taylor expansion to $\phi(X(\tau), Y(\tau))$ and $\phi(X_1, Y_1)$ at $(x^\top, y^\top)^\top = (x_1, \cdots, x_{d+1}, y_1, \cdots, y_{d+1})^\top$ and comparing the terms on both sides of (2.29), we choose $H_i^{(1)}$, $i = 0, \cdots, m$ such that

$$\frac{\partial H_i^{(1)}}{\partial x_{d+1}} = 0, \quad i = 0, \cdots, m,$$

$$\frac{\partial H_i^{(1)}}{\partial (y_1, \cdots, y_d)} = \frac{1}{2} \gamma e^{\gamma y_{d+1}} \sigma_i, \quad \frac{\partial H_i^{(1)}}{\partial (x_1, \cdots, x_d)} = \frac{1}{2} \sigma_i, \quad i = 1, \cdots, m,$$

$$\frac{\partial H_0^{(1)}}{\partial y_r} = \frac{1}{2} \sum_{j=1}^d \frac{\partial^2 F(y)}{\partial y_r \partial y_j} x_j + \frac{1}{2} \gamma e^{\gamma y_{d+1}} \frac{\partial F(y)}{\partial y_r}, \quad r = 1, \cdots, d,$$

$$\frac{\partial H_0^{(1)}}{\partial x_r} = \frac{1}{2} \frac{\partial F(y)}{\partial y_r} - \frac{1}{2} \gamma e^{-\gamma y_{d+1}} x_r, \quad r = 1, \cdots, d.$$

Then $\hat{S}$ is determined, based on which one can get the expression of the one-step approximation (2.28) as well as the numerical solution $\{(X_n, Y_n)\}_{n \in \mathbb{N}}$.

To deduce an approximation of the original system (2.6) based on scheme (2.28), we denote by X_n^- and Y_n^- the first d components of X_n and Y_n respectively, and denote

$$p_n = e^{-\gamma t_n} X_n^-, \quad q_n = Y_n^-.$$

We finally get the numerical scheme for (2.6)

$$\begin{cases} p_{n+1} = e^{-\gamma\tau} p_n - \dfrac{\tau^2}{2} \nabla^2 F(q_n) p_{n+1} - \tau \left(1 + \dfrac{\gamma\tau}{2}\right) e^{-\gamma\tau} \nabla F(q_n) \\ \qquad\quad - \left(1 + \dfrac{\gamma\tau}{2}\right) e^{-\gamma\tau} \sigma \Delta_{n+1} W, \\ q_{n+1} = q_n + \tau \left(1 - \dfrac{\gamma\tau}{2}\right) e^{\gamma\tau} p_{n+1} + \dfrac{\tau^2}{2} \nabla F(q_n) + \dfrac{\tau}{2} \sigma \Delta_{n+1} W \end{cases} \tag{2.30}$$

with $\triangle_{n+1} W = W(t_{n+1}) - W(t_n)$ and $t_n = n\tau$, $n \in \mathbb{N}$, which admits the discrete conformal symplectic structure and is of weak order two.

Theorem 2.8 *Scheme (2.30) possesses the discrete conformal symplectic conservation law, i.e,*

$$dp_{n+1} \wedge dq_{n+1} = e^{-\gamma\tau} dp_n \wedge dq_n.$$

Proof Note that

$$
\begin{aligned}
dp_{n+1} \wedge dq_{n+1} &= dp_{n+1} \wedge dq_n + \frac{\tau^2}{2} dp_{n+1} \wedge \left(\nabla^2 F(q_n)dq_n\right) \\
&= \left(I_d + \frac{\tau^2}{2}\nabla^2 F(q_n)\right) dp_{n+1} \wedge dq_n \\
&= d\left(e^{-\gamma\tau} p_n - \tau\left(1 + \frac{\gamma\tau}{2}\right)e^{-\gamma\tau}\nabla F(q_n)\right) \wedge dq_n \\
&= e^{-\gamma\tau} dp_n \wedge dq_n.
\end{aligned}
$$

It then completes the proof.   □

Theorem 2.9 *Assume that ∇F is globally Lipschitz continuous with linear growth. Then scheme (2.30) admits an invariant measure and has convergence order two in weak sense. More precisely, for a fixed $T > 0$,*

$$|\mathbb{E}\varphi(p(t_n), q(t_n)) - \mathbb{E}\varphi(p_n, q_n)| = O(\tau^2)$$

for all $\varphi \in \mathbf{C}_p^6(\mathbb{R}^{2d}, \mathbb{R})$ and $t_n \leq T$, $n \in \mathbb{N}$, where $\mathbf{C}_p^6(\mathbb{R}^{2d}, \mathbb{R})$ denotes the space of continuous functions whose derivatives are of polynomial growth up to the 6th order.

The readers are referred to [108] for the proof of above theorem. Moreover, the weak error plays an important role in approximating the invariant measure of the original system, which will be introduced in the following two sections.

2.4   Approximation of Invariant Measures via Ergodic Schemes

The error estimate for ergodic SDEs usually contains two aspects: the error between invariant measures and the error between the temporal average of the numerical solution and the ergodic limit. In this section, we concentrate on the first aspect. If the numerical solution is proved or assumed to be ergodic with a numerical invariant measure μ^τ, the error between the original invariant measure μ and the numerical one is defined as

$$e(\varphi) := \left| \int_{\mathbb{H}} \varphi d\mu - \int_{\mathbb{H}} \varphi d\mu^\tau \right|$$

for some kind of test functions φ. One can investigate the convergence order of $e(\varphi)$ to determine whether μ^τ is a proper approximation of μ (see [3, 140, 178, 182] and references therein).

Note that the ergodicity of the exact solution X and the numerical one $\{X_n\}_{n\in\mathbb{N}}$ ensures the following equations

$$\lim_{T\to\infty} \frac{1}{T} \int_0^T \mathbb{E}[\varphi(X(t))]dt = \int_{\mathbb{H}} \varphi d\mu,$$

$$\lim_{N\to\infty} \frac{1}{N} \sum_{n=0}^{N-1} \mathbb{E}[\varphi(X_n)] = \int_{\mathbb{H}} \varphi d\mu^\tau.$$

For a fixed time step-size τ, the difference between above equations shows that

$$e(\varphi) = \left| \lim_{T=N\tau\to\infty} \frac{1}{N\tau} \sum_{n=0}^{N-1} \int_{t_n}^{t_{n+1}} \left(\mathbb{E}[\varphi(X(t)] - \mathbb{E}[\varphi(X_n)]\right) dt \right|.$$

It indicates that the error between invariant measures is the same as the weak error if the weak error is shown to be independent of the time interval. However, the time-independent weak error is usually a difficult criterion to verify. The following theorem gives an error estimate of invariant measures via the solution of Kolmogorov equation avoiding proving that the weak error is time independent.

In Sect. 2.2, we show that both Eq. (2.1) and the Euler–Maruyama scheme (2.4) applied to it are ergodic under the assumptions in Theorem 2.3. Based on that result, the following theorem gives the error between invariant measures, whose proof is similar to that of Theorem 3.3 in [178]. We also refer to [178] for higher order approximations.

Theorem 2.10 *Under the same assumptions as in Theorem 2.3, the Euler–Maruyama scheme (2.4) possesses a numerical invariant measure μ^τ. The error between invariant measures μ and μ^τ is of order one, i.e.,*

$$e(\varphi) = O(\tau), \quad \forall \varphi \in \mathbf{C}_p^\infty(\mathbb{H}),$$

where $\mathbf{C}_p^\infty(\mathbb{H})$ denotes the space of smooth functions on $\mathbb{H}$ with all the derivatives having polynomial growth.

Proof We first claim that the numerical solution of (2.4) is uniformly bounded

$$\mathbb{E}\|X_n\|^p \le K_p(1 + \|x\|^p \exp(-\alpha_p t_n)) \tag{2.31}$$

with deterministic initial value $X_0 = x$ and positive constants K_p and α_p depending on p. In fact, we have known from the proof of Theorem 2.5 that

$$\mathbb{E}\|X_n\|^2 \le (1 - C\tau)\mathbb{E}\|X_{n-1}\|^2 + C\tau \le e^{-\alpha_2 t_n}\|x\|^2 + C$$

for some positive constant α_2, which verifies (2.31) with $p = 2$. For $p = 4$, note that

$$
\begin{aligned}
\mathbb{E}\|X_{n+1}\|^4 &= \mathbb{E}\Big[\|X_n\|^2 + 2\tau\langle X_n, b(X_n)\rangle + \tau^2\|b(X_n)\|^2 + \|\sigma(X_n)\triangle_{n+1}W\|^2 \\
&\quad + 2\langle X_n + b(X_n)\tau, \triangle_{n+1}W\rangle\Big]^2 \\
&= \mathbb{E}\Big[\|X_n\|^2 + 2\tau\langle X_n, b(X_n)\rangle + \tau^2\|b(X_n)\|^2\Big]^2 + \mathbb{E}\Big[\|\sigma(X_n)\triangle_{n+1}W\|^2\Big]^2 \\
&\quad + 2\mathbb{E}\Big[\Big(\|X_n\|^2 + 2\tau\langle X_n, b(X_n)\rangle + \tau^2\|b(X_n)\|^2\Big)\|\sigma(X_n)\triangle_{n+1}W\|^2\Big] \\
&\quad + 4\mathbb{E}\Big[\|\sigma(X_n)\triangle_{n+1}W\|^2\langle X_n + b(X_n)\tau, \triangle_{n+1}W\rangle\Big] \\
&\quad + 4\mathbb{E}\Big[\langle X_n + b(X_n)\tau, \triangle_{n+1}W\rangle\Big]^2 \\
&\leq (1 - C\tau)\mathbb{E}\|X_n\|^4 + C\tau^2 + C\tau(1 - C\tau)\mathbb{E}\|X_n\|^2,
\end{aligned}
$$

which, together with the estimate for $\mathbb{E}\|X_n\|^2$, verifies (2.31). Following this procedure, we then complete the claim by recursion.

In addition, (2.4) has been shown to be of weak order one

$$
|\mathbb{E}[\varphi(X(t_n))] - \mathbb{E}[\varphi(X_n)]| \leq C_n\tau
$$

with $t_n = n\tau$ and some constant C_n depending on n (see also [122, 144, 177]). Recall in Sect. 2.2 that $u(x, t) = \mathbb{E}[\varphi(X_x(t))]$ is the solution of the following Kolmogorov equation

$$
\begin{cases}
\dfrac{\partial}{\partial t}u(x, t) = \mathscr{L}u(x, t), \\
u(x, 0) = \varphi(x),
\end{cases}
$$

and the derivatives of $u(x, t)$ also have the boundedness similar to X_n (Theorem 3.4, [178])

$$
\|\partial_\iota u(x, t)\| \leq C_\iota(1 + \|x\|^{p_\iota})\exp(-\alpha_\iota t) \tag{2.32}
$$

with multi-index ι and positive constants p_ι, α_ι depending on ι. The Taylor expansion performed on $u(t_k, X_n)$ at X_{n-1} shows that

$$
\begin{aligned}
\mathbb{E}[u(t_k, X_n)] &= \mathbb{E}[u(t_k, X_{n-1})] + \mathbb{E}[\nabla u(t_k, X_{n-1})(X_n - X_{n-1})] \\
&\quad + \frac{1}{2}\mathbb{E}\big[\nabla^2 u(t_k, X_{n-1})(X_n - X_{n-1})^2\big] \\
&\quad + \frac{1}{3!}\mathbb{E}\big[\nabla^3 u(t_k, X_{n-1})(X_n - X_{n-1})^3\big]
\end{aligned}
$$

$$+ \frac{1}{4!} \mathbb{E}\left[\nabla^4 u(t_k, \theta(X_n - X_{n-1}) + X_{n-1})(X_n - X_{n-1})^4\right]$$

$$= \mathbb{E}[u(t_k, X_{n-1})] + \tau \mathbb{E}\left[\mathscr{L}u(t_k, X_{n-1})\right] + R_1^{k,n}\tau^2, \qquad (2.33)$$

where $\theta \in (0, 1)$ and $\nabla^p u(t, x)y^p := \nabla^p u(t, x)(y, \cdots, y)$ is the Fréchet derivative for some $p \in \mathbb{N}$. The remainder $R_1^{k,n}$ can be expressed by

$$R_1^{k,n} = \frac{1}{2} \mathbb{E}\left[\nabla^2 u(t_k, X_{n-1})b(X_{n-1})^2\right] + \frac{\tau}{6} \mathbb{E}\left[\nabla^3 u(t_k, X_{n-1})b(X_{n-1})^3\right]$$

$$+ \frac{1}{24\tau^2} \mathbb{E}\left[\nabla^4 u(t_k, \theta(X_n - X_{n-1}) + X_{n-1})(X_n - X_{n-1})^4\right].$$

According to (2.31) and (2.32), we have

$$|R_1^{k,n}| \le C(1 + \mathbb{E}\|X_n\|^p + \mathbb{E}\|X_{n-1}\|^p)\exp(-\xi_1 t_k) \le C(1 + \|x\|^p)\exp(-\xi_1 t_k)$$

for some p and ξ_1. In the same procedure as above, based on Itô's formula, we derive

$$\mathbb{E}[u(t_{k+1}, X_{n-1})] = \mathbb{E}[u(t_k, X_{n-1})] + \tau \mathbb{E}\left[\mathscr{L}u(t_k, X_{n-1})\right] + R_2^{k,n}\tau^2 \qquad (2.34)$$

with remainder $|R_2^{k,n}| \le C(1 + \|x\|^q)\exp(-\xi_2 t_k)$ for some positive q and ξ_2. The difference between (2.33) and (2.34) leads to

$$\mathbb{E}[u(t_k, X_n)] = \mathbb{E}[u(t_{k+1}, X_{n-1})] + (R_1^{k,n} - R_2^{k,n})\tau^2$$

$$= \mathbb{E}[u(t_{k+n}, x)] + \sum_{j=k}^{k+n}(R_1^{j,n+k-j} - R_2^{j,n+k-j})\tau^2.$$

Choosing $k = 0$ and taking the average of above terms for $n = 1, \cdots, N$, we obtain

$$\left|\frac{1}{N}\sum_{n=0}^{N-1}\mathbb{E}[u(0, X_n)] - \frac{1}{N}\sum_{n=0}^{N-1}\mathbb{E}[u(t_n, x)]\right| \le \frac{1}{N}\sum_{n=0}^{N-1}\sum_{j=0}^{\infty}\left|R_1^{j,n-j} - R_2^{j,n-j}\right|\tau^2$$

$$\le C(1 + \|x\|^p + \|x\|^q)\tau^2 \sum_{j=0}^{\infty}\exp(-(\xi_1 \wedge \xi_2)j\tau) \le C(1 + \|x\|^p + \|x\|^q)\tau,$$

since the series $\tau \sum_{j=0}^{\infty}\exp(-(\xi_1 \wedge \xi_2)j\tau)$ converges and is uniformly bounded with respect to τ. Let $N \to \infty$ in above inequality. The ergodicity of both X_n and $X(t_n)$ indicates that

$$\lim_{N\to\infty}\frac{1}{N}\sum_{n=0}^{N-1}\mathbb{E}[u(0, X_n)] = \lim_{N\to\infty}\frac{1}{N}\sum_{n=0}^{N-1}\mathbb{E}[\varphi(X_n)] = \int_{\mathbb{H}}\varphi d\mu^\tau.$$

and

$$\lim_{N\to\infty} \frac{1}{N} \sum_{n=0}^{N-1} \mathbb{E}[u(t_n, x)] = \lim_{N\to\infty} \frac{1}{N} \sum_{n=0}^{N-1} \mathbb{E}[\varphi(X(t_n))]$$

$$= \lim_{N\to\infty} \frac{1}{N\tau} \sum_{n=0}^{N-1} \int_{t_n}^{t_{n+1}} \mathbb{E}[\varphi(X(t_n))] ds$$

$$= \lim_{N\to\infty} \frac{1}{N\tau} \sum_{n=0}^{N-1} \left(\int_{t_n}^{t_{n+1}} \mathbb{E}[\varphi(X(s))] ds + O((1 + \|x\|^r)\tau^2) \right)$$

$$= \int_{\mathbb{H}} \varphi d\mu + O((1 + \|x\|^r)\tau),$$

where we have used the numerical integration and the polynomial growth of $\nabla\varphi$, more precisely, $\|\nabla\varphi(\theta(X_n - X(s)) + X(s))\| \le C(1 + \|x\|^r)$ for any $\theta \in (0, 1)$ and some $r \in \mathbb{N}$. It then completes the proof.   □

For ergodic processes or chains, their invariant measures are usually unknown. The error between invariant measures depends heavily on the uniform boundedness of the exact and numerical solutions. We next give an example, in which the invariant measures for both the exact solution and the numerical one have explicit forms. In this circumstance, the error between invariant measures is immediately obtained through the error between their densities.

Example 2.2 We still take the Ornstein–Uhlenbeck process with invariant measure $\mu = N(0, 1)$ in Example 1.1 as an instance. The Euler–Maruyama scheme is shown to be ergodic with a unique invariant measure $\mu^\tau = N(0, \frac{2}{2-\tau})$ in Example 2.1. Thus,

$$e(\varphi) = \left| \int_{\mathbb{R}} \varphi(x) \frac{1}{\sqrt{2\pi}} \left(e^{-\frac{x^2}{2}} - \left(\frac{2}{2-\tau} \right)^{-\frac{1}{2}} e^{-\frac{(2-\tau)x^2}{4}} \right) dx \right|$$

$$= \frac{1}{\sqrt{2\pi}} \left| \int_{\mathbb{R}} \varphi(x) e^{-\frac{(2-\tau)x^2}{4}} \left(e^{-\frac{\tau x^2}{4}} - \left(\frac{2-\tau}{2} \right)^{\frac{1}{2}} \right) dx \right|,$$

where

$$\left| e^{-\frac{\tau x^2}{4}} - \left(\frac{2-\tau}{2} \right)^{\frac{1}{2}} \right| = \left| \exp\left\{ -\frac{\tau x^2}{4} \right\} - \exp\left\{ \frac{1}{2} \ln\left(1 - \frac{\tau}{2} \right) \right\} \right|$$

$$\le \left| \frac{\tau x^2}{4} + \frac{1}{2} \ln\left(1 - \frac{\tau}{2} \right) \right|.$$

Hence, we finally get $e(\varphi) = O(\tau)$ for any $\varphi \in C_p^\infty(\mathbb{R})$ based on the fact that $\int_{\mathbb{R}} x^p e^{-\frac{(2-\tau)x^2}{4}} dx < \infty$ for $p \in \mathbb{N}$.

For SDEs with degenerate additive noise, such as the Langevin equation (2.6), and general numerical schemes in the form $X_{n+1} = F(X_n, \Delta_{n+1}W)$, the following theorem gives the error between invariant measures.

Theorem 2.11 (Theorem 7.3, [139]) *For SDEs with degenerate additive noise and satisfying Assumptions 2.1 and 2.2, if the numerical solution of scheme $X_{n+1} = F(X_n, \Delta_{n+1}W)$ satisfies assumptions in Proposition 2.2 and Assumption 2.2, then both the exact solution and the numerical one possess unique invariant measures μ and μ^τ, respectively. If, in addition,*

$$\mathbb{E}\|X(t_n) - X_n\|^2 \le C(1 + \|x\|^2)\tau^s$$

for any $n \in \mathbb{N}$ with initial value $X(0) = x$, then there exists some constant $\rho \in (0, \frac{1}{2})$ such that

$$e(\varphi) \le C\tau^{s\rho}, \quad \forall\, \varphi \in \mathbf{C}_b(\mathbb{H}).$$

We refer to [179] and references therein for the study of degenerate SDEs with multiplicative noises and the implicit Euler scheme. The author shows the ergodicity of the exact solution and the numerical one, and gives the error between invariant measures.

2.5 Approximation of the Ergodic Limit

In some other circumstances, the average of a broad class of empirical functions φ with respect to the invariant measure μ, i.e., the ergodic limit $\bar{\varphi} := \int_{\mathbb{H}} \varphi d\mu$, may be of more importance than the behavior of the solution itself. To approximate the ergodic limit, one usually construct a sequence of measures μ_N such that $\int_{\mathbb{H}} \varphi d\mu_N \to \int_{\mathbb{H}} \varphi d\mu$ in some sense as $N \to \infty$, while the numerical solution need not to be ergodic anymore.

Let X denote the exact solution of (2.1), and $\{X_n\}_{n\in\mathbb{N}}$ be a numerical solution of some proper numerical scheme. Denote the auxiliary measures μ_N, $N \in \mathbb{N}$, such that

$$\int_{\mathbb{H}} \varphi d\mu_N := \frac{1}{t_N} \sum_{n=1}^{N} \left[\tau_n \int_{\mathbb{H}} \varphi d\delta_{X_n} \right] = \frac{1}{t_N} \sum_{n=1}^{N} [\tau_n \varphi(X_n)]$$

with $t_n = \sum_{k=1}^{n} \tau_k$, $\tau_n = t_n - t_{n-1}$ and δ_{X_n} being the Dirac measure centered on the point X_n. When the step-size τ_n is decreasing, i.e., $\lim_{n\to\infty} \tau_n = 0$ and $\lim_{n\to\infty} t_n = \infty$, it has been shown that

$$\int_{\mathbb{H}} \varphi d\mu_N \to \int_{\mathbb{H}} \varphi d\mu =: \bar{\varphi}, \quad \text{a.s.} \tag{2.35}$$

We refer to [130, 131, 142, 159] and references therein for more details. However, if $\tau_n \equiv \tau$, we have $t_n = n\tau$ and

$$\int_{\mathbb{H}} \varphi d\mu_N = \frac{1}{N} \sum_{n=1}^{N} \varphi(X_n),$$

which together with the local weak error, also yields (2.35) (see e.g. [140, 148]) taking advantage of the Poisson equation (see e.g. [160–162, 175]). This approach relies especially on the local error of the numerical schemes, and is also applicable to the case with decreasing step-size τ_n.

This section contains a brief introduction of the approach mentioned above for (2.1) possessing a unique invariant measure μ under either uniform elliptic setting (see Theorem 2.3) or hypoelliptic setting (see Theorem 2.2).

We first give the Poisson equation on $\mathbb{H}$ associated to (2.1)

$$\mathscr{L}\Phi = \varphi - \bar{\varphi}, \tag{2.36}$$

where $\mathscr{L}$ is defined in (2.2) and the function $\varphi - \bar{\varphi}$ is centered in the sense $\int_{\mathbb{H}}(\varphi - \bar{\varphi})d\mu = 0$. The well-posedness of (2.36) has been studied in [160] for the elliptic setting and in [162] for the degenerate case. However, to gain the convergence rate of the temporal average $\frac{1}{N} \sum_{n=1}^{N} \varphi(X_n)$ when approximate the ergodic limit $\bar{\varphi}$, it is essential for Φ to be regular enough. Thus, in the following, we assume that the state space is compact, namely, $\mathbb{H} = \mathbb{T}^d$ for simplicity, where $\mathbb{T}^d$ denotes the torus in $\mathbb{R}^d$. In this case, the regularity of Φ is stated in the following theorem.

Theorem 2.12 (Theorem 4.1, [140]) *Assume that (2.1) possesses a unique invariant measure under the elliptic setting (resp. hypoelliptic setting) with smooth coefficients b and σ. For any $\varphi \in \mathbf{W}^{p,\infty}(\mathbb{H})$ (resp. $\varphi \in \mathbf{W}^{p+2,\infty}(\mathbb{H})$) with $p \in \mathbb{N}$, there exists a unique solution $\Phi \in \mathbf{W}^{p+2,\infty}(\mathbb{H})$ to (2.36).*

Assume that a numerical scheme could be solved explicitly by

$$X_{n+1} = X_n + \tau F(X_n, \tau) + G(X_n, \tau)\triangle_{n+1}W \tag{2.37}$$

with increments $\triangle_{n+1}W = W(t_{n+1}) - W(t_n)$ and, in addition, F and G satisfy that

$$\|b(x) - F(x, \tau)\| + \|\sigma(x) - G(x, \tau)\|_F \leq C\tau \tag{2.38}$$

for all $x \in \mathbb{H}$, sufficiently small τ and some constant C.

Note that there is an equivalent condition to (2.38), which reads that the local weak error between the exact solution and the numerical one is of order two, and is stated in the following proposition.

Proposition 2.3 *Consider (2.1) and (2.37) with smooth coefficients b and σ. For $X(0) = X_0$ and $\varphi \in \mathbf{W}^{4,\infty}(\mathbb{H})$, the following two conditions are equivalent:*

(i) $\|b(x) - F(x, \tau)\| + \|\sigma(x) - G(x, \tau)\|_F \leq C\tau, \forall x \in \mathbb{H}$,

(ii) $|\mathbb{E}[\varphi(X(\tau))] - \mathbb{E}[\varphi(X_1)]| = O(\tau^2)$.

Proof Based on Itô's formula and the uniform boundedness of b, σ and $\nabla^p \varphi$, $p = 1, \cdots, 4$, on a compact set, we have

$$\mathbb{E}[\varphi(X(\tau))] = \varphi(x) + \int_0^\tau \mathbb{E}[\mathscr{L}\varphi(X(s))]ds,$$

$$= \varphi(x) + \tau\mathscr{L}\varphi(x) + \int_0^\tau \int_0^s \mathbb{E}\left[\mathscr{L}^2\varphi(X(r))\right]drds$$

$$= \varphi(x) + \tau\mathscr{L}\varphi(x) + O(\tau^2). \tag{2.39}$$

According to the Taylor expansion, it shows

$$\mathbb{E}[\varphi(X_1)]$$
$$= \varphi(x) + \mathbb{E}\left[\nabla\varphi(x)\,(\tau F + G\triangle_1 W)\right] + \mathbb{E}\left[\frac{1}{2}\nabla^2\varphi(x)\,(\tau F + G\triangle_1 W)^2\right]$$
$$+ \mathbb{E}\left[\frac{1}{3!}\nabla^3\varphi(x)\,(\tau F + G\triangle_1 W)^3\right] + \mathbb{E}\left[\frac{1}{4!}\nabla^4\varphi(\theta x + (1-\theta)X_1)\,(\tau F + G\triangle_1 W)^4\right]$$
$$= \varphi(x) + \tau\left[\nabla\varphi(x)F + \frac{1}{2}\nabla^2\varphi(x)G^2\right] + \frac{\tau^2}{2}\left(\nabla^2\varphi(x)F^2 + \nabla^3\varphi(x)FG^2\right)$$
$$+ \frac{\tau^3}{6}\nabla^3\varphi(x)F^3 + \frac{1}{24}\mathbb{E}\left[\nabla^4\varphi(\theta x + (1-\theta)X_1)\,(\tau F + G\triangle_1 W)^4\right] \tag{2.40}$$

with F and G taking values at (x, τ) and $\theta \in (0, 1)$. Here, we write $\nabla^p\varphi(x)(y_1, \cdots, y_p)$ for the pth derivative evaluated in directions y_j, $j = 1, \cdots, p$, and $\nabla^p\varphi(x)y^p$ for short if all the directions are the same. The difference between (2.39) and (2.40) yields

$$\mathbb{E}[\varphi(X(\tau))] - \mathbb{E}[\varphi(X_1)]$$
$$= \tau\left[\nabla\varphi(x)(b(x) - F) + \frac{1}{2}\nabla^2\varphi(x)\left(\sigma(x)^2 - G^2\right)\right] - \frac{\tau^2}{2}\left(\nabla^2\varphi(x)F^2 + \nabla^3\varphi(x)FG^2\right)$$
$$- \frac{\tau^3}{6}\nabla^3\varphi(x)F^3 - \frac{1}{24}\mathbb{E}\left[\nabla^4\varphi(\theta x + (1-\theta)X_1)\,(\tau F + G\triangle_1 W)^4\right] + O(\tau^2),$$

from which one can get the equivalence of conditions (*i*) and (*ii*), since (*i*) could also imply the boundedness of F and G. □

It is worth mentioning that most numerical schemes can not be expressed as (2.37), but as

$$X_{n+1} = X_n + K(X_n, \tau, \triangle_{n+1}W) \tag{2.41}$$

for some function $K : \mathbb{H} \times (0, 1) \times \mathbb{R}^m \to \mathbb{H}$, according to the well-posedness of the scheme. The condition (2.38) is not available anymore in this circumstance. Thus, for general schemes, condition (ii) in Proposition 2.3 is frequently used to obtain the approximate error of the ergodic limit.

Theorem 2.13 *For any $\varphi \in \mathbf{W}^{4,\infty}(\mathbb{H})$, assume that the assumptions in Theorem 2.12 and condition (ii) in Proposition 2.3 hold for (2.1) and (2.41). Then*

$$\left| \frac{1}{N} \sum_{n=0}^{N-1} \mathbb{E}[\varphi(X_n)] - \bar{\varphi} \right| \leq C \left(\tau + \frac{1}{T} \right)$$

for $T = N\tau$ and $X_0 = X(0) = x$.

We refer to [140] for a similar version of this theorem, which is proved under the equivalent condition (2.38), and for a general version when the numerical scheme has local weak order $p + 1$ which is also introduced at the end of this section. The proof of Theorem 2.13 is given here for the readers' convenience.

Proof As (2.36) admits a unique solution $\Phi \in \mathbf{W}^{4,\infty}(\mathbb{H})$ according to Theorem 2.12, we derive that

$$\frac{1}{N} \sum_{n=0}^{N-1} \mathbb{E}[\varphi(X_n)] - \bar{\varphi} = \frac{1}{N} \sum_{n=0}^{N-1} \mathbb{E}[\mathscr{L}\Phi(X_n)]. \tag{2.42}$$

If we take X_n as an initial value and denote the solution at time τ as $X_{X_n}(\tau)$, then

$$\mathbb{E}[\Phi(X_{X_n}(\tau))] = \mathbb{E}[\Phi(X_n)] + \mathbb{E}\left[\int_0^\tau \mathscr{L}\Phi(X(s))ds \right]$$

$$= \mathbb{E}[\Phi(X_n)] + \tau\mathbb{E}[\mathscr{L}\Phi(X_n)] + \mathbb{E}\left[\int_0^\tau \int_0^s \mathscr{L}^2\Phi(X(r))drds \right]$$

$$= \mathbb{E}[\Phi(X_n)] + \tau\mathbb{E}[\mathscr{L}\Phi(X_n)] + O(\tau^2)$$

as $\Phi \in \mathbf{W}^{4,\infty}(\mathbb{H})$. On the other hand, since $\mathbb{E}[\varphi(X(\tau))] - \mathbb{E}[\varphi(X_1)] = O(\tau^2)$ for any initial value $x \in \mathbb{H}$ and any $\varphi \in \mathbf{W}^{4,\infty}(\mathbb{H})$, it leads to

$$\mathbb{E}[\Phi(X_{n+1})] - \mathbb{E}[\Phi(X_{X_n}(\tau))] = O(\tau^2).$$

Thus, we have

$$\frac{1}{N} \sum_{n=0}^{N-1} \mathbb{E}[\mathscr{L}\Phi(X_n)] = \frac{1}{N} \sum_{n=0}^{N-1} \frac{\mathbb{E}[\Phi(X_{X_n}(\tau)) - \Phi(X_n)]}{\tau} + O(\tau)$$

$$= \frac{1}{N} \sum_{n=0}^{N-1} \frac{\mathbb{E}[\Phi(X_{n+1}) - \Phi(X_n)]}{\tau} + O(\tau) = \frac{1}{T}(\mathbb{E}[\Phi(X_N)] - \Phi(x)) + O(\tau),$$

which, together with the uniform boundedness of Φ and (2.42), completes the proof. $\square$

Based on the results above, one can get the approximate error of the ergodic limit, which contains two parts intuitively: the weak error between the numerical solution and the exact one, and the error between the temporal average and the spatial one. More precisely, it can be separated into the following two parts

$$\frac{1}{N}\sum_{n=0}^{N-1}\mathbb{E}[\varphi(X_n)] - \bar{\varphi} = \frac{1}{T}\sum_{n=0}^{N-1}\int_{t_n}^{t_{n+1}}\Big(\mathbb{E}[\varphi(X_n)] - \mathbb{E}[\varphi(X(t))]\Big)dt$$

$$+ \frac{1}{T}\int_0^T\mathbb{E}[\varphi(X(t))]dt - \int_{\mathbb{H}}\varphi d\mu.$$

This approach is also applicable for high order cases, stated as follows.

Assumption 2.3 For all $\varphi \in \mathbf{W}^{2(p+2),\infty}(\mathbb{H})$ and sufficiently small τ

$$|\mathbb{E}[\varphi(X(\tau))] - \mathbb{E}[\varphi(X_1)]| \le C\tau^{p+1},$$

where the constant C depends on the initial value $X(0) = X_0 = x$, and is uniform over all φ with $\|\varphi\|_{\mathbf{W}^{2(p+2),\infty}(\mathbb{H})} \le 1$.

Theorem 2.14 (Theorem 5.6, [140]) *Assume that (2.1) and the numerical solution to (2.41) satisfy the assumptions in Theorem 2.12 and Assumption 2.3. Then for any $\varphi \in \mathbf{W}^{2(p+1),\infty}(\mathbb{H})$ with $\|\varphi\|_{\mathbf{W}^{2(p+1),\infty}(\mathbb{H})} \le 1$ and $X_0 = X(0) = x$,*

$$\left|\frac{1}{N}\sum_{n=0}^{N-1}\mathbb{E}[\varphi(X_n)] - \bar{\varphi}\right| \le C\left(\tau^p + \frac{1}{T}\right)$$

with $T = N\tau$.

We refer to [140] for the proof of above theorem, which can also be found in Chap. 6 for a finite dimensional approximation of stochastic Schrödinger equations.

Summary

This chapter mainly focuses on the existence and uniqueness of invariant measures for stochastic ordinary differential equations and numerical approximations of invariant measures. Non-degenerate SDEs and an important kind of degenerate SDEs–stochastic Langevin equations—are taken as the keystone to illustrate the procedure of approximating invariant measures via numerical schemes.

The stochastic Langevin equation introduced in Sect. 2.3 admits a unique invariant Boltzmann–Gibbs measure, which gives the probability that a system will be in a certain state and shows that states with lower energy will always have a higher probability of being occupied than the states with higher energy. We also refer to [27, 156, 166] for the study of Boltzmann–Gibbs measures for deterministic Korteweg–de Vries equations, and refer to [27, 28, 157, 181] for the study of deterministic nonlinear Schrödinger equations, which is also briefly introduced in Sect. 3.2.

Moreover, as a stochastic dissipative Hamiltonian system, the stochastic Langevin equation can be transformed to an equivalent autonomous stochastic Hamiltonian system. The generating function is then employed to construct conformal symplectic schemes for the stochastic Langevin equation, and the modified equation technique (inspired by the backward error analysis) is adopted to improve the accuracy of the proposed schemes and to release the simulation of the multiple integrals. There is plenty of work related to the construction of high order schemes for both deterministic and stochastic systems. We refer to [41, 122, 141, 164, 176, 195] for high order symplectic Runge–Kutta methods, to [7, 8, 83, 154, 185, 186, 188] for symplectic schemes constructed through generating functions, and to [1, 2, 46, 69, 126, 127, 163, 171, 180] for high order integrators based on modified equations.

Chapter 3
Invariant Measures for Stochastic Nonlinear Schrödinger Equations

This chapter focuses on the existence and uniqueness of invariant measures for stochastic nonlinear Schrödinger equations, and recalls several results related to the well-posedness and continuous dependence on the initial value as the starting point of this chapter.

Section 3.1 introduces some notations used throughout the following chapters. In addition, we recall some fundamental conserved quantities of deterministic nonlinear Schrödinger equations, which are essential in proving the well-posedness of Schrödinger equations.

Section 3.2 gives the expression of an invariant measure for the deterministic nonlinear Schrödinger equation under periodic setting, utilizing a Fourier finite dimensional truncation of the considered model.

Section 3.3 introduces the existence and uniqueness of solutions for stochastic nonlinear Schrödinger equations driven by additive or multiplicative noises.

Section 3.4 shows the pathwise continuous dependence on the initial data, which indicates that the transition semigroup generated by the solution is Markovian and Feller. These results form the basis of the study on invariant measures for stochastic Schrödinger equations.

Sections 3.5 and 3.6 are devoted to the study of invariant measures and ergodicity for stochastic Schrödinger equations with weak damping in different dimensions.

3.1 Preliminaries

We first introduce some notations used throughout the following chapters, and recall some classical results for deterministic nonlinear Schrödinger equations.

- Let $(\Omega, \mathscr{F}, \mathbb{P}, \{\mathscr{F}_t\}_{t \geq 0})$ be a filtered probability space.

© Springer Nature Singapore Pte Ltd. 2019
J. Hong and X. Wang, *Invariant Measures for Stochastic Nonlinear Schrödinger Equations*, Lecture Notes in Mathematics 2251,
https://doi.org/10.1007/978-981-32-9069-3_3

- Let $\{\beta_k\}_{k\in\mathbb{N}}$ be a family of mutually independent $\mathbb{R}$-valued standard Brownian motions on $(\Omega, \mathscr{F}, \mathbb{P}, \{\mathscr{F}_t\}_{t\geq 0})$.
- Denote by $\mathbf{L}^p(\mathscr{O})$ the classical Lebesgue space of $\mathbb{C}$-valued functions for $p \geq 1$. When $p = 2$, $\mathbf{L}^2(\mathscr{O})$ denotes the space of $\mathbb{C}$-valued square integrable functions with inner product

$$(u, v) = \int_{\mathscr{O}} u(x)\overline{v(x)}dx$$

  for $u, v \in \mathbf{L}^2(\mathscr{O})$. In particular, $\mathbf{L}^2(\mathscr{O}; \mathbb{R})$ denotes the space of $\mathbb{R}$-valued square integrable functions. Domain $\mathscr{O} \subset \mathbb{R}^d, d \geq 1$, may be different in the following sections, and will be pointed out at the beginning of each section.
- Denote by $\mathbf{H}^s(\mathscr{O}) := \mathbf{W}^{s,2}(\mathscr{O})$ the classical Sobolev space for $s \geq 0$, which is a Hilbert space endowed with inner product

$$(u, v)_{\mathbf{H}^s(\mathscr{O})} = \int_{\mathscr{O}} \sum_{|\alpha|\leq s} \left(D^\alpha u \overline{D^\alpha v}\right) dx, \quad u, v \in \mathbf{H}^s(\mathscr{O}).$$

- Denote by $\mathscr{L}_2^s := \mathscr{L}_2(\mathbf{L}^2(\mathscr{O}); \mathbf{H}^s(\mathscr{O})), s \geq 0$, the space of Hilbert–Schmidt operators from $\mathbf{L}^2(\mathscr{O})$ to $\mathbf{H}^s(\mathscr{O})$ equipped with the norm

$$\|\Phi\|_{\mathscr{L}_2^s} = \left(\sum_{k\in\mathbb{N}} \|\Phi e_k\|_{\mathbf{H}^s(\mathscr{O})}^2\right)^{\frac{1}{2}}, \quad \Phi \in \mathscr{L}_2^s.$$

- For a Banach space B, the space of γ-radonifying operators from $\mathbf{L}^2(\mathscr{O})$ to B is denoted by $R(\mathbf{L}^2(\mathscr{O}); B)$ with norm

$$\|\Psi\|_{R(\mathbf{L}^2(\mathscr{O}); B)} = \left(\tilde{\mathbb{E}}\left\|\sum_{k\in\mathbb{N}} \gamma_k \Psi e_k\right\|_B^2\right)^{\frac{1}{2}}, \quad \Psi \in R(\mathbf{L}^2(\mathscr{O}); B),$$

  where $\{\gamma_k\}_{k\in\mathbb{N}}$ is any sequence of mutually independent $\mathbb{R}$-valued normal random variables on another probability space $(\tilde{\Omega}, \tilde{\mathscr{F}}, \tilde{\mathbb{P}})$, and the norm is independent of $\{\gamma_k\}_{k\in\mathbb{N}}$ and $\{e_k\}_{k\in\mathbb{N}}$. In particular, if B is a Hilbert space, then

$$R(\mathbf{L}^2(\mathscr{O}); B) = \mathscr{L}_2(\mathbf{L}^2(\mathscr{O}); B).$$

The Schrödinger equation, originating in quantum mechanics, was first presented by E. Schrödinger in 1926. Physicists investigate the molecular structure and atomic structure of actual substance, and even subatomic and macroscopic systems, via solving time dependent evolution problems for the Schrödinger equation. It is central to the applications of quantum mechanics, statistical mechanics, plasma physics, as well as nonlinear optics.

Recall that the classical nonlinear Schrödinger equation

$$\begin{cases} du = \mathbf{i}\Delta u dt + \mathbf{i}\lambda|u|^{2\sigma} u dt, \\ u(0, x) = u_0(x), \ x \in \mathscr{O} \subset \mathbb{R}^d, \end{cases} \tag{3.1}$$

has an infinite number of invariant quantities. Among them, the following conservation laws are frequently considered in practical problems (see [29, 34, 81, 117] and references therein).

Charge conservation law. The charge of (3.1) is defined as

$$M(u(t)) := \|u(t)\|_{\mathbf{L}^2(\mathscr{O})}^2 = \int_{\mathscr{O}} |u(t, x)|^2 dx,$$

in which $|u(x, t)|^2$ represents the density for the particle to appear in state x at time t in quantum mechanics. The law of conservation of the charge reads

$$M(u(t)) = M(u_0)$$

for (3.1) under Dirichlet boundary condition.

Energy conservation law. The Hamiltonian

$$H(u(t)) := \frac{1}{2}\int_{\mathscr{O}} |\nabla u(t, x)|^2 dx - \frac{\lambda}{2\sigma + 2}\int_{\mathscr{O}} |u(t, x)|^{2\sigma+2} dx$$

is shown to be preserved for any time when the solution is well-posed and sufficiently smooth, i.e.,

$$H(u(t)) = H(u_0).$$

3.2 Invariant Measures for Deterministic Nonlinear Schrödinger Equations

In this section, we consider the invariant distributions of the following one dimensional nonlinear Schrödinger equation under periodic setting:

$$\begin{cases} \dot{u} = \mathbf{i}\Delta u + \mathbf{i}\lambda|u|^{2\sigma} u, \\ u(t, 0) = u(t, L), \quad t \in \mathbb{R}, \\ u(0, x) = u_0(x), \quad x \in (0, L). \end{cases} \tag{3.2}$$

Denote $\mathbf{L}^p := \mathbf{L}^p(0, L)$. We obtain from above that its Hamiltonian reads

$$H(u) = \frac{1}{2}\|\nabla u\|_{\mathbf{L}^2}^2 - \frac{\lambda}{2\sigma + 2}\|u\|_{\mathbf{L}^{2\sigma+2}}^{2\sigma+2},$$

based on which the authors in [132] define a formal and unnormalized Gibbs measure

$$e^{-\beta H(\phi)} \prod_{x \in [0,L]} d\phi(x) \tag{3.3}$$

$$= \exp\left(\frac{\lambda\beta}{2\sigma + 2}\|\phi\|_{L^p}^p\right)\left[\exp\left(-\frac{\beta}{2}\|\nabla\phi\|_{L^2}^2\right)\prod_{x \in [0,L]} d\phi(x)\right]$$

with $d\phi$ denoting the Lebesgue measure in the complex plane. The measure above is ill-defined since there is no analogue of Lebesgue measure on infinite dimensional Banach spaces. Even though it is possible to define Gaussian measure on infinite dimensional spaces, it can not be defined through a density in that case. The authors in [132] then modify the definition of this formal measure as

$$\exp\left(\frac{\lambda\beta}{2\sigma + 2}\|\phi\|_{L^p}^p\right) \mathbf{1}_{\left\{\phi \mid \|\phi\|_{L^2}^2 \leq K\right\}} d\mu_\beta(\phi)$$

with μ_β a Wiener measure on an appropriate statistical ensemble, such that it is well-defined and normalizable under the global well-posedness assumption of the equation. Since the Hamiltonian is not bounded below, the identity function based on the truncation of the L^2-norm of ϕ is added to ensure that the measure is normalizable.

Based on these investigations, the author in [27] proves further that the measure is invariant under the flow of (3.2) via a finite dimensional approximation of equation. More precisely, one can consider a finite dimensional approximation

$$\begin{cases} \dot{u}_N = \mathbf{i}\Delta u_N + \mathbf{i}\lambda\pi_N\left(|u_N|^{2\sigma} u_N\right), \\ u_N(0) = \pi_N u_0 \end{cases} \tag{3.4}$$

with π_N being the projection operator such that

$$\pi_N u(x,t) = \sum_{|n| \leq N} a_n(t) e^{2\pi \mathbf{i} n x}.$$

For the finite dimensional equation (3.4), it possesses the discrete charge and energy conservation laws:

$$M_N(a(t)) := \left(\sum_{|n| \leq N} |a_n(t)|^2\right)^{\frac{1}{2}} = M_N(a(0)),$$

$$H_N(a(t)) := 2\pi^2 \sum_{|n| \leq N} n^2 |a_n(t)|^2 - \frac{\lambda}{2\sigma + 2} \int_0^L \left|\sum_{|n| \leq N} a_n e^{2\pi \mathbf{i} n x}\right|^{2\sigma + 2} dx,$$

where $a(t) := (a_n)_{|n| \leq N} \in \mathbb{C}^{2N+1}$. Furthermore, (3.4) is also a Hamiltonian system

$$\dot{a} = -i \frac{\partial H_N}{\partial \overline{a}}$$

with $\overline{a}$ denoting the conjugate of the complex value a, or equivalently,

$$\dot{p}_N = \frac{\partial \hat{H}_N}{\partial q_N}, \quad \dot{q}_N = -\frac{\partial \hat{H}_N}{\partial p_N},$$

where $(p_N^\top, q_N^\top)^\top \in \mathbb{R}^{4N+2}$, $a = p_N + iq_N$ and $\hat{H}_N(p_N, q_N) = H_N(a)$. For this finite dimensional case, we can also define the measure similar to (3.3):

$$d\mu_N = e^{-H_N(a)} da$$

$$= \exp\left(\frac{\lambda}{2\sigma + 2} \int_0^L \left| \sum_{|n| \le N} a_n e^{2\pi i n x} \right|^{2\sigma + 2} dx \right) \cdot$$

$$\left[da_0 \otimes \exp\left(-2\pi^2 \sum_{|n| \le N, n \ne 0} n^2 |a_n(t)|^2 \right) d(a_n)_{|n| \le N, n \ne 0} \right]$$

$$=: \exp\left(\frac{\lambda}{2\sigma + 2} \int_0^L \left| \sum_{|n| \le N} a_n e^{2\pi i n x} \right|^{2\sigma + 2} dx \right) [da_0 \otimes d\rho_N],$$

where ρ_N is a measure on $\mathbb{C}^{2N}$.

Remark 3.1 The measure ρ_N is a Gaussian measure associated to a $2N$-dimensional Gaussian random variable $X(\omega) := \left(\frac{1}{2\pi n} \xi_n(\omega) \right)_{|n| \le N, n \ne 0}$ and of course can be normalizable, where $\{\xi_n\}_{|n| \le N, n \ne 0}$ is a family of independent normal random variables. In fact, for any Borel set $A \in \mathcal{B}(\mathbb{C}^{2N})$,

$$\mathbb{P}(X(\omega) \in A) = \int_A \prod_{|n| \le N, n \ne 0} \exp\left(-\frac{|a_n|^2}{2 \left(\frac{1}{2\pi n} \right)^2} \right) d(a_n)_{|n| \le N, n \ne 0}$$

$$= \int_A \exp\left(-2\pi^2 \sum_{|n| \le N, n \ne 0} n^2 |a_n(t)|^2 \right) d(a_n)_{|n| \le N, n \ne 0} = \rho_N(A).$$

The measure μ_N is well-defined (see [27, 132]) when choosing the space

$$\Omega_{N,K} := \left\{ a = (a_n)_{|n| \le N} \Big| \|a\| \le K \right\}$$

as the statistical ensemble, where $\| \cdot \|$ denotes the Euclidean norm in the corresponding space.

When considering the limits ρ and μ of ρ_N and μ_N, respectively, one can derive from Lemma 3.10 in [27] that ρ and μ are well-defined for $\sigma \le 2$. Moreover, Eq. (3.2) is globally well-posed μ-almost everywhere and admits an invariant measure with proper σ.

Theorem 3.1 ([27]) *For $\sigma \leq 2$, the initial value problem (3.2) is globally well-posed. If $1 \leq \sigma \leq 2$, the measure μ is invariant under the flow of (3.2).*

For the study of invariant measures for deterministic nonlinear Schrödinger equation in high dimensions, we refer to [26, 27, 30, 132] and references therein for more details.

In the following, we will turn our attention to the study of stochastic nonlinear Schrödinger equations, including their well-posedness, invariant measures, geometric structures (see Chap. 4) and numerical approximations (see Chaps. 4, 5 and 6).

3.3 Well-Posedness of Stochastic Nonlinear Schrödinger Equations

In this section, the local and global existence and uniqueness for the solution of stochastic nonlinear Schrödinger equations on $\mathscr{O} = \mathbb{R}^d$ are introduced, based on the evolution of charge and energy of the solution. Higher regularity of the solution is obtained for the one dimensional case $\mathscr{O} = (0, 1)$.

Let the noise in the considered stochastic nonlinear Schrödinger equation be a $\mathbb{C}$-valued Wiener process for the additive noise case, and be a $\mathbb{R}$-valued Wiener process for the multiplicative noise case. The precise assumption on the noise will be given separately for each case.

3.3.1 The Additive Noise Case

We first consider the nonlinear Schrödinger equation perturbed by an additive noise

$$\begin{cases} du = \mathbf{i}\big(\Delta u + \lambda |u|^{2\sigma} u\big)dt + dW(t), \\ u(0, x) = u_0(x), \ x \in \mathbb{R}^d, \end{cases} \tag{3.5}$$

where W is a Q-Wiener process on $\mathbf{L}^2(\mathbb{R}^d)$ with symmetric and positive definite covariance operator Q. More precisely, $W(t)$ has the following Karhunen–Loève expansion

$$W(t) = \sum_{k=1}^{\infty} Q^{\frac{1}{2}} e_k \beta_k(t)$$

with $\{e_k\}_{k\in\mathbb{N}}$ being an orthonormal basis for $\mathbf{L}^2(\mathbb{R}^d)$, and $\{\beta_k\}_{k\in\mathbb{N}}$ being a sequence of $\mathbb{R}$-valued mutually independent and identically distributed Brownian motions.

We give the well-posedness of the mild solution of the considered equation in this section. Process u is called a *mild solution* to (3.5) on $[0, T]$ if for any $t \in [0, T]$ it satisfies

$$u(t) = S(t)u_0 + i \int_0^t S(t-s)|u(s)|^{2\sigma} u(s)ds + \int_0^t S(t-s)dW(s) \qquad (3.6)$$

almost surely and each integral in the right hand side of above equation is well-defined, where $S(t) = e^{it\Delta}$ is the unitary group on $\mathbf{H}^s(\mathbb{R}^d)$ generated by the linear equation $\dot{u} = i\Delta u$.

The local existence of the solution of (3.6) is given first by defining a stopping time $\tau^*(u_0)$ such that

$$\tau^*(u_0) = +\infty \quad \text{or} \quad \lim_{t \to \tau^*(u_0)} \|u(t)\|_{\mathbf{H}^1(\mathbb{R}^d)} = +\infty \qquad (3.7)$$

holds almost surely.

Theorem 3.2 (Theorem 3.1, [64]) *Assume that $u_0 \in \mathbf{H}^1(\mathbb{R}^d)$ is $\mathscr{F}_0$ measurable, $Q^{\frac{1}{2}} \in \mathscr{L}_2^1$ and that*

$$\sigma \in I^d := \begin{cases} [0, +\infty) & \text{if } d = 1, 2, \\ \left[0, \dfrac{2}{d-2}\right] & \text{if } d \geq 3. \end{cases}$$

Then there exists a unique solution $u \in \mathbf{H}^1(\mathbb{R}^d)$ on $[0, \tau^*(u_0))$ such that $u(0) = u_0$.

To derive the global existence of the solution, that is $\tau^*(u_0) = +\infty$ almost surely, a priori estimates for the charge and energy of (3.5) are required. Different from the deterministic case (3.1), the charge and energy of (3.5) are not conserved any more, but satisfy the following evolution principles.

Proposition 3.1 ([64]) *Under the assumptions in Theorem 3.2, the charge and the energy of the solution of (3.5) satisfy*

$$M(u(\tau)) = M(u_0) + 2\Re\left[\sum_{k \in \mathbb{N}} \int_0^\tau (u(s), Q^{\frac{1}{2}} e_k)d\beta_k(s)\right] + \tau\|Q^{\frac{1}{2}}\|_{\mathscr{L}_2^0}^2$$

and

$$H(u(\tau)) = H(u_0) - \Re\left[\int_0^\tau (\Delta u + \lambda|u|^{2\sigma}u, dW(t))\right] + \frac{1}{2}\sum_{k \in \mathbb{N}} \int_0^\tau \|\nabla(Q^{\frac{1}{2}}e_k)\|_{L^2(\mathbb{R}^d)}^2 ds$$
$$- \frac{\lambda}{2}\sum_{k \in \mathbb{N}} \int_0^\tau \left[\left\||u|^\sigma Q^{\frac{1}{2}}e_k\right\|_{L^2(\mathbb{R}^d)}^2 + 2\sigma\left\||u|^{\sigma-1}\Re[\bar{u}Q^{\frac{1}{2}}e_k]\right\|_{L^2(\mathbb{R}^d)}^2\right]ds.$$

Moreover, for any $p \in \mathbb{N}$, the uniform estimate for the pth moment of $M(u(t))$ holds: there exists a constant $C_p > 0$ such that

$$\mathbb{E}\left[\sup_{t\in[0,\tau]} M^p(u(t))\right] \leq C_p\mathbb{E}\left[M^p(u_0)\right].$$

Based on the a priori estimate above, the global well-posedness of the solution is derived by showing the uniform boundedness of the second moment of $\|u(t)\|_{\mathbf{H}^1(\mathbb{R}^d)}$, i.e.,

$$\mathbb{E}\left[\sup_{t\leq\tau} \|u(t)\|_{\mathbf{H}^1(\mathbb{R}^d)}^2\right] \leq C(T, Q, u_0)$$

with any given $T > 0$ and stopping time $\tau < \inf\{T, \tau^*(u_0)\}$.

Theorem 3.3 (Theorem 3.4, [64]) *Assume that the assumptions in Theorem 3.2 hold, and that $\sigma < \frac{2}{d}$ or $\lambda = -1$. Then the solution given in Theorem 3.2 is global, i.e., $\tau^*(u_0) = +\infty$ almost surely.*

3.3.2   The Multiplicative Noise Case

For nonlinear Schrödinger equation perturbed by the linear multiplicative noise

$$\begin{cases} du = i(\Delta u + \lambda|u|^{2\sigma}u)dt + iu \circ dW(t), \\ u(0, x) = u_0(x), \ x \in \mathbb{R}^d, \end{cases} \tag{3.8}$$

the driven noise $W(t) = \sum\limits_{k=0}^{\infty} Q^{\frac{1}{2}}e_k\beta_k(t)$ is a Q-Wiener process on $\mathbf{L}^2(\mathbb{R}^d; \mathbb{R})$ with a symmetric and positive definite covariance operator Q. Here, $\{e_k\}_{k\in\mathbb{N}}$ is an orthonormal basis in $\mathbf{L}^2(\mathbb{R}^d; \mathbb{R})$, and $\{\beta_k\}_{k\in\mathbb{N}}$ as above. The stochastic integral arising on the right hand side of (3.8) holds in the Stratonovich sense, and the solution possesses the charge conservation law almost surely.

In this subsection, the global existence and uniqueness of the mild solution of (3.8) in $\mathbf{L}^2(\mathbb{R}^d)$, $\mathbf{H}^1(\mathbb{R}^d)$ and $\mathbf{H}^s(0, 1)$ for $s \geq 2$ are introduced.

Theorem 3.4 (Theorem 2.1, [61]) *Assume that $Q^{\frac{1}{2}} \in \mathscr{L}_2^0(\mathbb{R}^d) \cap R(\mathbf{L}^2(\mathbb{R}^d), \mathbf{L}^{2+\delta}(\mathbb{R}^d))$ for some $\delta > 2(d-1)$. Let parameters p, r and ρ satisfy*

$$\begin{cases} \max\left\{2\sigma + 2, 2\left(\dfrac{2}{\delta} + 1\right)\right\} \leq p \ \ \text{if} \ d = 1, \\ \max\left\{2\sigma + 2, 2\left(\dfrac{2}{\delta} + 1\right)\right\} \leq p \leq \dfrac{2d}{d-1} \ \ \text{if} \ d \geq 2, \end{cases}$$

$$\frac{2}{r} = d\left(\frac{1}{2} - \frac{1}{p}\right) \ \ \text{and} \ \ \rho \geq \max\left\{r, (2\sigma + 2)\left(\frac{4\sigma}{2 - d\sigma} + 1\right)\right\}.$$

Then for any $\mathscr{F}_0$-measurable initial datum $u_0 \in \mathbf{L}^\rho(\Omega; \mathbf{L}^2(\mathbb{R}^d))$ and $T > 0$, there is a unique solution

$$u \in \mathbf{L}^\rho(\Omega; \mathbf{C}([0, T]; \mathbf{L}^2(\mathbb{R}^d))) \cap \mathbf{L}^1(\Omega; \mathbf{L}^r([0, T]; \mathbf{L}^p(\mathbb{R}^d)))$$

of (3.8), which possesses the charge conservation law almost surely

$$M(u(t)) = M(u_0), \quad \forall t \geq 0.$$

To study the well-posedness of (3.8) in energy space $\mathbf{H}^1(\mathbb{R}^d)$, similar to the additive noise case, local existence and uniqueness is given first utilizing the stopping time $\tau^*(u_0)$ defined through (3.7).

Theorem 3.5 (Theorem 4.1, [64]) *Assume that $u_0 \in \mathbf{H}^1(\mathbb{R}^d)$ is $\mathscr{F}_0$-measurable, $Q^{\frac{1}{2}} \in \mathscr{L}_2^1 \cap R(\mathbf{L}^2(\mathbb{R}^d); \mathbf{W}^{1,\alpha}(\mathbb{R}^d))$ with $\alpha > 2d$ and that*

$$\sigma \in I^d := \begin{cases} (0, +\infty) & \text{if } d = 1, 2, \\ (0, 2) & \text{if } d = 3, \\ \left[\frac{1}{2}, \frac{2}{d-2}\right) \cup \left(0, \frac{1}{d-1}\right) & \text{if } d \geq 4. \end{cases}$$

Then there exists a unique solution

$$u \in \mathbf{C}([0, \tau]; \mathbf{H}^1(\mathbb{R}^d)) \cap \mathbf{L}^r([0, \tau]; \mathbf{W}^{1,p}(\mathbb{R}^d))$$

almost surely for any $\tau < \tau^*(u_0)$ and some $r \geq 2$ and p satisfying $\frac{2}{r} = d\left(\frac{1}{2} - \frac{1}{p}\right)$, where stopping time $\tau^*(u_0)$ is defined through (3.7).

The energy for the multiplicative noise case is not preserved, but satisfies the following evolution principle.

Proposition 3.2 (Proposition 4.5, [64]) *Assume that the assumptions in Theorem 3.5 hold. Then the solution u given in Theorem 3.5 satisfies*

$$H(u(\tau)) = H(u_0) - \Re\left[\sum_{k=0}^\infty \int_0^\tau \left(\nabla u, u\nabla(Q^{\frac{1}{2}}e_k)\right) d\beta_k(s)\right]$$
$$+ \frac{1}{2}\sum_{k=0}^\infty \int_0^\tau \|u\nabla(Q^{\frac{1}{2}}e_k)\|_{\mathbf{L}^2(\mathbb{R}^d)}^2 ds,$$

for any stopping time τ such that $\tau < \tau^*(u_0)$ almost surely, where $H(u)$ is the same as that in Sect. 3.1.

The global well-posedness of (3.8) in $\mathbf{H}^1(\mathbb{R}^d)$ is obtained based on the uniform estimate of the energy $H(u(t))$

$$\mathbb{E}\left[\sup_{t\le\tau}\|u(t)\|_{\mathbf{H}^1(\mathbb{R}^d)}^2\right] \le C(T, Q, u_0)$$

as stated in the additive noise case.

Theorem 3.6 (Theorem 4.6, [64]) *Assume that the assumptions in Theorem 3.5 hold and that either $\sigma < \frac{2}{d}$ or $\lambda = -1$. Then the solution given in Theorem 3.5 is global, i.e., $\tau^*(u_0) = +\infty$ almost surely.*

The global well-posedness result above is valid for the equation with a subcritical or defocusing nonlinearity. Here, the subcritical condition is given as $\sigma < \frac{2}{d}$, which is named as a comparison with the $\mathbf{H}^s(\mathbb{R}^d)$-critical condition $\sigma < \frac{2}{d-2s}$ proposed in the deterministic case (see e.g. [27, 44, 91]).

We refer the reader to [15, 16] for the existence and uniqueness of strong solutions in $\mathbf{L}^2(\mathbb{R}^d)$ and $\mathbf{H}^1(\mathbb{R}^d)$ of stochastic nonlinear Schrödinger equations with multiplicative noises, including both conservative and nonconservative cases.

For stochastic nonlinear Schrödinger equation (3.8) on bounded domain $\mathscr{O}$ with homogenous Dirichlet boundary condition, i.e., $u(t, x) = 0$ on $(0, T] \times \partial\mathscr{O}$ for some $T > 0$, the uniform boundedness of the solution in the energy space is also obtained for the defocusing case.

Proposition 3.3 (Corollary 1, [49]) *Let $p \ge 1$, $\lambda = -1$, $\sigma = 1$, $\mathscr{O} \subset \mathbb{R}^d$ be a bounded Lipschitz domain. The initial value u_0 is $\mathscr{F}_0$ measurable with homogenous Dirichlet boundary condition and $\mathbb{E}[H^p(u_0)] < \infty$. Assume that u is the solution of (3.8) in domain $(0, T] \times \mathscr{O}$ with homogenous Dirichlet boundary condition. Then there exists a constant $C = C(Q, T, u_0, p)$ such that*

$$\mathbb{E}\left[\sup_{0\le t\le T}\left(\|\nabla u(t)\|_{\mathbf{L}^2(\mathscr{O})}^{2p} + \|u(t)\|_{\mathbf{L}^4(\mathscr{O})}^{4p}\right)\right] \le C.$$

In particular, for the one dimensional case $\mathscr{O} = (0, 1)$, uniform boundedness of the solution in $\mathbf{H}^1(0, 1)$ and $\mathbf{H}^2(0, 1)$ is derived for both focusing and defocusing cases.

Proposition 3.4 ([49, 55]) *Let $\mathscr{O} = (0, 1)$, $\sigma = 1$ and $p \ge 1$. Assume that the initial value $u_0 \in \mathbf{H}^2(0, 1)$ is $\mathscr{F}_0$ measurable with homogenous Dirichlet boundary condition and $Q^{\frac{1}{2}} \in \mathscr{L}_2^2$. Then there exists a constant $C = C(Q, T, u_0, p)$ such that*

$$\mathbb{E}\left[\sup_{0\le t\le T}\|u(t)\|_{\mathbf{H}^2(0,1)}^{2p}\right] \le C.$$

The regularity of the solution can be further improved for the one dimensional case $\mathscr{O} = (0, 1)$ utilizing the Lyapunov functional

$$f(u) := \|\nabla^s u\|_{\mathbf{L}^2}^2 - \lambda\Re\left((-\Delta)^{s-1}u, |u|^2u\right), \quad u \in \mathbf{H}^s(0, 1)$$

for $s \ge 2$. We refer to [55] for more details.

3.4 Continuous Dependence of the Solutions on the Initial Data

The continuous dependence of the solutions on the initial data plays an important role in proving the existence of invariant measures. In fact, the Krylov–Bogoliubov theorem (see Theorem 1.2) is frequently used in proving the existence of invariant measures for time homogenous Markov processes (see e.g. [31, 76]). It requires the Feller property of the semigroup generated through the solution, which is a consequence of the continuous dependence on the initial data (see e.g. [60, 76]).

The continuous dependence on the initial data and noise is established in [64] by setting $z(t) = \int_0^t S(t - s)dW(s)$ and $v(t) = u(t) - z(t)$. If $v(t)$ satisfies

$$v(t) = S(t)u_0 + i\lambda \int_0^t S(t - s)|v(s) + z(s)|^{2\sigma}(v(s) + z(s))ds, \qquad (3.9)$$

then $u(t)$ is the mild solution of (3.5). We denote by $v(z, u_0, \cdot)$ the solution of (3.9) if exists.

Theorem 3.7 (Proposition 3.5, [64]) *Assume that* $u_0^* \in \mathbf{H}^1(\mathbb{R}^d)$

$$z^* \in \mathbf{C}([0, T]; \mathbf{H}^1(\mathbb{R}^d)) \cap \mathbf{L}^r([0, T]; \mathbf{W}^{1,2\sigma+2}(\mathbb{R}^d))$$

with $r = \frac{4(\sigma+1)}{d\sigma}$ *and solution* $v(z^*, u_0^*, \cdot)$ *to* (3.9) *exists in* $\mathbf{C}([0, T]; \mathbf{H}^1(\mathbb{R}^d))$. *Then there exist neighborhoods* $B_{z^*} \subset \mathbf{C}([0, T]; \mathbf{H}^1(\mathbb{R}^d)) \cap \mathbf{L}^r([0, T]; \mathbf{W}^{1,2\sigma+2}(\mathbb{R}^d))$ *of* z^* *and* $B_{u_0^*} \subset \mathbf{C}([0, T]; \mathbf{H}^1(\mathbb{R}^d))$ *of* u_0^* *such that for arbitrary* $(z, u_0) \in B_{z^*} \times B_{u_0^*}$ *there exists a unique solution*

$$v(z, u_0, \cdot) \in \mathbf{C}([0, T]; \mathbf{H}^1(\mathbb{R}^d)) \cap \mathbf{L}^r([0, T]; \mathbf{W}^{1,2\sigma+2}(\mathbb{R}^d))$$

to (3.9). *In addition, the mapping*

$$f : B_{z^*} \times B_{u_0^*} \to \mathbf{C}([0, T]; \mathbf{H}^1(\mathbb{R}^d))$$
$$(z, u_0) \mapsto v(z, u_0, \cdot)$$

is continuous.

Based on above results, one can get the large deviation results: for $\varepsilon > 0$, the solution u^ε of the following equation

$$du^\varepsilon = i\left(\Delta u^\varepsilon + \lambda|u^\varepsilon|^{2\sigma}u^\varepsilon\right)dt + \varepsilon dW(t)$$

converges to the solution u^0 of the deterministic equation ($\varepsilon = 0$) in $\mathbf{C}([0, T]; \mathbf{H}^1(\mathbb{R}^d))$ almost surely as $\varepsilon \to 0$. More details can be found in [63, 64].

For the multiplicative noise case with $\mathcal{O} = (0, 1)$ and $\sigma = 1$, the continuous dependence of solutions to (3.8) on initial data and noises is derived by showing the exponential integrability of the solution.

Lemma 3.1 ([54, 55]) *Let* $\mathcal{O} = (0, 1)$, $\sigma = 1$, $q \geq 1$ *and* $Q^{\frac{1}{2}} \in \mathcal{L}_2^2$. *Assume that the initial value* u_0 *is* $\mathcal{F}_0$ *measurable and satisfies*

$$\mathbb{E}\left[\exp(H(u_0))\right] + \mathbb{E}\left[\exp\left(\frac{\|u_0\|_{\mathbf{L}^2(0,1)}^6}{2} + 4q^2 T^2 \|u_0\|_{\mathbf{L}^2(0,1)}^2 \exp(C(u_0))\right)\right] < \infty$$

for some $T > 0$, *where*

$$C(u_0) = 2L^2 T \|Q^{\frac{1}{2}}\|_{\mathcal{L}_2^2}^2 \|u_0\|_{\mathbf{L}^2(0,1)}^2$$

and L *is a constant such that* $\|f\|_{\mathbf{L}^\infty(0,1)} \leq L\|f\|_{\mathbf{H}^1(0,1)}$ *for any* $f \in \mathbf{H}^1(0, 1)$. *Then there exists a constant* $C = C(Q, T, u_0, q)$ *such that*

$$\mathbb{E}\left[\exp\left(q\int_0^T \|u_0\|_{\mathbf{L}^2(0,1)}\|\nabla u(t)\|_{\mathbf{L}^2(0,1)}^2 dt\right)\right] \leq C.$$

Exponential integrability of exact and numerical solutions is important in the numerical analysis to obtain the strong convergence rate of numerical schemes especially when the equation has a non-globally Lipschitz continuous nonlinearity (see e.g. [54, 55, 112]). As an application of the exponential integrability, the continuous dependence on both initial value and noises are given in [55].

Theorem 3.8 (Corollary 3.1, [55]) *Assume that the assumptions in Lemma 3.1 hold for two different initial values* u_0 *and* v_0 *with* $q = 4p$ *for* $p = 2$ *or* $p \geq 4$. *Then there exists a constant* $C = C(T, Q, p, u_0, v_0)$ *such that*

$$\left(\mathbb{E}\left[\sup_{t \in [0,T]} \|u(t) - v(t)\|_{\mathbf{L}^2(0,1)}^p\right]\right)^{\frac{1}{p}} \leq C\left(\mathbb{E}\left[\|u_0 - v_0\|_{\mathbf{L}^2(0,1)}^{2p}\right]\right)^{\frac{1}{2p}},$$

where u *and* v *denote the solution of* (3.8) *with initial values* u_0 *and* v_0 *respectively.*

Furthermore, the large deviation result also holds for the multiplicative noise case. For $\varepsilon > 0$, the solution u^ε of the following equation

$$du^\varepsilon = \mathbf{i}(\Delta u^\varepsilon + \lambda|u^\varepsilon|^2 u^\varepsilon)dt + \mathbf{i}\varepsilon u^\varepsilon \circ dW(t)$$

satisfies

$$\left(\mathbb{E}\left[\sup_{t \in [0,T]} \|u^\varepsilon(t) - u^0(t)\|_{\mathbf{L}^2(0,1)}^p\right]\right)^{\frac{1}{p}} \leq C\varepsilon$$

under the assumptions in Theorem 3.8, which is proved in [55].

When studying the microscopic dynamical trajectory of an individual macroscopic system, it is more convenient to study appropriate ensembles in the state space instead. It then turns our focus on the study of invariant measures for the considered system.

The following sections are devoted to showing the invariant distributions of stochastic Schrödinger equations with damping (see [70, 76, 109]).

3.5 Stochastic Linear Schrödinger Equation with Weak Damping

This section is devoted to studying the distribution of the following linear equation

$$du = (\mathbf{i}\Delta u - \alpha u + \mathbf{i}\lambda u)dt + Q^{\frac{1}{2}}dW \tag{3.10}$$

with $\alpha > 0$, $\lambda \in \mathbb{R}$, and operator Q commuting with Δ. Moreover, W is a cylindrical Wiener process on $\mathbf{L}^2(\mathscr{O})$ with $\mathscr{O} = (0, 1)$ such that

$$W = \sum_{m=1}^{\infty} e_m \beta_m$$

with $\{e_m\}_{m \in \mathbb{N}}$ being an orthogonal basis for $\mathbf{L}^2(\mathscr{O}; \mathbb{R})$ and $\{\beta_m\}_{m \in \mathbb{N}}$ being a sequence of mutually independent and identically distributed $\mathbb{C}$-valued Brownian motions.

Rewriting above equation through its components $u^m := \langle u, e_m \rangle$, we obtain

$$du^m = (-\lambda_m^\alpha + \mathbf{i}\lambda)u^m dt + \sum_{i=1}^{\infty} \langle Q^{\frac{1}{2}} e_i, e_m \rangle d\beta_i, \quad m \in \mathbb{N}.$$

Its solution is given by an Ornstein–Uhlenbeck process

$$u^m(t) = e^{(-\lambda_m^\alpha + \mathbf{i}\lambda)t} u^m(0) + \sum_{i=1}^{\infty} \int_0^t e^{(-\lambda_m^\alpha + \mathbf{i}\lambda)(t-s)} \langle Q^{\frac{1}{2}} e_i, e_m \rangle d\beta_i(s),$$

where $u^m(0) = \langle u_0, e_m \rangle$ and λ_m^α are the eigenvalues for the linear operator $-\mathbf{i}\Delta + \alpha Id$ with Id being the identity operator.

Note that $\{u^m(t)\}_{t \geq 0}$ satisfies a complex Gaussian distribution $\mathscr{N}(\mathbf{m}, \mathbf{C}, \mathbf{R})$ defined by its mean $\mathbf{m}$, covariance $\mathbf{C}$ and relation $\mathbf{R}$:

$$\mathbf{m}\left(u^m(t)\right) := \mathbb{E}\left[u^m(t)\right] = e^{(-\lambda_m^\alpha + \mathbf{i}\lambda)t} \mathbf{m}\left[u^m(0)\right],$$

$$\mathbf{C}\left(u^m(t)\right) := \mathbb{E}\left|u^m(t) - \mathbf{m}\left(u^m(t)\right)\right|^2 = e^{-2\alpha t} \mathbf{C}\left(u^m(0)\right) + \frac{1 - e^{-2\alpha t}}{\alpha} \|Q^{\frac{1}{2}} e_m\|^2,$$

$$\mathbf{R}\left(u^m(t)\right) := \mathbb{E}\left(u^m(t) - \mathbf{m}\left(u^m(t)\right)\right)^2 = e^{2(-\lambda_m^\alpha + \mathbf{i}\lambda)t} \mathbf{R}\left(u^m(0)\right).$$

We use the notation $\mu_t^m := \mathcal{N}(\mathbf{m}(u^m(t)), \mathbf{C}(u^m(t)), \mathbf{R}(u^m(t)))$ for simplicity.

Remark 3.2 We consider a one-dimensional $\mathbb{C}$-valued Gaussian random variable $Z = \mathbf{a} + \mathbf{ib}$ with $\mathbf{a}$ and $\mathbf{b}$ being two $\mathbb{R}$-valued Gaussian random variables. If its relation vanishes, i.e.,

$$\mathbf{R}(Z) = \mathbb{E}|\mathbf{a} - \mathbb{E}\mathbf{a}|^2 - \mathbb{E}|\mathbf{b} - \mathbb{E}\mathbf{b}|^2 + 2\mathbf{i}(\mathbb{E}[\mathbf{ab}] - \mathbb{E}\mathbf{a}\mathbb{E}\mathbf{b}) = 0,$$

it implies $\mathbb{E}|\mathbf{a} - \mathbb{E}\mathbf{a}|^2 = \mathbb{E}|\mathbf{b} - \mathbb{E}\mathbf{b}|^2$ and $\mathbb{E}[\mathbf{ab}] = \mathbb{E}\mathbf{a}\mathbb{E}\mathbf{b}$. Since $\mathbf{a}$ and $\mathbf{b}$ are both Gaussian, we obtain equivalently that $\mathbf{a}$ and $\mathbf{b}$ are independent with the same covariance.

Remark 3.3 The characteristic function of a one-dimensional complex Gaussian variable Z with distribution $\nu = \mathcal{N}(\mathbf{m}, \mathbf{C}, \mathbf{R})$ reads (see e.g. [6])

$$\hat{\nu}(c) := \mathbb{E}[\exp\{i\Re(\overline{c}Z)\}] = \int_{\mathbb{C}} \exp\{i\Re(\overline{c}z)\}\nu(dz)$$

$$= \exp\left\{i\Re(\overline{c}\mathbf{m}) - \frac{1}{4}(\overline{c}\mathbf{C}c + \Re(\overline{c}\mathbf{R}\overline{c}))\right\}, \quad c \in \mathbb{C}.$$

It can be generalized for the infinite dimensional case utilizing inner product in H:

$$\hat{\nu}(w) := \exp\left\{i\Re\langle \overline{w}, \mathbf{m}\rangle - \frac{1}{4}(\langle \mathbf{C}\overline{w}, w\rangle + \Re\langle \mathbf{R}\overline{w}, \overline{w}\rangle)\right\}, \quad w \in H.$$

Hence, we get that the unique invariant measure of (3.10) is a complex Gaussian distribution, which is stated in the following theorem. We refer to [70, 76] and references therein for the existence of invariant measures for the nonlinear case, and refer to [32, 60] and references therein for other types of SPDEs.

Theorem 3.9 *Assume that Q is a nonnegative and symmetric trace operator. The solution u in (3.10) possesses a unique invariant measure*

$$\mu_\infty = \mathcal{N}\left(0, \frac{1}{\alpha}Q, 0\right).$$

Proof Based on Remark 3.2, we define

$$u_\infty^m = \frac{\|Q^{\frac{1}{2}}e_m\|}{\sqrt{2\alpha}}(\xi_m + \mathbf{i}r_m)$$

with $\{\xi_m, r_m\}_{m\in\mathbb{N}}$ being independent standard $\mathbb{R}$-valued normal random variables, i.e., $\xi_m, r_m \sim \mathcal{N}(0, 1)$. Apparently,

$$u_\infty^m \sim \mathcal{N}\left(0, \frac{\|Q^{\frac{1}{2}}e_m\|^2}{\alpha}, 0\right) =: \mu_\infty^m.$$

We claim that the following random variable has the distribution μ_∞:

$$u_\infty := \sum_{m=1}^{\infty} u_\infty^m e_m = \sum_{m=1}^{\infty} \frac{\|Q^{\frac{1}{2}}e_m\|}{\sqrt{2\alpha}}(\xi_m + \mathbf{i}r_m)e_m.$$

Compared with $u(t) = \sum_{m=1}^{\infty} u^m(t)e_m$, it then suffices to show that the distribution μ_t^m of $u^m(t)$ converges to μ_∞^m. As a result of Remark 3.3, the characteristic function of μ_t^m is

$$\hat{\mu}_t^m(c) = \exp\left\{ \mathbf{i}\Re\left(\overline{c}e^{(-\lambda_m^\alpha + \mathbf{i}\lambda)t}\mathbb{E}\left[u^m(0)\right]\right) - \frac{1}{4}\Re\left(e^{2(-\lambda_m^\alpha + \mathbf{i}\lambda)t}\mathbf{R}\left(u^m(0)\right)\overline{c}^2\right)\right.$$
$$\left. - \frac{1}{4}\left(e^{-2\alpha t}\mathbf{C}\left(u^m(0)\right) + \frac{1 - e^{-2\alpha t}}{\alpha}\|Q^{\frac{1}{2}}e_m\|^2|c|^2\right)\right\}$$

and $\hat{\mu}_t^m(c) \to \exp\{-\frac{\|Q^{\frac{1}{2}}e_m\|^2}{4\alpha}|c|^2\} = \hat{\mu}_\infty^m(c)$.   □

3.6 Stochastic Nonlinear Schrödinger Equation with Weak Damping

In this section, we consider the stochastic damped nonlinear Schrödinger equation in the following form

$$\begin{cases} du = (\mathbf{i}\Delta u - \alpha u + \mathbf{i}\lambda|u|^{2\sigma}u)dt + dW \\ u(0) = u_0, \quad x \in \mathscr{O} \end{cases} \tag{3.11}$$

with $\alpha > 0$ and W being a Q-Wiener process on $\mathbf{L}^2(\mathscr{O})$ with a symmetric and positive definite covariance operator Q such that

$$Q^{\frac{1}{2}}e_k = \sqrt{\eta_k}e_k, \quad \eta_k > 0, \ k \in \mathbb{N}.$$

We still denote the linear operator $A_\alpha := -\mathbf{i}\Delta + \alpha Id$.

3.6.1 One Dimensional Case

Let $\mathscr{O} = (0, 1)$. Note that the real parts of eigenvalues of the operator $-A_\alpha$ are negative. This model is called a weakly damped stochastic nonlinear Schrödinger equation in [70]. This establishes the ergodicity and even polynomial mixing for (3.11) with a more general assumption: there exists some $N_* \in \mathbb{N}_+$ such that $\eta_k > 0$

for all $k \leq N_*$. The geometric structure and numerical approximations for this model are considered in Sect. 4.5 and Chap. 5, respectively.

Let $\{P_t\}_{t\geq 0}$ be the Markov semigroup generated by the solution u of the considered model.

Theorem 3.10 (Theorem 1.1, [70]) *Assume that* $\|Q^{\frac{1}{2}}\|_{\mathscr{L}_2^3} < \infty$. *There exists a unique stationary probability measure* μ *of* $\{P_t\}_{t\geq 0}$ *on* $\dot{\mathbf{H}}^1(\mathcal{O})$ *with*

$$C_p := \int_{\dot{\mathbf{H}}^1(\mathcal{O})} \|y\|_1^{2p} d\mu(y) < \infty$$

for any $p \in \mathbb{N}_+$. *Moreover, the measures converge to equilibrium with polynomial speed at any order, i.e.,*

$$\|P_t\nu - \mu\|_W \leq \tilde{C}_p(1+t)^{-p}(1+C_1)$$

for some $\tilde{C}_p > 0$ *and any* $\nu \in \mathscr{P}(\dot{\mathbf{H}}^1(\mathcal{O}))$.

Here, $\|\cdot\|_W$ is a Wasserstein type norm, and we refer to [70] for more details.

3.6.2 High Dimensional Case

Let $\mathcal{O} = \mathbb{R}^d$. The existence of invariant measures and ergodic measures is established in $\mathbf{H}^1(\mathcal{O})$ by [76].

Theorem 3.11 *Assume that* $u_0 \in \mathbf{H}^1(\mathcal{O})$ *is* $\mathscr{F}_0$-*measurable,* $Q^{\frac{1}{2}} \in \mathscr{L}_2^1$ *and that*

$$\sigma \in I_{\alpha,d} := \begin{cases} [0, \dfrac{2}{d}) & \text{if } \lambda = 1, \\[2mm] [0, +\infty) & \text{if } \lambda = -1, \ d = 1, 2, \\[2mm] [0, \dfrac{2}{d-2}) & \text{if } \lambda = -1, \ d \geq 3. \end{cases}$$

Then Eq. (3.11) *possesses an invariant measure.*

We also refer to [27, 28, 30, 70, 76, 121, 155] for more details on invariant measures and ergodicity for Schrödinger equations in both the deterministic and the stochastic case.

It is not difficult to find out that the ergodicity of (3.11) can be obtained based on the above results. Thus, the temporal average converge to the ergodic limit almost surely as a result of the mixing property:

$$\lim_{t \to \infty} \mathbb{E}[f(u(t, u_0))] = \int_{\dot{\mathbf{H}}^1(\mathcal{O})} f(y) d\mu(y)$$

with $f \in \mathbf{C}_b$ and $u(t, u_0)$ denoting the solution starting from u_0. It also leads to the ergodicity

$$\lim_{T \to \infty} \frac{1}{T} \int_0^T \mathbb{E}[f(u(t, u_0))]dt = \int_{\dot{\mathbf{H}}^1(\mathscr{O})} f(y)d\mu(y), \qquad (3.12)$$

which will be considered in Chaps. 5 and 6.

Summary

In this chapter, the invariant measures for nonlinear Schrödinger equations, together with the well-posedness and continuous dependence on the initial value as the preliminaries, are introduced.

The existence of an invariant Gibbs measure for deterministic nonlinear Schrödinger equations, which are also Hamiltonian partial differential equations, has been investigated by many authors taking advantage of the Hamiltonian, see [20, 27, 28, 40, 132, 157, 173, 183, 184, 192–194] and references therein.

For stochastic Schrödinger equations, the Hamiltonian is not conserved anymore in general. Hence, dissipative terms, which have applications in forced and damped quantum, are usually involved to ensure the existence and uniqueness of an invariant measure for stochastic nonlinear Schrödinger equations. We refer to [70, 76, 121, 150] for more details. Numerical approximation of invariant measures for stochastic nonlinear Schrödinger equations will be studied in Chaps. 5 and 6.

There are lots of profound researches on invariant measures for other kinds of SPDEs, see e.g. [17, 37, 60, 89, 128, 137, 152, 158, 165], and we refer to [31, 32] for numerical approximations of invariant measures. In particular, the study of invariant measures for stochastic Navier–Stokes equations can be found in [5, 33, 38, 59, 73, 84, 96] and references therein. It still remains to be investigated whether stochastic Hamiltonian partial differential equations admit a unique invariant measure.

Chapter 4
Geometric Structures and Numerical Schemes for Nonlinear Schrödinger Equations

The stochastic (conformal) symplectic structure in the finite dimension case has been introduced briefly in Chap. 1 for the stochastic Kubo oscillator and stochastic dissipative Hamiltonian systems. In this chapter, we turn to considering the geometric structures—stochastic symplectic and multi-symplectic structures—in the infinite dimension case for stochastic nonlinear Schrödinger equations as well as their numerical approximations.

Section 4.1 shows the symplectic and multi-symplectic structures for deterministic nonlinear Schrödinger equations, based on which Sect. 4.2 gives several numerical approximations which could inherit the symplecticity or multi-symplecticity. Sections 4.3 and 4.4 generalize these definitions to SPDEs, and show that stochastic nonlinear Schrödinger equations with either additive or linear multiplicative noise preserve the symplectic and multi-symplectic conservation laws. Numerical approximations which could inherit these properties are also given in these sections, which are also used in Chap. 6. Section 4.5 focuses on stochastic nonlinear Schrödinger equations with weak damping. The conformal multi-symplectic conservation law for the considered equation is introduced, which is used in the numerical analysis of invariant measures in Chap. 5.

4.1 Preliminaries

It is introduced for the finite dimensional case in Sect. 4.1 that the Lagrange equation and the Hamiltonian system are equivalent based on the unique Legendre transform. In infinite dimensional case, the Lagrange–Hamiltonian duality is no longer uniquely defined. One can define the symplectic structure for infinite dimensional Hamiltonian systems or the multi-symplectic structure when regarding the considered model as a Hamiltonian partial differential equation.

© Springer Nature Singapore Pte Ltd. 2019
J. Hong and X. Wang, *Invariant Measures for Stochastic Nonlinear Schrödinger Equations*, Lecture Notes in Mathematics 2251,
https://doi.org/10.1007/978-981-32-9069-3_4

We now recall the geometric structures for the deterministic nonlinear Schrödinger equation

$$\begin{cases} du = i\Delta u dt + i\lambda |u|^{2\sigma} u dt, \\ u(0, x) = u_0(x), \ x \in \mathcal{O} \subset \mathbb{R}^d, \\ u(t, x) = 0, \ x \in \partial\mathcal{O}, \ t \geq 0. \end{cases} \tag{4.1}$$

It is known that (4.1) can be regarded as an infinite dimensional Hamiltonian system. Therefore its geometric structures could be investigated in a similar way to the finite dimensional case by taking average with respect to the spatial variable.

Denoting by p and q the real and imaginary parts of solution u, which satisfy

$$\begin{cases} \dot{p} = -\Delta q - \lambda \left(p^2 + q^2\right)^\sigma q, \\ \dot{q} = \Delta p + \lambda \left(p^2 + q^2\right)^\sigma p, \end{cases} \tag{4.2}$$

we can transform (4.1) into

$$\dot{p} = \frac{\delta \hat{H}}{\delta q}(p, q), \quad \dot{q} = -\frac{\delta \hat{H}}{\delta p}(p, q),$$

where δ denotes the variation and

$$\hat{H}(p, q) := H(u) = \frac{1}{2} \int_{\mathcal{O}} \left(|\nabla p|^2 + |\nabla q|^2\right) dx - \frac{\lambda}{2\sigma + 2} \int_{\mathcal{O}} \left(p^2 + q^2\right)^{\sigma+1} dx.$$

Theorem 4.1 *The symplectic conservation law holds for* (4.2): *the averaged differential 2-form $\int_{\mathcal{O}} dp(t) \wedge dq(t) dx$ is invariant at any time, i.e.,*

$$\int_{\mathcal{O}} dp(t) \wedge dq(t) dx = \int_{\mathcal{O}} dp_0 \wedge dq_0 dx, \quad \forall t \geq 0$$

with $p(0) = p_0$ and $q(0) = q_0$.

Proof Denoting

$$\varpi(t) := \int_{\mathcal{O}} dp(t) \wedge dq(t) dx$$
$$= \int_{\mathcal{O}} \left(\frac{\partial p}{\partial p_0} \frac{\partial q}{\partial q_0} - \frac{\partial p}{\partial q_0} \frac{\partial q}{\partial p_0}\right) dp_0 \wedge dq_0 dx,$$

it then suffices to show that

$$\frac{d}{dt}\varpi(t) = \int_{\mathcal{O}} \frac{d}{dt}\left(\frac{\partial p}{\partial p_0} \frac{\partial q}{\partial q_0} - \frac{\partial p}{\partial q_0} \frac{\partial q}{\partial p_0}\right) dp_0 \wedge dq_0 dx = 0.$$

In fact, by denoting

$$F(p, q) := \frac{\lambda}{2\sigma + 2} \left(p^2 + q^2 \right)^{\sigma+1}$$

and calculating

$$\frac{d}{dt} \left(\frac{\partial p}{\partial p_0} \right) = \frac{\partial}{\partial p_0} \left[-\Delta q - \lambda \left(p^2 + q^2 \right)^{\sigma} q \right] = \frac{\partial}{\partial p_0} \left[-\Delta q - \frac{\partial F}{\partial q} \right]$$

$$= -\Delta \left(\frac{\partial q}{\partial p_0} \right) - \frac{\partial^2 F}{\partial p \partial q} \left(\frac{\partial p}{\partial p_0} \right) - \frac{\partial^2 F}{\partial q^2} \left(\frac{\partial q}{\partial p_0} \right),$$

$$\frac{d}{dt} \left(\frac{\partial p}{\partial q_0} \right) = -\Delta \left(\frac{\partial q}{\partial q_0} \right) - \frac{\partial^2 F}{\partial p \partial q} \left(\frac{\partial p}{\partial q_0} \right) - \frac{\partial^2 F}{\partial q^2} \left(\frac{\partial q}{\partial q_0} \right),$$

$$\frac{d}{dt} \left(\frac{\partial q}{\partial p_0} \right) = \Delta \left(\frac{\partial p}{\partial p_0} \right) + \frac{\partial^2 F}{\partial p^2} \left(\frac{\partial p}{\partial p_0} \right) + \frac{\partial^2 F}{\partial q \partial p} \left(\frac{\partial q}{\partial p_0} \right),$$

$$\frac{d}{dt} \left(\frac{\partial q}{\partial q_0} \right) = \Delta \left(\frac{\partial p}{\partial q_0} \right) + \frac{\partial^2 F}{\partial p^2} \left(\frac{\partial p}{\partial q_0} \right) + \frac{\partial^2 F}{\partial q \partial p} \left(\frac{\partial q}{\partial q_0} \right),$$

we finally get

$$\frac{d}{dt} \varpi(t) = \int_{\mathcal{O}} \left[-\Delta \left(\frac{\partial q}{\partial p_0} \right) \left(\frac{\partial q}{\partial q_0} \right) + \Delta \left(\frac{\partial q}{\partial q_0} \right) \left(\frac{\partial q}{\partial p_0} \right) \right.$$

$$\left. + \Delta \left(\frac{\partial p}{\partial q_0} \right) \left(\frac{\partial p}{\partial p_0} \right) - \Delta \left(\frac{\partial p}{\partial p_0} \right) \left(\frac{\partial p}{\partial q_0} \right) \right] dp_0 \wedge dq_0 dx$$

$$= -\int_{\mathcal{O}} d(\Delta q) \wedge dq + d(\Delta p) \wedge dp \, dx$$

$$= -\int_{\mathcal{O}} \nabla \left[d(\nabla q) \wedge dq + d(\nabla p) \wedge dp \right] dx = 0$$

due to the homogenous boundary condition. □

Next we turn to investigating the geometric structures for (4.1) in the view of the Hamiltonian partial differential equation. In this case, the derivatives of u with respect to the spatial variable should also be taken as a conjugate variable when defining the Legendre transform introduced in Sect. 1.5.

Definition 4.1 A deterministic partial differential equation is called a Hamiltonian partial differential equation if it can be written in the form

$$M z_t + K z_x = \nabla S(z), \quad z \in \mathbb{R}^n, \ n \geq 3,$$

where M and K are skew-symmetric matrices and S is a smooth function of the state variable z.

For simplicity, let $d = 1$ and $\mathscr{O} = (0, 1)$. Rewriting (4.1) utilizing the real part p and imaginary part q of u, we obtain

$$\begin{cases} p_t = -q_{xx} - \lambda(p^2 + q^2)^\sigma q, \\ q_t = p_{xx} + \lambda(p^2 + q^2)^\sigma p. \end{cases}$$

Denoting auxiliary variables $v = p_x$, $w = q_x$, we transform above system into the following form

$$\begin{cases} p_t + w_x = -\lambda \left(p^2 + q^2\right)^\sigma q, \\ q_t - v_x = \lambda \left(p^2 + q^2\right)^\sigma p, \\ \quad p_x = v, \\ \quad q_x = w \end{cases} \tag{4.3}$$

with Hamiltonian function

$$\tilde{H}(p, q, v, w) := \frac{1}{2} \left(v^2 + w^2\right) - \frac{\lambda}{2\sigma + 2} \left(p^2 + q^2\right)^{\sigma+1}.$$

Denoting further $z = (p, q, v, w)^\top$, two skew symmetric matrices

$$M_4 = \begin{pmatrix} 0 & 1 & 0 & 0 \\ -1 & 0 & 0 & 0 \\ 0 & 0 & 0 & 0 \\ 0 & 0 & 0 & 0 \end{pmatrix}, \quad K_4 = \begin{pmatrix} 0 & 0 & -1 & 0 \\ 0 & 0 & 0 & -1 \\ 1 & 0 & 0 & 0 \\ 0 & 1 & 0 & 0 \end{pmatrix}$$

and $S(z) = \frac{\lambda}{2\sigma+2}(p^2 + q^2)^{\sigma+1} + \frac{1}{2}(v^2 + w^2)$, we obtain its equivalent Hamiltonian partial differential equation form

$$M_4 z_t + K_4 z_x = \nabla S(z).$$

Theorem 4.2 *The nonlinear Schrödinger equation (4.1) possesses the multisymplectic conservation law:*

$$\frac{\partial}{\partial t} \omega(t, x) + \frac{\partial}{\partial x} \kappa(t, x) = 0$$

with $\omega = \frac{1}{2}dz \wedge M_4 dz$ and $\kappa = \frac{1}{2}dz \wedge K_4 dz$. More precisely,

$$\frac{\partial}{\partial t}\left(dp \wedge dq\right) - \frac{\partial}{\partial x}\left(dp \wedge dv + dq \wedge dw\right) = 0.$$

Proof Performing partial derivative to ω with respect to t yields

$$\frac{\partial}{\partial t}\omega = \frac{1}{2}\Big[dz_t \wedge M_4 dz + dz \wedge M_4 dz_t\Big]$$

$$= dz \wedge d(M_4 z_t),$$

where we used the fact that $M_4^\top = -M_4$ and

$$dz_t \wedge M_4 dz = M_4^\top dz_t \wedge dz = dz \wedge M_4 dz_t.$$

Similarly, we have

$$\frac{\partial}{\partial x}\kappa = dz \wedge d(K_4 z_x).$$

As a result,

$$\frac{\partial}{\partial t}\omega + \frac{\partial}{\partial x}\kappa = dz \wedge d(M_4 z_t + K_4 z_x) = dz \wedge d\nabla S(z) = dz \wedge \nabla^2 S(z) dz = 0$$

due to the symmetry of $\nabla^2 S(z)$. $\qquad\qquad\qquad\qquad\qquad\qquad\qquad\qquad\qquad\Box$

4.2 Symplectic and Multi-symplectic Methods for Deterministic Schrödinger Equations

This section is devoted to introducing several symplectic temporal semi-discretizations or multi-symplectic full discretizations for the deterministic nonlinear Schrödinger equation (4.1).

4.2.1 Symplectic Temporal Semi-discretizations

Since the midpoint scheme possesses the discrete symplectic conservation law when applied to finite dimensional Hamiltonian systems (see e.g. [95]), we consider the midpoint scheme as the temporal semi-discretization for the nonlinear Schrödinger equation. We apply the midpoint scheme to (4.2) and get

$$\begin{cases} p^{n+1} = p^n - \tau \Delta q^{n+\frac{1}{2}} - \tau\lambda\left[\left(p^{n+\frac{1}{2}}\right)^2 + \left(q^{n+\frac{1}{2}}\right)^2\right]^\sigma q^{n+\frac{1}{2}}, \\[2ex] q^{n+1} = q^n + \tau \Delta p^{n+\frac{1}{2}} + \tau\lambda\left[\left(p^{n+\frac{1}{2}}\right)^2 + \left(q^{n+\frac{1}{2}}\right)^2\right]^\sigma p^{n+\frac{1}{2}}, \end{cases} \qquad (4.4)$$

where τ is the uniform time step-size, $p^{n+\frac{1}{2}} := \frac{1}{2}(p^{n+1} + p^n)$ and $q^{n+\frac{1}{2}} := \frac{1}{2}(q^{n+1} + q^n)$. Scheme (4.4) is equivalent to the midpoint scheme applied to (4.1) directly, that is, it can be written in the following form

$$u^{n+1} = u^n + \mathbf{i}\Delta u^{n+\frac{1}{2}}\tau + \mathbf{i}\lambda |u^{n+\frac{1}{2}}|^{2\sigma} u^{n+\frac{1}{2}}\tau$$

with $u^{n+\frac{1}{2}} := \frac{1}{2}(u^{n+1} + u^n)$ and $u^n = p^n + \mathbf{i}q^n$ for $n \in \mathbb{N}$.

Theorem 4.3 *The temporal semi-discretization based on the midpoint scheme for nonlinear Schrödinger equation* (4.1) *possesses the discrete symplectic conservation law, i.e.,*

$$\int_\mathcal{O} \mathrm{d}p^{n+1} \wedge \mathrm{d}q^{n+1} dx = \int_\mathcal{O} \mathrm{d}p^n \wedge \mathrm{d}q^n dx.$$

Proof We use the same notation

$$F(p, q) := \frac{\lambda}{2\sigma + 2}(p^2 + q^2)^{\sigma+1}$$

as that in the proof of Theorem 4.1. Then according to scheme (4.4), we have

$$\begin{cases} \mathrm{d}p^{n+1} - \mathrm{d}p^n = -\tau\Delta\left(\mathrm{d}q^{n+\frac{1}{2}}\right) - \tau\frac{\partial^2 F}{\partial p\partial q}\mathrm{d}p^{n+\frac{1}{2}} - \tau\frac{\partial^2 F}{\partial q^2}\mathrm{d}q^{n+\frac{1}{2}}, \\ \mathrm{d}q^{n+1} - \mathrm{d}q^n = \tau\Delta\left(\mathrm{d}p^{n+\frac{1}{2}}\right) + \tau\frac{\partial^2 F}{\partial p^2}\mathrm{d}p^{n+\frac{1}{2}} + \tau\frac{\partial^2 F}{\partial q\partial p}\mathrm{d}q^{n+\frac{1}{2}}. \end{cases}$$

Performing the wedge product between $\mathrm{d}p^{n+1} - \mathrm{d}p^n$ and $\mathrm{d}q^{n+\frac{1}{2}}$ and taking integral with respect to x, we get

$$\int_\mathcal{O} \left(\mathrm{d}p^{n+1} - \mathrm{d}p^n\right) \wedge \mathrm{d}q^{n+\frac{1}{2}} dx$$

$$= \int_\mathcal{O} \left(-\tau\Delta\left(\mathrm{d}q^{n+\frac{1}{2}}\right) - \tau\frac{\partial^2 F}{\partial p\partial q}\mathrm{d}p^{n+\frac{1}{2}} - \tau\frac{\partial^2 F}{\partial q^2}\mathrm{d}q^{n+\frac{1}{2}}\right) \wedge \mathrm{d}q^{n+\frac{1}{2}} dx$$

$$= -\tau\int_\mathcal{O} \frac{\partial^2 F}{\partial p\partial q}\mathrm{d}p^{n+\frac{1}{2}} \wedge \mathrm{d}q^{n+\frac{1}{2}} dx$$

and similarly

$$\int_\mathcal{O} \mathrm{d}p^{n+\frac{1}{2}} \wedge \left(\mathrm{d}q^{n+1} - \mathrm{d}q^n\right) dx = \tau\int_\mathcal{O} \mathrm{d}p^{n+\frac{1}{2}} \wedge \frac{\partial^2 F}{\partial q\partial p}\mathrm{d}q^{n+\frac{1}{2}} dx.$$

Adding the above two equations, we finally complete the proof.   □

More generally, we apply the s-stage Runge–Kutta method to (4.2) in the temporal direction and get

$$\begin{cases} P_i = p^n - \tau \sum_{j=1}^{s} a_{ij} \left(\Delta Q_j + \lambda (P_j^2 + Q_j^2)^\sigma Q_j \right), \\[2mm] Q_i = q^n + \tau \sum_{j=1}^{s} a_{ij} \left(\Delta P_j + \lambda (P_j^2 + Q_j^2)^\sigma P_j \right), \\[2mm] p^{n+1} = p^n - \tau \sum_{i=1}^{s} b_i \left(\Delta Q_i + \lambda (P_i^2 + Q_i^2)^\sigma Q_i \right), \\[2mm] q^{n+1} = q^n + \tau \sum_{i=1}^{s} b_i \left(\Delta P_i + \lambda (P_i^2 + Q_i^2)^\sigma P_i \right) \end{cases} \tag{4.5}$$

with parameters a_{ij} and b_i for $i, j = 1, \cdots, s$.

Theorem 4.4 *If the coefficients a_{ij} and b_i satisfy*

$$b_i b_j = b_i a_{ij} + b_j a_{ji}, \quad \forall\, i, j = 1, \cdots, s, \tag{4.6}$$

then the temporal semi-discretization (4.5) has a solution, and it possesses the discrete symplectic conservation law

$$\int_{\mathscr{O}} \mathrm{d} p^{n+1} \wedge \mathrm{d} q^{n+1} dx = \int_{\mathscr{O}} \mathrm{d} p^n \wedge \mathrm{d} q^n dx.$$

Condition (4.6) is named the symplectic condition for Runge–Kutta methods. The proof of Theorem 4.4 is a special case of that of Theorem 4.7 for the stochastic case, and thus is omitted here.

It is not hard to find out that the midpoint scheme (4.4) is a 1-stage Runge–Kutta method with coefficients $a_{11} = \frac{1}{2}$ and $b_1 = 1$, which satisfy the symplectic condition apparently.

4.2.2 Multi-symplectic Full Discretizations

When regarding (4.1) as a partial differential equation instead of an evolution equation in time, discretizations in the spatial direction will be taken into consideration.

We discretize (4.3) utilizing the midpoint scheme in both temporal and spatial directions. Based on the notations $v = p_x$ and $w = q_x$, the full discretization reads

$$\begin{cases} \dfrac{p_{j+\frac{1}{2}}^{n+1} - p_{j+\frac{1}{2}}^{n}}{\tau} + \dfrac{w_{j+1}^{n+\frac{1}{2}} - w_{j}^{n+\frac{1}{2}}}{h} = -\lambda \left[\left(p_{j+\frac{1}{2}}^{n+\frac{1}{2}} \right)^2 + \left(q_{j+\frac{1}{2}}^{n+\frac{1}{2}} \right)^2 \right]^{\sigma} q_{j+\frac{1}{2}}^{n+\frac{1}{2}}, \\[3mm] \dfrac{q_{j+\frac{1}{2}}^{n+1} - q_{j+\frac{1}{2}}^{n}}{\tau} - \dfrac{v_{j+1}^{n+\frac{1}{2}} - v_{j}^{n+\frac{1}{2}}}{h} = \lambda \left[\left(p_{j+\frac{1}{2}}^{n+\frac{1}{2}} \right)^2 + \left(q_{j+\frac{1}{2}}^{n+\frac{1}{2}} \right)^2 \right]^{\sigma} p_{j+\frac{1}{2}}^{n+\frac{1}{2}}, \\[3mm] \dfrac{p_{j+1}^{n+\frac{1}{2}} - p_{j}^{n+\frac{1}{2}}}{h} = v_{j+\frac{1}{2}}^{n+\frac{1}{2}}, \\[3mm] \dfrac{q_{j+1}^{n+\frac{1}{2}} - q_{j}^{n+\frac{1}{2}}}{h} = w_{j+\frac{1}{2}}^{n+\frac{1}{2}}, \end{cases} \tag{4.7}$$

which is known as the Preissman scheme (see e.g. [35]). Here,

$$p_{j}^{n+\frac{1}{2}} := \frac{1}{2} \left(p_{j}^{n+1} + p_{j}^{n} \right), \quad p_{j+\frac{1}{2}}^{n} := \frac{1}{2} \left(p_{j+1}^{n} + p_{j}^{n} \right),$$

and τ and h denote the uniform time and space step-sizes, respectively.

Theorem 4.5 *The Preissman scheme* (4.7) *preserves the discrete multi-symplectic structure, that is,*

$$\begin{aligned} &\frac{1}{\tau} \left(dp_{j+\frac{1}{2}}^{n+1} \wedge dq_{j+\frac{1}{2}}^{n+1} - dp_{j+\frac{1}{2}}^{n} \wedge dq_{j+\frac{1}{2}}^{n} \right) \\ &- \frac{1}{h} \left(dp_{j+1}^{n+\frac{1}{2}} \wedge dv_{j+1}^{n+\frac{1}{2}} - dp_{j}^{n+\frac{1}{2}} \wedge dv_{j}^{n+\frac{1}{2}} \right) \\ &- \frac{1}{h} \left(dq_{j+1}^{n+\frac{1}{2}} \wedge dw_{j+1}^{n+\frac{1}{2}} - dq_{j}^{n+\frac{1}{2}} \wedge dw_{j}^{n+\frac{1}{2}} \right) = 0. \end{aligned}$$

Proof Utilizing the Hamiltonian function $\tilde{H}(p, q, v, w)$ introduced for (4.3) in Sect. 4.1, scheme (4.7) can be expressed as

$$\begin{cases} \dfrac{p_{j+\frac{1}{2}}^{n+1} - p_{j+\frac{1}{2}}^{n}}{\tau} + \dfrac{w_{j+1}^{n+\frac{1}{2}} - w_{j}^{n+\frac{1}{2}}}{h} = \dfrac{\partial}{\partial q} \tilde{H}_{j+\frac{1}{2}}^{n+\frac{1}{2}}, \\[3mm] \dfrac{q_{j+\frac{1}{2}}^{n+1} - q_{j+\frac{1}{2}}^{n}}{\tau} - \dfrac{v_{j+1}^{n+\frac{1}{2}} - v_{j}^{n+\frac{1}{2}}}{h} = -\dfrac{\partial}{\partial p} \tilde{H}_{j+\frac{1}{2}}^{n+\frac{1}{2}}, \\[3mm] \dfrac{p_{j+1}^{n+\frac{1}{2}} - p_{j}^{n+\frac{1}{2}}}{h} = \dfrac{\partial}{\partial v} \tilde{H}_{j+\frac{1}{2}}^{n+\frac{1}{2}}, \\[3mm] \dfrac{q_{j+1}^{n+\frac{1}{2}} - q_{j}^{n+\frac{1}{2}}}{h} = \dfrac{\partial}{\partial w} \tilde{H}_{j+\frac{1}{2}}^{n+\frac{1}{2}}, \end{cases} \tag{4.8}$$

where

$$\tilde{H}_{j+\frac{1}{2}}^{n+\frac{1}{2}} := \tilde{H} \left(p_{j+\frac{1}{2}}^{n+\frac{1}{2}}, q_{j+\frac{1}{2}}^{n+\frac{1}{2}}, v_{j+\frac{1}{2}}^{n+\frac{1}{2}}, w_{j+\frac{1}{2}}^{n+\frac{1}{2}} \right).$$

Taking the exterior derivative of the first equation in (4.8) and performing the wedge product between the obtained equation and $dq_{j+\frac{1}{2}}^{n+\frac{1}{2}}$, we obtain

$$
\frac{1}{\tau}\left[d\left(p_{j+\frac{1}{2}}^{n+1}-p_{j+\frac{1}{2}}^{n}\right)\wedge dq_{j+\frac{1}{2}}^{n+\frac{1}{2}}\right]-\frac{1}{h}\left[dq_{j+\frac{1}{2}}^{n+\frac{1}{2}}\wedge d\left(w_{j+1}^{n+\frac{1}{2}}-w_{j}^{n+\frac{1}{2}}\right)\right]
$$

$$
=\frac{1}{2\tau}\left[dp_{j+\frac{1}{2}}^{n+1}\wedge dq_{j+\frac{1}{2}}^{n+1}+dp_{j+\frac{1}{2}}^{n+1}\wedge dq_{j+\frac{1}{2}}^{n}-dp_{j+\frac{1}{2}}^{n}\wedge dq_{j+\frac{1}{2}}^{n+1}-dp_{j+\frac{1}{2}}^{n}\wedge dq_{j+\frac{1}{2}}^{n}\right]
$$

$$
-\frac{1}{2h}\left[dq_{j+1}^{n+\frac{1}{2}}\wedge dw_{j+1}^{n+\frac{1}{2}}-dq_{j+1}^{n+\frac{1}{2}}\wedge dw_{j}^{n+\frac{1}{2}}+dq_{j}^{n+\frac{1}{2}}\wedge dw_{j+1}^{n+\frac{1}{2}}-dq_{j}^{n+\frac{1}{2}}\wedge dw_{j}^{n+\frac{1}{2}}\right]
$$

$$
=\left(\frac{\partial^{2}}{\partial p\partial q}\tilde{H}_{j+\frac{1}{2}}^{n+\frac{1}{2}}\right)dp_{j+\frac{1}{2}}^{n+\frac{1}{2}}\wedge dq_{j+\frac{1}{2}}^{n+\frac{1}{2}}+\left(\frac{\partial^{2}}{\partial q^{2}}\tilde{H}_{j+\frac{1}{2}}^{n+\frac{1}{2}}\right)dq_{j+\frac{1}{2}}^{n+\frac{1}{2}}\wedge dq_{j+\frac{1}{2}}^{n+\frac{1}{2}}
$$

$$
+\left(\frac{\partial^{2}}{\partial v\partial q}\tilde{H}_{j+\frac{1}{2}}^{n+\frac{1}{2}}\right)dv_{j+\frac{1}{2}}^{n+\frac{1}{2}}\wedge dq_{j+\frac{1}{2}}^{n+\frac{1}{2}}+\left(\frac{\partial^{2}}{\partial w\partial q}\tilde{H}_{j+\frac{1}{2}}^{n+\frac{1}{2}}\right)dw_{j+\frac{1}{2}}^{n+\frac{1}{2}}\wedge dq_{j+\frac{1}{2}}^{n+\frac{1}{2}}
$$

$$
=\left(\frac{\partial^{2}}{\partial p\partial q}\tilde{H}_{j+\frac{1}{2}}^{n+\frac{1}{2}}\right)dp_{j+\frac{1}{2}}^{n+\frac{1}{2}}\wedge dq_{j+\frac{1}{2}}^{n+\frac{1}{2}}.
$$

Similarly, for the second equation in (4.8), we take the exterior derivative, perform the wedge product with $dp_{j+\frac{1}{2}}^{n+\frac{1}{2}}$ and obtain

$$
\frac{1}{2\tau}\left[dp_{j+\frac{1}{2}}^{n+1}\wedge dq_{j+\frac{1}{2}}^{n+1}-dp_{j+\frac{1}{2}}^{n+1}\wedge dq_{j+\frac{1}{2}}^{n}+dp_{j+\frac{1}{2}}^{n}\wedge dq_{j+\frac{1}{2}}^{n+1}-dp_{j+\frac{1}{2}}^{n}\wedge dq_{j+\frac{1}{2}}^{n}\right]
$$

$$
-\frac{1}{2h}\left[dp_{j+1}^{n+\frac{1}{2}}\wedge dv_{j+1}^{n+\frac{1}{2}}-dp_{j+1}^{n+\frac{1}{2}}\wedge dv_{j}^{n+\frac{1}{2}}+dp_{j}^{n+\frac{1}{2}}\wedge dv_{j+1}^{n+\frac{1}{2}}-dp_{j}^{n+\frac{1}{2}}\wedge dv_{j}^{n+\frac{1}{2}}\right]
$$

$$
=-dp_{j+\frac{1}{2}}^{n+\frac{1}{2}}\wedge\left(\frac{\partial^{2}}{\partial q\partial p}\tilde{H}_{j+\frac{1}{2}}^{n+\frac{1}{2}}\right)dq_{j+\frac{1}{2}}^{n+\frac{1}{2}}.
$$

Adding the above two equations and utilizing the fact that

$$
\left(\frac{\partial^{2}}{\partial p\partial q}\tilde{H}_{j+\frac{1}{2}}^{n+\frac{1}{2}}\right)dp_{j+\frac{1}{2}}^{n+\frac{1}{2}}\wedge dq_{j+\frac{1}{2}}^{n+\frac{1}{2}}=dp_{j+\frac{1}{2}}^{n+\frac{1}{2}}\wedge\left(\frac{\partial^{2}}{\partial q\partial p}\tilde{H}_{j+\frac{1}{2}}^{n+\frac{1}{2}}\right)dq_{j+\frac{1}{2}}^{n+\frac{1}{2}},
$$

we derive

$$
\frac{1}{\tau}\left[dp_{j+\frac{1}{2}}^{n+1}\wedge dq_{j+\frac{1}{2}}^{n+1}-dp_{j+\frac{1}{2}}^{n}\wedge dq_{j+\frac{1}{2}}^{n}\right]
$$

$$
-\frac{1}{2h}\left[dq_{j+1}^{n+\frac{1}{2}}\wedge dw_{j+1}^{n+\frac{1}{2}}-dq_{j+1}^{n+\frac{1}{2}}\wedge dw_{j}^{n+\frac{1}{2}}+dq_{j}^{n+\frac{1}{2}}\wedge dw_{j+1}^{n+\frac{1}{2}}-dq_{j}^{n+\frac{1}{2}}\wedge dw_{j}^{n+\frac{1}{2}}\right]
$$

$$
-\frac{1}{2h}\left[dp_{j+1}^{n+\frac{1}{2}}\wedge dv_{j+1}^{n+\frac{1}{2}}-dp_{j+1}^{n+\frac{1}{2}}\wedge dv_{j}^{n+\frac{1}{2}}+dp_{j}^{n+\frac{1}{2}}\wedge dv_{j+1}^{n+\frac{1}{2}}-dp_{j}^{n+\frac{1}{2}}\wedge dv_{j}^{n+\frac{1}{2}}\right]
$$

$$
=0.
$$

We use a similar argument for the last two equations in (4.8). Taking their exterior derivatives, performing the wedge product with $dv_{j+\frac{1}{2}}^{n+\frac{1}{2}}$ and $dw_{j+\frac{1}{2}}^{n+\frac{1}{2}}$ respectively and adding the two derived equations together, we get

$$
\frac{1}{2h}\left[dp_{j+1}^{n+\frac{1}{2}} \wedge dv_{j+1}^{n+\frac{1}{2}} + dp_{j+1}^{n+\frac{1}{2}} \wedge dv_{j}^{n+\frac{1}{2}} - dp_{j}^{n+\frac{1}{2}} \wedge dv_{j+1}^{n+\frac{1}{2}} - dp_{j}^{n+\frac{1}{2}} \wedge dv_{j}^{n+\frac{1}{2}}\right]
$$
$$
+ \frac{1}{2h}\left[dq_{j+1}^{n+\frac{1}{2}} \wedge dw_{j+1}^{n+\frac{1}{2}} + dq_{j+1}^{n+\frac{1}{2}} \wedge dw_{j}^{n+\frac{1}{2}} - dq_{j}^{n+\frac{1}{2}} \wedge dw_{j+1}^{n+\frac{1}{2}} - dq_{j}^{n+\frac{1}{2}} \wedge dw_{j}^{n+\frac{1}{2}}\right]
$$
$$
= 0.
$$

We finally conclude the required result in the theorem based on above two equations. □

Note that there is an alternative way to show the multi-symplecticity of scheme (4.7) since it can be rewritten into the following form

$$
M_4 \delta_t z_{j+\frac{1}{2}}^n + K_4 \delta_x z_j^{n+\frac{1}{2}} = \nabla S\left(z_{j+\frac{1}{2}}^{n+\frac{1}{2}}\right)
$$

with the notation $z_j^n = (p_j^n, q_j^n, v_j^n, w_j^n)^\top$,

$$
\delta_t z_j^n := \frac{z_j^{n+1} - z_j^n}{\tau} \quad \text{and} \quad \delta_x z_j^n := \frac{z_{j+1}^n - z_j^n}{h}.
$$

Some more details will be given in the proof of the stochastic Preissman scheme in Sect. 4.4. This scheme is also called the centered cell scheme (see e.g. [117]) or the central box scheme (see e.g. [107]).

As for the nonlinear Schrödinger equations, we concern mainly on the value of u, equivalently p and q. As a result, it is more convenient to eliminate auxiliary variables v and w and get an equivalent numerical scheme for (4.1).

Corollary 4.1 *The Preissman scheme (4.7) has the following equivalent form*

$$
\frac{\left(u_{j+1}^{n+1} + 2u_j^{n+1} + u_{j-1}^{n+1}\right) - \left(u_{j+1}^n + 2u_j^n + u_{j-1}^n\right)}{2\tau} - i\frac{u_{j+1}^{n+\frac{1}{2}} - 2u_j^{n+\frac{1}{2}} + u_{j-1}^{n+\frac{1}{2}}}{h^2}
$$
$$
= \frac{1}{2}i\lambda\left(\left|u_{j+\frac{1}{2}}^{n+\frac{1}{2}}\right|^{2\sigma} u_{j+\frac{1}{2}}^{n+\frac{1}{2}} + \left|u_{j-\frac{1}{2}}^{n+\frac{1}{2}}\right|^{2\sigma} u_{j-\frac{1}{2}}^{n+\frac{1}{2}}\right)
$$

as a numerical scheme for (4.1).

Proof For simplicity, we consider the equivalent form (4.8) of (4.7) based on the Hamiltonian function $\tilde{H}$. According to the first equation in scheme (4.8), we have

$$
\frac{p_{j+\frac{1}{2}}^{n+1} - p_{j+\frac{1}{2}}^n}{\tau} + \frac{w_{j+1}^{n+\frac{1}{2}} - w_j^{n+\frac{1}{2}}}{h} = \frac{\partial}{\partial q}\tilde{H}_{j+\frac{1}{2}}^{n+\frac{1}{2}},
$$

$$\frac{p_{j-\frac{1}{2}}^{n+1} - p_{j-\frac{1}{2}}^{n}}{\tau} + \frac{w_{j}^{n+\frac{1}{2}} - w_{j-1}^{n+\frac{1}{2}}}{h} = \frac{\partial}{\partial q} \tilde{H}_{j-\frac{1}{2}}^{n+\frac{1}{2}},$$

whose summation shows

$$\frac{\left(p_{j+1}^{n+1} + 2p_{j}^{n+1} + p_{j-1}^{n+1} \right) - \left(p_{j+1}^{n} + 2p_{j}^{n} + p_{j-1}^{n} \right)}{2\tau} + \frac{w_{j+1}^{n+\frac{1}{2}} - w_{j-1}^{n+\frac{1}{2}}}{h}$$

$$= \frac{\partial}{\partial q} \tilde{H}_{j+\frac{1}{2}}^{n+\frac{1}{2}} + \frac{\partial}{\partial q} \tilde{H}_{j-\frac{1}{2}}^{n+\frac{1}{2}}.$$

Furthermore, the last equation in (4.8) leads to

$$\frac{q_{j+1}^{n+\frac{1}{2}} - q_{j}^{n+\frac{1}{2}}}{h} - \frac{q_{j}^{n+\frac{1}{2}} - q_{j-1}^{n+\frac{1}{2}}}{h} = \frac{\partial}{\partial w} \tilde{H}_{j+\frac{1}{2}}^{n+\frac{1}{2}} - \frac{\partial}{\partial w} \tilde{H}_{j-\frac{1}{2}}^{n+\frac{1}{2}} = \frac{1}{2} \left(w_{j+1}^{n+\frac{1}{2}} - w_{j-1}^{n+\frac{1}{2}} \right).$$

Combining the above two equations together, we derive

$$\frac{\left(p_{j+1}^{n+1} + 2p_{j}^{n+1} + p_{j-1}^{n+1} \right) - \left(p_{j+1}^{n} + 2p_{j}^{n} + p_{j-1}^{n} \right)}{2\tau} + \frac{2\left(q_{j+1}^{n+\frac{1}{2}} - 2q_{j}^{n+\frac{1}{2}} + q_{j-1}^{n+\frac{1}{2}} \right)}{h^{2}}$$

$$= \frac{\partial}{\partial q} \tilde{H}_{j+\frac{1}{2}}^{n+\frac{1}{2}} + \frac{\partial}{\partial q} \tilde{H}_{j-\frac{1}{2}}^{n+\frac{1}{2}}.$$

The same procedure applied to the second and third equations in (4.8) yields

$$\frac{\left(q_{j+1}^{n+1} + 2q_{j}^{n+1} + q_{j-1}^{n+1} \right) - \left(q_{j+1}^{n} + 2q_{j}^{n} + q_{j-1}^{n} \right)}{2\tau} - \frac{2\left(p_{j+1}^{n+\frac{1}{2}} - 2p_{j}^{n+\frac{1}{2}} + p_{j-1}^{n+\frac{1}{2}} \right)}{h^{2}}$$

$$= -\left(\frac{\partial}{\partial p} \tilde{H}_{j+\frac{1}{2}}^{n+\frac{1}{2}} + \frac{\partial}{\partial p} \tilde{H}_{j-\frac{1}{2}}^{n+\frac{1}{2}} \right).$$

Then utilizing the notation $u_{j}^{n} := p_{j}^{n} + \mathbf{i}q_{j}^{n}$, we finally get

$$\frac{\left(u_{j+1}^{n+1} + 2u_{j}^{n+1} + u_{j-1}^{n+1} \right) - \left(u_{j+1}^{n} + 2u_{j}^{n} + u_{j-1}^{n} \right)}{2\tau} - 2\mathbf{i}\frac{u_{j+1}^{n+\frac{1}{2}} - 2u_{j}^{n+\frac{1}{2}} + u_{j-1}^{n+\frac{1}{2}}}{h^{2}}$$

$$= \mathbf{i}\lambda \left(\left| u_{j+\frac{1}{2}}^{n+\frac{1}{2}} \right|^{2\sigma} u_{j+\frac{1}{2}}^{n+\frac{1}{2}} + \left| u_{j-\frac{1}{2}}^{n+\frac{1}{2}} \right|^{2\sigma} u_{j-\frac{1}{2}}^{n+\frac{1}{2}} \right),$$

which completes the proof. $\qquad\qquad\square$

4.3   Stochastic Symplectic Geometric Structure and Numerical Schemes

The symplecticity and multi-symplecticity are preserved by stochastic nonlinear Schrödinger equations with additive or linear multiplicative noises, similar to the deterministic case. We show these properties in this section and the following one based on the variational principle with a stochastic forcing.

The classical Hamilton's principle seeks the extremal $q(t)$ such that the action functional $\int_0^T L(t, q, \dot{q}) dt$ is stationary with $\delta q(0) = \delta q(T) = 0$ under variation, where L is the Lagrangian of the deterministic Hamiltonian system. Based on the Legendre transform $H(p, q, t) = p\dot{q} - L(t, q, \dot{q})$ with $p = \frac{\delta L}{\delta \dot{q}}$, the author in [92] gives the modified Hamilton's principle

$$\delta \int_0^T p\dot{q} - H(p, q, t) dt \equiv 0.$$

However, for Hamiltonian systems influenced by an additional nonconservative force, i.e., an external noise, its Hamiltonian energy turns out to be $H_0(p, q, t) + H_1(p, q, t) \circ \dot{\chi}$ instead of $H(p, q, t)$. Here, $H_1(p, q, t) \circ \dot{\chi}$ denotes the work done by a spatio-temporal noise $\dot{\chi}$, and $\dot{\chi} = \frac{dW(t)}{dt}$ is a formal time derivative of an $\mathbb{R}$-valued Wiener process W. In this case, especially when p, q depending on the spatial variable $x \in \mathbb{R}^d$, by introducing the generalized action functional

$$G(p, q) := \int_0^T \left[\int_{\mathbb{R}^d} p\dot{q} dx - H_0(p, q, t) - H_1(p, q, t) \circ \dot{\chi} \right] dt,$$

the generalized modified Hamilton's principle reads

$$
\begin{aligned}
\delta G &= \delta \int_0^T \left[\int_{\mathbb{R}^d} p\dot{q} dx - H_0(p, q, t) - H_1(p, q, t) \circ \dot{\chi} \right] dt \\
&= \int_0^T \int_{\mathbb{R}^d} \left[p\delta\dot{q} + \dot{q}\delta p - \frac{\delta H_0}{\delta p}\delta p - \frac{\delta H_0}{\delta q}\delta q - \frac{\delta H_1}{\delta p} \circ \dot{\chi}\delta p - \frac{\delta H_1}{\delta q} \circ \dot{\chi}\delta q \right] dx dt \\
&\equiv 0
\end{aligned}
\tag{4.9}
$$

under the condition of fixed endpoints, that is, the increments at time 0 and T are zeros: $\delta q(0) = \delta q(T) = 0$. The principle above, together with the fact that $\int_0^T p\delta\dot{q} dt = -\int_0^T \dot{p}\delta q dt$, yields the generalized Hamiltonian equations of the motion with noise:

$$
\begin{cases}
\dot{p} = -\dfrac{\delta H_0}{\delta q} - \dfrac{\delta H_1}{\delta q} \circ \dot{\chi}, \\[2mm]
\dot{q} = \dfrac{\delta H_0}{\delta p} + \dfrac{\delta H_1}{\delta p} \circ \dot{\chi},
\end{cases}
$$

which is rigorously interpreted by the following SPDE

$$\begin{cases} dp = -\dfrac{\delta H_0}{\delta q} dt - \dfrac{\delta H_1}{\delta q} \circ dW, \\[3mm] dq = \dfrac{\delta H_0}{\delta p} dt + \dfrac{\delta H_1}{\delta p} \circ dW, \end{cases} \tag{4.10}$$

where the variation of H_0 is defined by

$$\delta H_0(p, q) = \left. \frac{d}{d\varepsilon} \right|_{\varepsilon=0} H_0(p + \varepsilon \delta p, q + \varepsilon \delta q),$$

and similarly for H_1. For (4.10) with a fixed initial value $(p(0), q(0))^\top = (p_0, q_0)^\top$, we refer to [90, 120] for the differentiability of the solution with respect to initial data.

Theorem 4.6 *If the solution of* (4.10) *is differentiable with respect to the initial value. Then the phase flow of system* (4.10) *preserves the symplectic structure almost surely, that is,*

$$\varpi(t) := \int_{\mathcal{O}} dp(t) \wedge dq(t) dx = \int_{\mathcal{O}} dp_0 \wedge dq_0 dx = \varpi(0).$$

Next, we focus on the symplectic structure and numerical approximations for the stochastic nonlinear Schrödinger equation (3.8) with an $\mathbb{R}$-valued linear multiplicative noise. Denoting by p and q the real and imaginary parts of solution u, we have that (3.8) is equivalent to

$$\begin{cases} dp = -\left(\Delta q + \lambda (p^2 + q^2)^\sigma q \right) dt - q \circ dW, \\[2mm] dq = \left(\Delta p + \lambda (p^2 + q^2)^\sigma p \right) dt + p \circ dW. \end{cases} \tag{4.11}$$

It is not hard to check that (4.11) can be rewritten in the form (4.10) with functionals (see also [47])

$$H_0(p, q) = -\frac{1}{2} \int_{\mathcal{O}} |\nabla p|^2 + |\nabla q|^2 dx + \frac{\lambda}{2\sigma + 2} \int_{\mathcal{O}} (p^2 + q^2)^{\sigma+1} dx$$

and

$$H_1(p, q) = \frac{1}{2} \int_{\mathcal{O}} (p^2 + q^2) dx.$$

To clear up ambiguities, we give an explanation about the notation $\frac{\delta H_0}{\delta p}$. It is easy to calculate the variation of H_0:

$$\delta H_0(p, q) = \frac{d}{d\varepsilon}\Big|_{\varepsilon=0} H_0(p + \varepsilon\delta p, q + \varepsilon\delta q)$$

$$= -\int_{\mathscr{O}} [\nabla p \nabla(\delta p) + \nabla q \nabla(\delta q)]\, dx + \lambda \int_{\mathscr{O}} \left(p^2 + q^2\right)^{\sigma} (p\delta p + q\delta q)dx$$

$$= \int_{\mathscr{O}} \left\{\left[\Delta p + \lambda \left(p^2 + q^2\right)^{\sigma} p\right]\delta p + \left[\Delta q + \lambda \left(p^2 + q^2\right)^{\sigma} q\right]\delta q\right\} dx$$

for any increments δp and δq. Compared with (4.9), we just denote

$$\frac{\delta H_0}{\delta p} = \Delta p + \lambda \left(p^2 + q^2\right)^{\sigma} p.$$

Similarly, we get

$$\frac{\delta H_0}{\delta q} = \Delta q + \lambda \left(p^2 + q^2\right)^{\sigma} q,$$

$\frac{\delta H_1}{\delta p} = p$ and $\frac{\delta H_1}{\delta q} = q$.

Denote $V := \lambda \left(p^2 + q^2\right)^{\sigma}$. Then the variations of $\frac{\delta H_0}{\delta p}$ and $\frac{\delta H_0}{\delta q}$ are

$$\delta\left(\frac{\delta H_0}{\delta p}\right) = \left(\Delta + V_p p + V\right)\delta p + V_q p\delta q,$$

$$\delta\left(\frac{\delta H_0}{\delta q}\right) = V_p q\delta p + \left(\Delta + V_q q + V\right)\delta q$$

with $V_p := \frac{\partial V}{\partial p}$ and $V_q := \frac{\partial V}{\partial q}$, such that

$$\frac{\delta^2 H_0}{\delta p^2} = \Delta + V_p p + V, \quad \frac{\delta^2 H_0}{\delta p\delta q} = V_q p, \quad \frac{\delta^2 H_0}{\delta q\delta p} = V_p q, \quad \frac{\delta^2 H_0}{\delta q^2} = \Delta + V_q q + V,$$

and similarly,

$$\frac{\delta^2 H_1}{\delta p^2} = 1, \quad \frac{\delta^2 H_1}{\delta p\delta q} = \frac{\delta^2 H_1}{\delta q\delta p} = 0, \quad \frac{\delta^2 H_1}{\delta q^2} = 1.$$

Hence, we derive the following equalities related to the directional derivatives of p and q in the directions p_0 and q_0:

$$d\left(\frac{\partial p}{\partial p_0}\right) = -\left(\frac{\delta^2 H_0}{\delta q\delta p}\frac{\partial p}{\partial p_0} + \frac{\delta^2 H_0}{\delta q^2}\frac{\partial q}{\partial p_0}\right) dt - \left(\frac{\delta^2 H_1}{\delta q\delta p}\frac{\partial p}{\partial p_0} + \frac{\delta^2 H_1}{\delta q^2}\frac{\partial q}{\partial p_0}\right) \circ dW$$

$$= -\left[(\Delta + V_q q + V)\frac{\partial q}{\partial p_0} + V_p q \frac{\partial p}{\partial p_0}\right] dt - \frac{\partial q}{\partial p_0} \circ dW,$$

$$d\left(\frac{\partial p}{\partial q_0}\right) = -\left[(\Delta + V_q q + V)\frac{\partial q}{\partial q_0} + V_p q \frac{\partial p}{\partial q_0}\right] dt - \frac{\partial q}{\partial q_0} \circ dW,$$

$$d\left(\frac{\partial q}{\partial p_0}\right) = \left[(\Delta + V_p p + V)\frac{\partial p}{\partial p_0} + V_q p \frac{\partial q}{\partial p_0}\right] dt + \frac{\partial p}{\partial p_0} \circ dW$$

and

$$d\left(\frac{\partial q}{\partial q_0}\right) = \left[(\Delta + V_p p + V)\frac{\partial p}{\partial q_0} + V_q p \frac{\partial q}{\partial q_0}\right] dt + \frac{\partial p}{\partial q_0} \circ dW.$$

Based on above equalities, one can check that

$$d\left(\frac{\partial p}{\partial p_0}\frac{\partial q}{\partial q_0} - \frac{\partial p}{\partial q_0}\frac{\partial q}{\partial p_0}\right) = \left[-\Delta\left(\frac{\partial q}{\partial p_0}\right)\frac{\partial q}{\partial q_0} + \frac{\partial p}{\partial p_0}\Delta\left(\frac{\partial p}{\partial q_0}\right)\right.$$
$$\left. + \Delta\left(\frac{\partial q}{\partial q_0}\right)\frac{\partial q}{\partial p_0} - \frac{\partial p}{\partial q_0}\Delta\left(\frac{\partial p}{\partial p_0}\right)\right] dt, \qquad (4.12)$$

where we have used the fact that $V_p q = V_q p$. Now we define the averaged differential 2-form $\varpi(t) := \int_{\mathcal{O}} dp(t) \wedge dq(t) dx$ similar to the deterministic case. Then, based on (4.12), the derivative with respect to time gives the stochastic symplectic conservation law (see also [47])

$$d\varpi(t) := d\int_{\mathcal{O}} dp \wedge dq dx = \int_{\mathcal{O}} d\left(\frac{\partial p}{\partial p_0}\frac{\partial q}{\partial q_0} - \frac{\partial p}{\partial q_0}\frac{\partial q}{\partial p_0}\right) dp_0 \wedge dq_0 dx$$

$$= -\int_{\mathcal{O}} [d(\Delta p) \wedge dp + d(\Delta q) \wedge dq] dx dt$$

$$= -\int_{\mathcal{O}} \nabla [d(\nabla p) \wedge dp + d(\nabla q) \wedge dq] dx dt$$

$$= 0,$$

where the homogenous boundary condition is used in the last step. It then shows the symplecticity of (4.11).

For stochastic nonlinear Schrödinger equations, their temporal semi-discretizations based on the s-stage Runge–Kutta methods with certain conditions could inherit the symplecticity of the original systems as in the case of deterministic nonlinear Schrödinger equations (Theorems 4.3 and 4.4).

The s-stage Runge–Kutta method applied to (4.11) in temporal direction reads

$$\begin{cases} P_i = p^n - \tau \sum_{j=1}^{s} a_{ij}^{(0)} \left(\Delta Q_j + \lambda(P_j^2 + Q_j^2)^\sigma Q_j \right) - \sum_{j=1}^{s} a_{ij}^{(1)} Q_j \delta_{n+1} W, \\[2ex] Q_i = q^n + \tau \sum_{j=1}^{s} a_{ij}^{(0)} \left(\Delta P_j + \lambda(P_j^2 + Q_j^2)^\sigma P_j \right) + \sum_{j=1}^{s} a_{ij}^{(1)} P_j \delta_{n+1} W, \\[2ex] p^{n+1} = p^n - \tau \sum_{i=1}^{s} b_i^{(0)} \left(\Delta Q_i + \lambda(P_i^2 + Q_i^2)^\sigma Q_i \right) - \sum_{i=1}^{s} b_i^{(1)} Q_i \delta_{n+1} W, \\[2ex] q^{n+1} = q^n + \tau \sum_{i=1}^{s} b_i^{(0)} \left(\Delta P_i + \lambda(P_i^2 + Q_i^2)^\sigma P_i \right) + \sum_{i=1}^{s} b_i^{(1)} P_i \delta_{n+1} W, \end{cases} \tag{4.13}$$

where $t_n = n\tau$ and $\delta_{n+1} W := W(t_{n+1}) - W(t_n)$ denotes the increment of the Wiener process. Two classes of parameters $\{a_{ij}^{(0)}, b_i^{(0)}\}_{i,j=1,\cdots,s}$ and $\{a_{ij}^{(1)}, b_i^{(1)}\}_{i,j=1,\cdots,s}$ may be different.

Scheme (4.13) can be expressed as an approximation for the original system (3.8) by denoting $U_i := P_i + \mathbf{i} Q_i$ and $u^n := p^n + \mathbf{i} q^n$:

$$\begin{cases} U_i = u^n + \mathbf{i}\tau \sum_{j=1}^{s} a_{ij}^{(0)} \left(\Delta U_j + \lambda|U_j|^{2\sigma} U_j \right) + \mathbf{i} \sum_{j=1}^{s} a_{ij}^{(1)} U_j \delta_{n+1} W, \\[2ex] u^{n+1} = u^n + \mathbf{i}\tau \sum_{i=1}^{s} b_i^{(0)} \left(\Delta U_i + \lambda|U_i|^{2\sigma} U_i \right) + \mathbf{i} \sum_{i=1}^{s} b_i^{(1)} U_i \delta_{n+1} W. \end{cases} \tag{4.14}$$

Theorem 4.7 ([47]) *Assume that the coefficients satisfy*

$$b_i^{(0)} b_j^{(0)} = b_i^{(0)} a_{ij}^{(0)} + b_j^{(0)} a_{ji}^{(0)},$$
$$b_i^{(0)} b_j^{(1)} = b_i^{(0)} a_{ij}^{(1)} + b_j^{(1)} a_{ji}^{(0)},$$
$$b_i^{(1)} b_j^{(1)} = b_i^{(1)} a_{ij}^{(1)} + b_j^{(1)} a_{ji}^{(1)}.$$

The temporal semi-discretization (4.13) possesses a solution in $\mathbf{L}^2(\mathcal{O})$, which preserves both the discrete charge conservation law

$$\|u^{n+1}\|_{\mathbf{L}^2(\mathcal{O})}^2 = \|u^n\|_{\mathbf{L}^2(\mathcal{O})}^2$$

and the discrete symplectic conservation law

$$\int_{\mathcal{O}} \mathrm{d}p^{n+1} \wedge \mathrm{d}q^{n+1} dx = \int_{\mathcal{O}} \mathrm{d}p^n \wedge \mathrm{d}q^n dx, \quad \forall\, n \in \mathbb{N}$$

almost surely.

Proof **Step 1**. We first show the existence of the solution, which follows from a standard Galerkin method and Brouwer's theorem. In fact, we first consider the following finite dimensional approximation

$$d\tilde{u} = \mathbf{i}\Delta\tilde{u}dt + \mathbf{i}\lambda\pi_M\left(|\tilde{u}|^{2\sigma}\tilde{u}\right)dt + \mathbf{i}\pi_M\left(\tilde{u}\circ dW(t)\right), \tag{4.15}$$

which is obtained based on the spectral Galerkin approximation. Here, π_M is the projection operator from $\mathbf{L}^2(\mathcal{O})$ to an M-dimensional space V_M. The well-posedness for (4.15) follows from the argument of the well-posedness for (3.8). Scheme (4.14) applied to (4.15) yields

$$\begin{cases} \tilde{U}_i = \tilde{u}^n + \mathbf{i}\tau\sum_{j=1}^{s}a_{ij}^{(0)}\tilde{A}_j + \mathbf{i}\sum_{j=1}^{s}a_{ij}^{(1)}\tilde{B}_j, \\ \tilde{u}^{n+1} = \tilde{u}^n + \mathbf{i}\tau\sum_{i=1}^{s}b_i^{(0)}\tilde{A}_i + \mathbf{i}\sum_{i=1}^{s}b_i^{(1)}\tilde{B}_i, \end{cases}$$

where we use the notations

$$\tilde{A}_i := \Delta\tilde{U}_i + \lambda\pi_M\left(|\tilde{U}_i|^{2\sigma}\tilde{U}_i\right) \quad \text{and} \quad \tilde{B}_i := \pi_M\left(\tilde{U}_i\delta_{n+1}W\right).$$

Based on the scheme above, we have

$$\tilde{u}^{n+\frac{1}{2}} = \tilde{u}^n + \frac{\mathbf{i}\tau}{2}\sum_{i=1}^{s}b_i^{(0)}\tilde{A}_i + \frac{\mathbf{i}}{2}\sum_{i=1}^{s}b_i^{(1)}\tilde{B}_i$$

and

$$\begin{aligned} \tilde{u}^{n+\frac{1}{2}} &= \frac{\tilde{u}^{n+1} - \tilde{u}^n}{2} + \tilde{u}^n \\ &= \frac{\mathbf{i}\tau}{2}\sum_{i=1}^{s}b_i^{(0)}\tilde{A}_i + \frac{\mathbf{i}}{2}\sum_{i=1}^{s}b_i^{(1)}\tilde{B}_i + \tilde{U}_i - \mathbf{i}\tau\sum_{j=1}^{s}a_{ij}^{(0)}\tilde{A}_j - \mathbf{i}\sum_{j=1}^{s}a_{ij}^{(1)}\tilde{B}_j. \end{aligned}$$

Defining the continuous function

$$g(\tilde{u}^{n+\frac{1}{2}}) := \tilde{u}^{n+\frac{1}{2}} - \tilde{u}^n - \frac{\mathbf{i}\tau}{2}\sum_{i=1}^{s}b_i^{(0)}\tilde{A}_i - \frac{\mathbf{i}}{2}\sum_{i=1}^{s}b_i^{(1)}\tilde{B}_i,$$

we obtain based on the assumption on the coefficients that

$$\Re\left(g(\tilde{u}^{n+\frac{1}{2}}),\tilde{u}^{n+\frac{1}{2}}\right)=\Re\left(\tilde{u}^{n+\frac{1}{2}}-\tilde{u}^n,\tilde{u}^{n+\frac{1}{2}}\right)$$

$$+\frac{\tau^2}{4}\sum_{i,j=1}^{s}\left(b_i^{(0)}b_j^{(0)}-b_i^{(0)}a_{ij}^{(0)}-b_j^{(0)}a_{ji}^{(0)}\right)\Re(\tilde{A}_i,\tilde{A}_j)$$

$$-\frac{\tau}{2}\sum_{i,j=1}^{s}\left(b_i^{(0)}b_j^{(1)}-b_i^{(0)}a_{ij}^{(1)}-b_j^{(1)}a_{ji}^{(0)}\right)\Re(\tilde{A}_i,\tilde{B}_j)$$

$$+\frac{1}{4}\sum_{i,j=1}^{s}\left(b_i^{(1)}b_j^{(1)}-b_i^{(1)}a_{ij}^{(1)}-b_j^{(1)}a_{ji}^{(1)}\right)\Re(\tilde{B}_i,\tilde{B}_j)$$

$$-\Re\left(\frac{\mathbf{i}\tau}{2}\sum_{i=1}^{s}b_i^{(0)}\tilde{A}_i+\frac{\mathbf{i}}{2}\sum_{i=1}^{s}b_i^{(1)}\tilde{B}_i,\tilde{U}_i\right)$$

$$=\left\|\tilde{u}^{n+\frac{1}{2}}\right\|_{\mathbf{L}^2(\mathcal{O})}^2-\Re\left(\tilde{u}^n,\tilde{u}^{n+\frac{1}{2}}\right)$$

$$\geq\left\|\tilde{u}^{n+\frac{1}{2}}\right\|_{\mathbf{L}^2(\mathcal{O})}\left(\left\|\tilde{u}^{n+\frac{1}{2}}\right\|_{\mathbf{L}^2(\mathcal{O})}-\left\|\tilde{u}^n\right\|_{\mathbf{L}^2(\mathcal{O})}\right).$$

Then utilizing the $\mathbb{C}$-valued version of Brouwer's fixed-point result (see e.g. [4]), we get the existence for the solution of the finite dimensional approximation. The existence of (4.13) (or equivalently (4.14)) then follows from the convergence analysis of the spectral Galerkin approximation in [57].

Step 2. Now we show that the numerical solution is charge-preserved. We consider the second equation in scheme (4.14)

$$u^{n+1}-u^n=\mathbf{i}\tau\sum_{i=1}^{s}b_i^{(0)}A_i+\mathbf{i}\sum_{i=1}^{s}b_i^{(1)}B_i$$

with notations

$$A_i:=\Delta U_i+\lambda|U_i|^{2\sigma}U_i\quad\text{and}\quad B_i:=U_i\delta_{n+1}W.$$

Then we can conclude

$$\frac{1}{2}\left(\left\|u^{n+1}\right\|_{\mathbf{L}^2(\mathcal{O})}^2-\left\|u^n\right\|_{\mathbf{L}^2(\mathcal{O})}^2\right)=\Re\left(u^{n+1}-u^n,u^{n+\frac{1}{2}}\right)=0$$

following the same procedure in **Step 1**.

Step 3. We finally prove the symplecticity of the numerical scheme. By denoting

$$A_i^Q := \Delta Q_i + \lambda (P_i^2 + Q_i^2)^\sigma Q_i,$$
$$A_i^P := \Delta P_i + \lambda (P_i^2 + Q_i^2)^\sigma P_i,$$
$$B_i^Q := Q_i \delta_{n+1} W,$$
$$B_i^P := P_i \delta_{n+1} W,$$

we rewrite scheme (4.13) as

$$\begin{cases} P_i = p^n - \tau \sum_{j=1}^{s} a_{ij}^{(0)} A_j^Q - \sum_{j=1}^{s} a_{ij}^{(1)} B_j^Q, \\[2mm] Q_i = q^n + \tau \sum_{j=1}^{s} a_{ij}^{(0)} A_j^P + \sum_{j=1}^{s} a_{ij}^{(1)} B_j^P, \\[2mm] p^{n+1} = p^n - \tau \sum_{i=1}^{s} b_i^{(0)} A_i^Q - \sum_{i=1}^{s} b_i^{(1)} B_i^Q, \\[2mm] q^{n+1} = q^n + \tau \sum_{i=1}^{s} b_i^{(0)} A_i^P + \sum_{i=1}^{s} b_i^{(1)} B_i^P. \end{cases}$$

Performing the wedge product between the last two equations above and noting that

$$\begin{cases} dp^n = dP_i + d\left(\tau \sum_{j=1}^{s} a_{ij}^{(0)} A_j^Q + \sum_{j=1}^{s} a_{ij}^{(1)} B_j^Q \right), \\[3mm] dq^n = dQ_i - d\left(\tau \sum_{j=1}^{s} a_{ij}^{(0)} A_j^P + \sum_{j=1}^{s} a_{ij}^{(1)} B_j^P \right), \end{cases}$$

we obtain

$$\begin{aligned} dp^{n+1} \wedge dq^{n+1} = {} & dp^n \wedge dq^n + dP_i \wedge d\left(\tau \sum_{i=1}^{s} b_i^{(0)} A_i^P + \sum_{i=1}^{s} b_i^{(1)} B_i^P \right) \\ & - d\left(\tau \sum_{i=1}^{s} b_i^{(0)} A_i^Q + \sum_{i=1}^{s} b_i^{(1)} B_i^Q \right) \wedge dQ_i \\ & - \tau^2 \sum_{i,j=1}^{s} \left(b_i^{(0)} b_j^{(0)} - b_i^{(0)} a_{ij}^{(0)} - b_j^{(0)} a_{ji}^{(0)} \right) dA_i^Q \wedge dA_j^P \\ & - \tau \sum_{i,j=1}^{s} \left(b_i^{(0)} b_j^{(1)} - b_i^{(0)} a_{ij}^{(1)} - b_j^{(1)} a_{ji}^{(0)} \right) dA_i^Q \wedge dB_j^P \end{aligned}$$

$$- \tau \sum_{i,j=1}^{s} \left(b_i^{(1)} b_j^{(0)} - b_i^{(1)} a_{ij}^{(0)} - b_j^{(0)} a_{ji}^{(1)} \right) \mathrm{d}B_i^Q \wedge \mathrm{d}A_j^P$$

$$- \sum_{i,j=1}^{s} \left(b_i^{(1)} b_j^{(1)} - b_i^{(1)} a_{ij}^{(1)} - b_j^{(1)} a_{ji}^{(1)} \right) \mathrm{d}B_i^Q \wedge \mathrm{d}B_j^P.$$

Note that

$$\mathrm{d}P_i \wedge \mathrm{d}B_i^P = 0, \quad \mathrm{d}B_i^Q \wedge \mathrm{d}Q_i = 0$$

and

$$\int_{\mathcal{O}} \mathrm{d}P_i \wedge \mathrm{d}A_i^P - \mathrm{d}A_i^Q \wedge \mathrm{d}Q_i \mathrm{d}x$$

$$= \int_{\mathcal{O}} \left[-\mathrm{d}(\nabla P_i) \wedge \mathrm{d}(\nabla P_i) + \mathrm{d}(\nabla Q_i) \wedge \mathrm{d}(\nabla Q_i) + \mathrm{d}P_i \wedge \left(\frac{\partial^2 F}{\partial p^2} \mathrm{d}P_i \right) \right.$$

$$\left. - \frac{\partial^2 F}{\partial q^2} \mathrm{d}Q_i \wedge \mathrm{d}Q_i + \mathrm{d}P_i \wedge \left(\frac{\partial^2 F}{\partial q \partial p} \right) \mathrm{d}Q_i - \frac{\partial^2 F}{\partial p \partial q} \mathrm{d}P_i \wedge \mathrm{d}Q_i \right] \mathrm{d}x$$

$$= 0$$

with

$$F(p, q) := \frac{\lambda}{2\sigma + 2} \left(p^2 + q^2 \right)^{\sigma+1}$$

and the partial derivatives of F taking value at (P_i, Q_i). We then conclude that

$$\int_{\mathcal{O}} \mathrm{d}p^{n+1} \wedge \mathrm{d}q^{n+1} \mathrm{d}x = \int_{\mathcal{O}} \mathrm{d}p^n \wedge \mathrm{d}q^n \mathrm{d}x$$

based on the assumption on the coefficients.   □

As a special case of the 1-stage stochastic Runge–Kutta methods, the midpoint scheme with $a_{11}^{(0)} = a_{11}^{(1)} = \frac{1}{2}$ and $b_1^{(0)} = b_1^{(1)} = 1$ also possesses the discrete symplectic conservation law.

4.4 Stochastic Multi-symplectic Geometric Structure and Numerical Schemes

The stochastic symplectic structure investigated above is obtained when regarding (3.8) as an infinite dimensional evolution equation in time. However, when the spatial variation is of interest for a Hamiltonian partial differential equation, such as spatially periodic waves, the multi-symplectic structure is usually involved, which is first investigated in [34] for the deterministic case. For simplicity, we show the stochastic multi-symplecticity for (3.8) on the domain $[0, 1] \subset \mathbb{R}$ with homogeneous boundary condition $u(t, 0) = u(t, 1) = 0$.

In this point of view, in addition of the conjugate momenta $p = \frac{\partial L}{\partial q_t}$, the Lagrangian $L = L(t, x, q, q_t(p, q), q_x, p_x)$ also depends on space derivatives q_x and p_x. Defining the conjugate coordinates $v = \frac{\partial L}{\partial p_x}$ and $w = \frac{\partial L}{\partial q_x}$, the generalized Hamilton's principle shows

$$\delta G(p, q, v, w) = \delta \int_0^T \int_0^1 pq_t + vp_x + wq_x - S_0(p, q, v, w) - S_1(p, q, v, w) \circ \dot{\chi} dx dt$$

$$= \int_0^T \int_0^1 \left[q_t - v_x - \frac{\partial S_0}{\partial p} - \frac{\partial S_1}{\partial p} \circ \dot{\chi} \right] \delta p + \left[-p_t - w_x - \frac{\partial S_0}{\partial q} - \frac{\partial S_1}{\partial q} \circ \dot{\chi} \right] \delta q$$

$$+ \left[p_x - \frac{\partial S_0}{\partial v} - \frac{\partial S_1}{\partial v} \circ \dot{\chi} \right] \delta v + \left[q_x - \frac{\partial S_0}{\partial w} - \frac{\partial S_1}{\partial w} \circ \dot{\chi} \right] \delta w dx dt \equiv 0.$$

By introducing $S_0 = \frac{\lambda}{2\sigma+2} \left(p^2 + q^2 \right)^{\sigma+1} + \frac{1}{2} \left(v^2 + w^2 \right)$ and $S_1 = \frac{1}{2} \left(p^2 + q^2 \right)$, the equation above shows

$$q_t - v_x = \lambda \left(p^2 + q^2 \right)^\sigma p + p \circ \dot{\chi},$$

$$-p_t - w_x = \lambda \left(p^2 + q^2 \right)^\sigma q + q \circ \dot{\chi},$$

$$p_x = v,$$

$$q_x = w.$$

It gives the equivalent multi-symplectic form of (4.11):

$$M_4 dz + K_4 z_x dt = \nabla S_0 dt + \nabla S_1 \circ dW$$

with $z = (p, q, v, w)^\top$ and skew-symmetric matrices M_4, K_4 being the same as those in Sect. 4.1.

For the numerical approximation of the above equation, we introduce the stochastic Preissman scheme similar to the deterministic case

$$M_4 \left(\frac{z_{j+\frac{1}{2}}^{n+1} - z_{j+\frac{1}{2}}^n}{\tau} \right) + K_4 \left(\frac{z_{j+1}^{n+\frac{1}{2}} - z_j^{n+\frac{1}{2}}}{h} \right)$$

$$= \nabla S_0(z_{j+\frac{1}{2}}^{n+\frac{1}{2}}) + \nabla S_1(z_{j+\frac{1}{2}}^{n+\frac{1}{2}}) \frac{\delta_{n+1} W_{j+\frac{1}{2}}}{\tau}, \tag{4.16}$$

where

$$\delta_{n+1} W_{j+\frac{1}{2}} = W(t_{n+1}, x_{j+\frac{1}{2}}) - W(t_n, x_{j+\frac{1}{2}})$$

denotes the increments of W at $x_{j+\frac{1}{2}} = \frac{1}{2}(x_{j+1} + x_j)$. Scheme (4.16) has the following equivalent form

$$\begin{cases} \dfrac{p_{j+\frac{1}{2}}^{n+1} - p_{j+\frac{1}{2}}^{n}}{\tau} + \dfrac{w_{j+1}^{n+\frac{1}{2}} - w_{j}^{n+\frac{1}{2}}}{h} = -\lambda\left[(p_{j+\frac{1}{2}}^{n+\frac{1}{2}})^2 + (q_{j+\frac{1}{2}}^{n+\frac{1}{2}})^2\right]^{\sigma} q_{j+\frac{1}{2}}^{n+\frac{1}{2}} - q_{j+\frac{1}{2}}^{n+\frac{1}{2}} \dfrac{\delta_{n+1}W_{j+\frac{1}{2}}}{\tau}, \\[4mm] \dfrac{q_{j+\frac{1}{2}}^{n+1} - q_{j+\frac{1}{2}}^{n}}{\tau} - \dfrac{v_{j+1}^{n+\frac{1}{2}} - v_{j}^{n+\frac{1}{2}}}{h} = \lambda\left[(p_{j+\frac{1}{2}}^{n+\frac{1}{2}})^2 + (q_{j+\frac{1}{2}}^{n+\frac{1}{2}})^2\right]^{\sigma} p_{j+\frac{1}{2}}^{n+\frac{1}{2}} + p_{j+\frac{1}{2}}^{n+\frac{1}{2}} \dfrac{\delta_{n+1}W_{j+\frac{1}{2}}}{\tau}, \\[4mm] \dfrac{p_{j+1}^{n+\frac{1}{2}} - p_{j}^{n+\frac{1}{2}}}{h} = v_{j+\frac{1}{2}}^{n+\frac{1}{2}}, \\[4mm] \dfrac{q_{j+1}^{n+\frac{1}{2}} - q_{j}^{n+\frac{1}{2}}}{h} = w_{j+\frac{1}{2}}^{n+\frac{1}{2}}. \end{cases}$$

Now we show that the stochastic Preissman scheme preserves the stochastic multi-symplectic structure.

Theorem 4.8 ([119]) *The full discretization* (4.16) *preserves the stochastic multisymplectic conservation law, i.e.,*

$$\frac{\omega_{j+\frac{1}{2}}^{n+1} - \omega_{j+\frac{1}{2}}^{n}}{\tau} + \frac{\kappa_{j+1}^{n+\frac{1}{2}} - \kappa_{j}^{n+\frac{1}{2}}}{h} = 0$$

with $\omega_{j+\frac{1}{2}}^{n} := \frac{1}{2}dz_{j+\frac{1}{2}}^{n} \wedge M_4 dz_{j+\frac{1}{2}}^{n}$ *and* $\kappa_{j}^{n+\frac{1}{2}} := \frac{1}{2}dz_{j}^{n+\frac{1}{2}} \wedge K_4 dz_{j}^{n+\frac{1}{2}}$.

Proof Taking the exterior derivative in the phase space on both sides of (4.16), we get

$$M_4 \left(\frac{dz_{j+\frac{1}{2}}^{n+1} - dz_{j+\frac{1}{2}}^{n}}{\tau} \right) + K_4 \left(\frac{dz_{j+1}^{n+\frac{1}{2}} - dz_{j}^{n+\frac{1}{2}}}{h} \right)$$
$$= \left(\nabla^2 S_0(z_{j+\frac{1}{2}}^{n+\frac{1}{2}}) + \nabla^2 S_1(z_{j+\frac{1}{2}}^{n+\frac{1}{2}}) \frac{\delta_{n+1}W_{j+\frac{1}{2}}}{\tau} \right) dz_{j+\frac{1}{2}}^{n+\frac{1}{2}}. \qquad (4.17)$$

Then performing the wedge product between

$$dz_{j+\frac{1}{2}}^{n+\frac{1}{2}} = \frac{dz_{j+\frac{1}{2}}^{n+1} + dz_{j+\frac{1}{2}}^{n}}{2} = \frac{dz_{j+1}^{n+\frac{1}{2}} + dz_{j}^{n+\frac{1}{2}}}{2}$$

and (4.17) yields

$$\left(\frac{dz_{j+\frac{1}{2}}^{n+1} + dz_{j+\frac{1}{2}}^{n}}{2} \right) \wedge M_4 \left(\frac{dz_{j+\frac{1}{2}}^{n+1} - dz_{j+\frac{1}{2}}^{n}}{\tau} \right)$$
$$+ \left(\frac{dz_{j+1}^{n+\frac{1}{2}} + dz_{j}^{n+\frac{1}{2}}}{2} \right) \wedge K_4 \left(\frac{dz_{j+1}^{n+\frac{1}{2}} - dz_{j}^{n+\frac{1}{2}}}{h} \right)$$

$$= dz_{j+\frac{1}{2}}^{n+\frac{1}{2}} \wedge \left(\nabla^2 S_0 + \nabla^2 S_1 \frac{\delta_{n+1} W_{j+\frac{1}{2}}}{\tau} \right) dz_{j+\frac{1}{2}}^{n+\frac{1}{2}} = 0 \qquad (4.18)$$

due to the symmetry of $\nabla^2 S_0$ and $\nabla^2 S_1$.

Moreover, according to the fact that matrices M_4 and K_4 are skew-symmetric, we get

$$dz_{j+\frac{1}{2}}^{n} \wedge M_4 dz_{j+\frac{1}{2}}^{n+1} - dz_{j+\frac{1}{2}}^{n+1} \wedge M_4 dz_{j+\frac{1}{2}}^{n} = 0,$$

$$dz_{j}^{n+\frac{1}{2}} \wedge K_4 dz_{j+1}^{n+\frac{1}{2}} - dz_{j+1}^{n+\frac{1}{2}} \wedge K_4 dz_{j}^{n+\frac{1}{2}} = 0,$$

which, together with (4.18), completes the proof. $\square$

4.5 Conformal Multi-symplectic Structure for the Damped Case

Some modifications to the canonical Schrödinger equation arise as mathematical models to suppress the loss of the soliton signal caused by the Gordon–Haus effect (see e.g. [93]). Several types of modifications with damped terms and noises are investigated in [70, 76, 79, 80] and references therein. As one of the modifications above, the following one-dimensional damped stochastic nonlinear Schrödinger equation (see also [50, 70])

$$\begin{cases} du = \left(\mathrm{i} \Delta u - \alpha u + \mathrm{i} \lambda |u|^2 u \right) dt + \varepsilon dW \\ u(t, 0) = u(t, 1) = 0, \ t \geq 0 \\ u(0, x) = u_0(x), \ x \in \mathcal{O} := (0, 1) \end{cases} \qquad (4.19)$$

with the absorption coefficient $\alpha > 0$ will be mainly studied in the following sections.

This section is devoted to introducing some internal properties of (4.19), including geometric structures and the charge evolution law, with $\alpha > 0$, $\lambda = \pm 1$ and the noise intensity ε. Note that the solution u is a $\mathbb{C}$-valued random field on a probability space $(\Omega, \mathscr{F}, \mathbb{P})$. We consider $W = W_1 + \mathrm{i} W_2$ as a Q-Wiener process on $\mathbf{L}^2(\mathcal{O})$ with W_1 and W_2 being its real and imaginary parts, respectively. The covariance operator Q is assumed to be a symmetric, positive definite and trace operator such that W has the following Karhunen–Loève expansion

$$W = \sum_{k=1}^{\infty} Q^{\frac{1}{2}} e_k \beta_k = \sum_{k=1}^{\infty} \sqrt{\eta_k} e_k \beta_k, \quad \eta_k > 0 \ \text{ and } \ \eta := \sum_{k=1}^{\infty} \eta_k < \infty,$$

where $\{e_k\}_{k \in \mathbb{N}}$ is an orthogonal basis for $\mathbf{L}^2(\mathcal{O}; \mathbb{R})$, $\{\beta_k = \beta_k^1 + \mathrm{i} \beta_k^2\}_{k \in \mathbb{N}}$ is associated to a filtration $\{\mathscr{F}_t\}_{t \geq 0}$, and $\{\beta_k^i\}_{k \in \mathbb{N}}^{i=1,2}$ is a family of independent and identically distributed $\mathbb{R}$-valued Wiener processes. In addition, we assume that Q commutes with the Laplace operator, and hence $\{e_k\}_{k\in\mathbb{N}}$ is also the eigenbasis of the Dirichlet Laplacian. Actually, the assumption on Q can be generalized to the degenerate case, which will be explained in Remark 5.1.

We set the linear operator $A_\alpha := -i\Delta + \alpha Id$, and the semigroup $S(t) := e^{-tA_\alpha} = e^{t(i\Delta - \alpha Id)}$ is generated by $-A_\alpha$. The mild solution of (4.19) exists globally and can be written as

$$u(t) = S(t)u_0 + i\lambda \int_0^T S(t-s)|u(s)|^2 u(s)ds + \varepsilon \int_0^T S(t-s)dW(s).$$

Let $\{\lambda_k^\alpha\}_{k\in\mathbb{N}_+}$ be an increasing sequence of eigenvalues of A_α with $1 \leq |\lambda_k^\alpha| \to +\infty$.

Definition 4.2 For all $s \in \mathbb{N}$, we define the normed linear space

$$\dot{\mathbf{H}}^s := \text{Dom}(A_\alpha^{\frac{s}{2}}) = \left\{ u \,\Big|\, u = \sum_{k=1}^\infty (u, e_k)e_k \in \mathbf{L}^2(\mathscr{O}) \text{ s.t. } \sum_{k=1}^\infty |(u, e_k)|^2 |\lambda_k^\alpha|^s < \infty \right\},$$

endowed with the s-norm

$$\|u\|_s := \left(\sum_{k=1}^\infty |(u, e_k)|^2 |\lambda_k^\alpha|^s \right)^{\frac{1}{2}}.$$

In particular, $\|u\|_0 = \|u\|_{\mathbf{L}^2(\mathscr{O})}$ for any $u \in \dot{\mathbf{H}}^0$.

In the sequel, we use the notation $\mathbf{L}^2 := \mathbf{L}^2(\mathscr{O})$ and $\mathbf{H}^s := \mathbf{H}^s(\mathscr{O})$. It is easy to check that the above norms satisfy $\|u\|_r \leq \|u\|_s$ for any $0 \leq r \leq s$, and $\|u\|_s \cong \|u\|_{\mathbf{H}^s}$ ($s = 0, 1, 2$) for any $u \in \dot{\mathbf{H}}^s$, where $\|u\|_{\mathbf{H}^s}$ denotes the natural Sobolev norm.

The operator norm is defined as

$$\|E\|_{\mathscr{L}(\dot{\mathbf{H}}^s, \dot{\mathbf{H}}^r)} = \sup_{u \in \dot{\mathbf{H}}^s} \frac{\|Eu\|_r}{\|u\|_s}, \quad \forall\, r, s \in \mathbb{N}$$

for any operator E. Hence, for $0 \leq r \leq s$, we have

$$\|S(t)\|_{\mathscr{L}(\dot{\mathbf{H}}^s, \dot{\mathbf{H}}^r)} = \sup_{u \in \dot{\mathbf{H}}^s} \frac{\left(\sum_{k=1}^\infty |(e^{t(i\Delta - \alpha Id)}u, e_k)|^2 |\lambda_k^\alpha|^r \right)^{\frac{1}{2}}}{\|u\|_s} = \sup_{u \in \dot{\mathbf{H}}^s} \frac{e^{-\alpha t}\|u\|_r}{\|u\|_s} \leq e^{-\alpha t}.$$

In the following, operator $Q^{\frac{1}{2}}$ is assumed to be a Hilbert–Schmidt operator from $\mathbf{L}^2$ to $\dot{\mathbf{H}}^s$ with norm

$$\|Q^{\frac{1}{2}}\|_{\mathscr{L}_2^s}^2 := \sum_{k=1}^\infty \|Q^{\frac{1}{2}}e_k\|_s^2 = \sum_{k=1}^\infty |\lambda_k^\alpha|^s \eta_k < \infty.$$

We define the spatio-temporal noise $\dot\chi = \frac{dW}{dt}$, set $u = p + \mathbf{i}q$, $\dot\chi = \dot\chi_1 + \mathbf{i}\dot\chi_2$ with $p, q, \dot\chi_1 = \frac{dW_1}{dt}$ and $\dot\chi_2 = \frac{dW_2}{dt}$, and rewrite (4.19) as

$$\begin{cases} p_t + q_{xx} + \alpha p + \lambda(p^2 + q^2)q = \varepsilon\dot\chi_1, \\ -q_t + p_{xx} - \alpha q + \lambda(p^2 + q^2)p = -\varepsilon\dot\chi_2. \end{cases} \tag{4.20}$$

Denoting $v = p_x$, $w = q_x$, $z = (p, q, v, w)^\top$, Eq. (4.20) can be transformed into a compact form

$$M_4 d_t z + K_4 \partial_x z dt = -\alpha M_4 z dt + \nabla S_0(z)dt + \nabla S_1(z) \circ dW_1 + \nabla S_2(z) \circ dW_2, \tag{4.21}$$

where M_4, K_4 are the same as those in Sect. 4.1 and

$$S_0(z) = \frac{\lambda}{4}(p^2 + q^2)^2 + \frac{1}{2}(v^2 + w^2), \quad S_1(z) = -\varepsilon q, \quad S_2(z) = \varepsilon p.$$

Without pointing it out, the equations below hold in the sense $\mathbb{P}$-a.s. We now prove that (4.19) possesses the stochastic conformal multi-symplectic conservation law, whose definition is also given in the following theorem.

Theorem 4.9 *Equation* (4.19) *is a stochastic conformal multi-symplectic Hamiltonian system, and preserves the stochastic conformal multi-symplectic conservation law*

$$d_t\omega(t, x) + \partial_x\kappa(t, x)dt = -\alpha\omega(t, x)dt,$$

which means

$$\int_{x_0}^{x_1} \omega(t_1, x)dx - \int_{x_0}^{x_1} \omega(t_0, x)dx + \int_{t_0}^{t_1} \kappa(t, x_1)dt - \int_{t_0}^{t_1} \kappa(t, x_0)dt$$
$$= -\int_{x_0}^{x_1}\int_{t_0}^{t_1} \alpha\omega(t, x)dtdx, \tag{4.22}$$

where $\omega = \frac{1}{2}dz \wedge M_4 dz$ and $\kappa = \frac{1}{2}dz \wedge K_4 dz$ are two differential 2-forms associated with two skew-symmetric matrices M_4 and K_4.

Proof To simplify the proof, we denote $(z_1, z_2, z_3, z_4) := (p, q, v, w) = z^\top$ and $(z_l)_t^x := z_l(t, x)$ for $l = 1, 2, 3, 4$. Noting that $\omega = dz_2 \wedge dz_1$ and $\kappa = dz_1 \wedge dz_3 + dz_2 \wedge dz_4$, we have

$$\int_{x_0}^{x_1} \omega(t_1, x)dx - \int_{x_0}^{x_1} \omega(t_0, x)dx$$

$$= \int_{x_0}^{x_1} \left[d(z_2)_{t_1}^x \wedge d(z_1)_{t_1}^x - d(z_2)_{t_0}^x \wedge d(z_1)_{t_0}^x \right] dx$$

$$= \int_{x_0}^{x_1} \left[\left(\sum_{l=1}^{4} \frac{\partial (z_2)_{t_1}^x}{\partial (z_l)_{t_0}^{x_0}} d(z_l)_{t_0}^{x_0} \right) \wedge \left(\sum_{i=1}^{4} \frac{\partial (z_1)_{t_1}^x}{\partial (z_i)_{t_0}^{x_0}} d(z_i)_{t_0}^{x_0} \right) \right.$$

$$\left. - \left(\sum_{l=1}^{4} \frac{\partial (z_2)_{t_0}^x}{\partial (z_l)_{t_0}^{x_0}} d(z_l)_{t_0}^{x_0} \right) \wedge \left(\sum_{i=1}^{4} \frac{\partial (z_1)_{t_0}^x}{\partial (z_i)_{t_0}^{x_0}} d(z_i)_{t_0}^{x_0} \right) \right] dx$$

$$= \sum_{l=1}^{4} \sum_{i=1}^{4} \left[\int_{x_0}^{x_1} \left(\frac{\partial (z_2)_{t_1}^x}{\partial (z_l)_{t_0}^{x_0}} \frac{\partial (z_1)_{t_1}^x}{\partial (z_i)_{t_0}^{x_0}} - \frac{\partial (z_2)_{t_0}^x}{\partial (z_l)_{t_0}^{x_0}} \frac{\partial (z_1)_{t_0}^x}{\partial (z_i)_{t_0}^{x_0}} \right) dx \right] d(z_l)_{t_0}^{x_0} \wedge d(z_i)_{t_0}^{x_0}$$

$$=: \sum_{l=1}^{4} \sum_{i=1}^{4} \mathscr{H}_{l,i}(t_1, x_1) d(z_l)_{t_0}^{x_0} \wedge d(z_i)_{t_0}^{x_0} \tag{4.23}$$

with $\mathscr{H}_{l,i}(t_1, x_1) = \int_{x_0}^{x_1} \left(\frac{\partial (z_2)_{t_1}^x}{\partial (z_l)_{t_0}^{x_0}} \frac{\partial (z_1)_{t_1}^x}{\partial (z_i)_{t_0}^{x_0}} - \frac{\partial (z_2)_{t_0}^x}{\partial (z_l)_{t_0}^{x_0}} \frac{\partial (z_1)_{t_0}^x}{\partial (z_i)_{t_0}^{x_0}} \right) dx$. Similarly,

$$\int_{t_0}^{t_1} \kappa(t, x_1)dt - \int_{t_0}^{t_1} \kappa(t, x_0)dt$$

$$= \sum_{l=1}^{4} \sum_{i=1}^{4} \left[\int_{t_0}^{t_1} \left(- \frac{\partial (z_1)_t^{x_1}}{\partial (z_i)_{t_0}^{x_0}} \frac{\partial (z_3)_t^{x_1}}{\partial (z_l)_{t_0}^{x_0}} + \frac{\partial (z_1)_t^{x_0}}{\partial (z_i)_{t_0}^{x_0}} \frac{\partial (z_3)_t^{x_0}}{\partial (z_l)_{t_0}^{x_0}} \right. \right.$$

$$\left. \left. - \frac{\partial (z_2)_t^{x_1}}{\partial (z_l)_{t_0}^{x_0}} \frac{\partial (z_4)_t^{x_1}}{\partial (z_i)_{t_0}^{x_0}} + \frac{\partial (z_2)_t^{x_0}}{\partial (z_l)_{t_0}^{x_0}} \frac{\partial (z_4)_t^{x_0}}{\partial (z_i)_{t_0}^{x_0}} \right) dt \right] d(z_l)_{t_0}^{x_0} \wedge d(z_i)_{t_0}^{x_0}$$

$$=: \sum_{l=1}^{4} \sum_{i=1}^{4} \mathscr{M}_{l,i}(t_1, x_1) d(z_l)_{t_0}^{x_0} \wedge d(z_i)_{t_0}^{x_0} \tag{4.24}$$

and

$$\int_{x_0}^{x_1} \int_{t_0}^{t_1} \alpha \omega(t, x)dtdx$$

$$= 2\alpha \sum_{l=1}^{4} \sum_{i=1}^{4} \left[\int_{x_0}^{x_1} \int_{t_0}^{t_1} \left(\frac{\partial (z_2)_t^x}{\partial (z_l)_{t_0}^{x_0}} \frac{\partial (z_1)_t^x}{\partial (z_i)_{t_0}^{x_0}} \right) dtdx \right] d(z_l)_{t_0}^{x_0} \wedge d(z_i)_{t_0}^{x_0}$$

$$=: 2\alpha \sum_{l=1}^{4} \sum_{i=1}^{4} \mathscr{N}_{l,i}(t_1, x_1) d(z_l)_{t_0}^{x_0} \wedge d(z_i)_{t_0}^{x_0}. \tag{4.25}$$

Adding (4.23), (4.24) and (4.25) together, we get that (4.22) holds if

$$\mathcal{H}_{l,i}(t_1, x_1) + \mathcal{M}_{l,i}(t_1, x_1) + 2\alpha \mathcal{N}_{l,i}(t_1, x_1) = 0 \tag{4.26}$$

for any $l, i \in \{1, 2, 3, 4\}$, $t_1 \in \mathbb{R}_+$ and $x_1 \in \mathbb{R}$. In fact, rewritting (4.20) as

$$\begin{cases} d_t z_1 = -\partial_x(z_4)dt - \alpha z_1 dt + \dfrac{\partial S_0(z)}{\partial z_2}dt + \varepsilon dW_1, \\[2mm] d_t z_2 = \partial_x(z_3)dt - \alpha z_2 dt - \dfrac{\partial S_0(z)}{\partial z_1}dt + \varepsilon dW_2, \end{cases}$$

and taking partial derivatives with respect to $(z_i)_{t_0}^{x_0}$ and $(z_l)_{t_0}^{x_0}$ respectively, we have

$$\begin{cases} d_t \dfrac{\partial(z_1)_t^x}{\partial(z_i)_{t_0}^{x_0}} = -\dfrac{\partial}{\partial x}\dfrac{\partial(z_4)_t^x}{\partial(z_i)_{t_0}^{x_0}}dt - \alpha\dfrac{\partial(z_1)_t^x}{\partial(z_i)_{t_0}^{x_0}}dt + \sum_{l=1}^{4}\dfrac{\partial S_1(z)}{\partial(z_2)_t^x\partial(z_l)_t^x}\dfrac{\partial(z_l)_t^x}{\partial(z_i)_{t_0}^{x_0}}dt, \\[3mm] d_t \dfrac{\partial(z_2)_t^x}{\partial(z_l)_{t_0}^{x_0}} = \dfrac{\partial}{\partial x}\dfrac{\partial(z_3)_t^x}{\partial(z_l)_{t_0}^{x_0}}dt - \alpha\dfrac{\partial(z_2)_t^x}{\partial(z_l)_{t_0}^{x_0}}dt - \sum_{i=1}^{4}\dfrac{\partial S_1(z)}{\partial(z_1)_t^x\partial(z_i)_t^x}\dfrac{\partial(z_i)_t^x}{\partial(z_l)_{t_0}^{x_0}}dt. \end{cases}$$

Furthermore,

$$\begin{aligned} d_{t_1}\mathcal{H}_{l,i}(t_1, x_1) &= \int_{x_0}^{x_1}\dfrac{\partial(z_1)_{t_1}^x}{\partial(z_i)_{t_0}^{x_0}}d_{t_1}\left(\dfrac{\partial(z_2)_{t_1}^x}{\partial(z_l)_{t_0}^{x_0}}\right) + \dfrac{\partial(z_2)_{t_1}^x}{\partial(z_l)_{t_0}^{x_0}}d_{t_1}\left(\dfrac{\partial(z_1)_{t_1}^x}{\partial(z_i)_{t_0}^{x_0}}\right)dx \\[2mm] &= \int_{x_0}^{x_1}\dfrac{\partial(z_1)_{t_1}^x}{\partial(z_i)_{t_0}^{x_0}}\left(\dfrac{\partial}{\partial x}\dfrac{\partial(z_3)_{t_1}^x}{\partial(z_l)_{t_0}^{x_0}} - \alpha\dfrac{\partial(z_2)_{t_1}^x}{\partial(z_l)_{t_0}^{x_0}}\right)dt_1 dx \\[2mm] &\quad - \int_{x_0}^{x_1}\dfrac{\partial(z_2)_{t_1}^x}{\partial(z_l)_{t_0}^{x_0}}\left(\dfrac{\partial}{\partial x}\dfrac{\partial(z_4)_{t_1}^x}{\partial(z_i)_{t_0}^{x_0}} + \alpha\dfrac{\partial(z_1)_{t_1}^x}{\partial(z_i)_{t_0}^{x_0}}\right)dt_1 dx \\[2mm] &= \dfrac{\partial(z_1)_{t_1}^{x_1}}{\partial(z_i)_{t_0}^{x_0}}\dfrac{\partial(z_3)_{t_1}^{x_1}}{\partial(z_l)_{t_0}^{x_0}} - \dfrac{\partial(z_1)_{t_1}^{x_0}}{\partial(z_i)_{t_0}^{x_0}}\dfrac{\partial(z_3)_{t_1}^{x_0}}{\partial(z_l)_{t_0}^{x_0}} - \dfrac{\partial(z_2)_{t_1}^{x_1}}{\partial(z_l)_{t_0}^{x_0}}\dfrac{\partial(z_4)_{t_1}^{x_1}}{\partial(z_i)_{t_0}^{x_0}} \\[2mm] &\quad + \dfrac{\partial(z_2)_{t_1}^{x_0}}{\partial(z_l)_{t_0}^{x_0}}\dfrac{\partial(z_4)_{t_1}^{x_0}}{\partial(z_i)_{t_0}^{x_0}} - 2\alpha\int_{x_0}^{x_1}\dfrac{\partial(z_1)_{t_1}^x}{\partial(z_i)_{t_0}^{x_0}}\dfrac{\partial(z_2)_{t_1}^x}{\partial(z_l)_{t_0}^{x_0}}dx dt_1 \\[2mm] &= -d_{t_1}\mathcal{M}_{l,i}(t_1, x_1) - 2\alpha d_{t_1}\mathcal{N}_{l,i}(t_1, x_1), \end{aligned}$$

which together with the fact that $\mathcal{H}_{l,i}(t_0, x_1) + \mathcal{M}_{l,i}(t_0, x_1) + 2\alpha\mathcal{N}_{l,i}(t_0, x_1) = 0$ yields (4.26). $\qquad\square$

Next, we show that the charge of the solution $u(t)$, although is not conserved anymore, satisfies an exponential type evolution law.

Proposition 4.1 *Assume that $M(u_0) < \infty$, then the solution of (4.19) is uniformly bounded with*

$$\mathbb{E}M(u(t)) = e^{-2\alpha t}\mathbb{E}M(u_0) + \frac{\varepsilon^2\eta}{\alpha}(1 - e^{-2\alpha t}), \tag{4.27}$$

where $M(u) = \|u\|_{\mathbf{L}^2}^2$.

Proof Itô's formula applied to $M(u(t))$ yields

$$dM(u(t)) = -2\alpha M(u(t))dt + 2\varepsilon\Re\left[\int_0^1 \bar{u}dxdW\right] + 2\varepsilon^2\eta dt,$$

where $\Re[\cdot]$ denotes the real part of a complex value. Taking the expectation on both sides of the above equation and solving the ordinary differential equation, we derive

$$\mathbb{E}M(u(t)) = e^{-2\alpha t}\left(\int_0^t 2\varepsilon^2\eta e^{2\alpha s}ds + \mathbb{E}M(u_0)\right)$$

$$= e^{-2\alpha t}\mathbb{E}M(u_0) + \frac{\varepsilon^2\eta}{\alpha}(1 - e^{-2\alpha t}).$$

$\square$

We hence conclude that the damped stochastic nonlinear Schrödinger equation (4.19) possesses the stochastic conformal multi-symplectic conservation law (4.22) with its solution being uniformly bounded with (4.27).

Summary

The phase flow of Schrödinger equations in both deterministic and stochastic cases preserves the symplectic structure when the equations are regarded as infinite dimensional Hamiltonian systems, and possesses the multi-symplectic conservation law when the equations are interpreted as Hamiltonian partial differential equations.

Symplectic temporal semi-discretization and multi-symplectic full discretizations are proposed to inherit the geometric structures of the original system in this chapter. We refer to [11, 117, 124] and [11, 34, 105, 117] for the study of symplectic and multi-symplectic schemes of deterministic Schrödinger equations, respectively. Symplectic and multi-symplectic schemes for stochastic Schrödinger equations can be found in [47, 48, 56] and [56, 107, 110, 119], respectively.

Moreover, Maxwell equations and wave equations are also Hamiltonian partial differential equations with symplectic and multi-symplectic structures. We refer to [51, 101, 103, 125] for the design of structure-preserving methods for both deterministic and stochastic Maxwell equations, and refer to [104, 135, 164] for those of wave equations.

Chapter 5
Numerical Invariant Measures for Damped Stochastic Nonlinear Schrödinger Equations

Stochastic nonlinear Schrödinger equations (NLSEs) arise in many applications, for instance, to model the nonlinear dispersive waves with perturbation due to random media or thermal fluctuations. In this chapter and Chap. 6, we investigate the construction of ergodic numerical approximations and the analysis of the error between invariant measures or error to the ergodic limit for stochastic NLSEs. This chapter mainly focuses on the stochastic NLSE with additive noise introduced in Sects. 3.3.1 and 4.5.

5.1 Ergodic Approximation and Numerical Invariant Measures

This section is devoted to constructing a full discretization of

$$\begin{cases} du = \left(\mathbf{i}\Delta u - \alpha u + \mathbf{i}\lambda |u|^2 u\right)dt + \varepsilon dW, \\ u(t,0) = u(t,1) = 0, \ t \geq 0, \\ u(0,x) = u_0(x), \ x \in \mathscr{O} := (0,1) \end{cases} \tag{5.1}$$

with W the same as that in (4.19), and to solving the first problem mentioned at the beginning of Chap. 2. To this end, an ergodic semi-discretization and an ergodic full discretization are given in Sects. 5.1.1 and 5.1.2, respectively. In addition, the weak error, as well as the error between invariant measures, is given in Sect. 5.1.3. In particular, we choose the noise intensity $\varepsilon = 1$ throughout this section.

Throughout this chapter, the s-norm $\|\cdot\|_s, s \in \mathbb{N}$, defined in Sect. 4.5 will be used.

© Springer Nature Singapore Pte Ltd. 2019
J. Hong and X. Wang, *Invariant Measures for Stochastic Nonlinear Schrödinger Equations*, Lecture Notes in Mathematics 2251,
https://doi.org/10.1007/978-981-32-9069-3_5

5.1.1 Spectral Semi-discretization

The M-dimensional spectral space with finite $M \in \mathbb{N}_+$ is defined as

$$V_M := \text{span}\{e_k\}_{k=1}^M.$$

Let $\pi_M : \dot{\mathbf{H}}^0 \to V_M$ be a projection operator, which is defined as

$$\pi_M u = \sum_{k=1}^M (u, e_k) e_k, \quad \forall u = \sum_{k=1}^{\infty} (u, e_k) e_k \in \dot{\mathbf{H}}^0.$$

We use u_M as an approximation to the original solution u, and the spatial semi-discrete scheme is expressed as

$$\begin{cases} du_M = \left(\mathrm{i}\Delta u_M - \alpha u_M + \mathrm{i}\lambda \pi_M \left(|u_M|^2 u_M\right)\right) dt + \pi_M dW \\ u_M(0, x) = \pi_M u_0(x), \end{cases} \tag{5.2}$$

where $\pi_M dW = \sum_{k=1}^M \sqrt{\eta_k} e_k d\beta_k$, and the projection operator π_M is bounded

$$\|\pi_M\|_{\mathscr{L}(\dot{\mathbf{H}}^s, \mathbf{L}^2)} \leq 1, \quad \forall s \in \mathbb{N}.$$

Theorem 5.1 *Let u_M be the solution of Eq. (5.2), then u_M possesses a unique invariant measure, denoted by μ_M. Thus, u_M is ergodic.*

Proof Following from Theorems 1.3 and 1.5, we need to show three properties of u_M, "strong Feller", "irreducibility" and the "Lyapunov condition", in order to show the ergodicity of u_M. Thus the proof is divided into three parts as follows.

Part 1. Strong Feller. We transform (5.2) into an equivalent finite-dimensional SDE. Denoting $a_k(t) = \left(u_M(t, x), e_k(x)\right)$, we have

$$u_M(t, x) = \sum_{k=1}^M a_k(t) e_k(x).$$

Itô's formula applied to $a_k(t)$ leads to

$$da_k(t) = \left[-\lambda_k^\alpha a_k(t) + \left(\mathrm{i}\lambda \pi_M \left(|u_M|^2 u_M\right), e_k\right)\right] dt + \sqrt{\eta_k} d\beta_k(t) \tag{5.3}$$

for $1 \leq k \leq M$. We decompose the above equation into its real and imaginary parts by denoting $a_k = a_k^1 + \mathrm{i}a_k^2$, $\lambda_k^\alpha = \lambda_k^1 + \mathrm{i}\lambda_k^2$ and $\beta_k = \beta_k^1 + \mathrm{i}\beta_k^2$, where $\{\beta_k^i\}_{1 \leq k \leq M}^{i=1,2}$ is a family of independent $\mathbb{R}$-valued Wiener processes and the superscripts 1 and 2 mean the real and imaginary parts of a complex number, respectively, and obtain

$$\begin{cases} da_k^1 = \left[-\lambda_k^1 a_k^1 + \lambda_k^2 a_k^2 + \Re\left(i\lambda\pi_M\left(|u_M|^2 u_M\right), e_k\right) \right]dt + \sqrt{\eta_k}d\beta_k^1(t), \\ da_k^2 = \left[-\lambda_k^2 a_k^1 - \lambda_k^1 a_k^2 + \Im\left(i\lambda\pi_M\left(|u_M|^2 u_M\right), e_k\right) \right]dt + \sqrt{\eta_k}d\beta_k^2(t). \end{cases}$$

With notations $X(t) = (a_1^1(t), a_1^2(t), \cdots, a_M^1(t), a_M^2(t))^\top$, $\beta = (\beta_1^1, \beta_1^2, \cdots, \beta_M^1, \beta_M^2)^\top \in \mathbb{R}^{2M}$, $F = \text{diag}\{\Lambda_1, \cdots, \Lambda_M\}$,

$$\Lambda_i = \begin{pmatrix} -\lambda_i^1 & \lambda_i^2 \\ -\lambda_i^2 & -\lambda_i^1 \end{pmatrix}, \quad G(X(t)) = \begin{pmatrix} \Re\left(i\lambda\pi_M\left(|u_M|^2 u_M\right), e_1\right) \\ \Im\left(i\lambda\pi_M\left(|u_M|^2 u_M\right), e_1\right) \\ \vdots \\ \Re\left(i\lambda\pi_M\left(|u_M|^2 u_M\right), e_M\right) \\ \Im\left(i\lambda\pi_M\left(|u_M|^2 u_M\right), e_M\right) \end{pmatrix}$$

and

$$Z = \begin{pmatrix} \sqrt{\eta_1} & & & & \\ & \sqrt{\eta_1} & & & \\ & & \ddots & & \\ & & & \sqrt{\eta_M} & \\ & & & & \sqrt{\eta_M} \end{pmatrix} := (Z_1^1, Z_1^2 \cdots, Z_M^1, Z_M^2),$$

we get an equivalent form of (5.2)

$$dX(t) = \left[FX(t) + G(X(t)) \right]dt + \sum_{k=1}^{M}\sum_{i=1}^{2} Z_k^i d\beta_k^i := Y(X(t))\, dt + \sum_{k=1}^{M}\sum_{i=1}^{2} Z_k^i d\beta_k^i.$$

It is obvious that

$$\text{span}\{Z_1^1, Z_1^2, \cdots, Z_M^1, Z_M^2\} = \mathbb{R}^{2M},$$

which means that the Hörmander condition holds. According to Theorem 2.2 (see also [111]), $X(t)$ is a strong Feller process.

Part 2. Irreducibility. By using the same notations as above, we have

$$dX = Y(X)dt + Zd\beta, \tag{5.4}$$

with $X = X(t) \in \mathbb{R}^{2M}$, $X(0) = y$ and Z being invertible. Using a similar technique as [139], we consider the associated control problem

$$d\overline{X} = Y(\overline{X})dt + ZdU, \tag{5.5}$$

with $\overline{X} = \overline{X}(t)$ and a smooth control function $U \in \mathbf{C}^1(0, T)$. For any fixed $T > 0$, $y \in \mathbb{R}^{2M}$ and $y^+ \in \mathbb{R}^{2M}$, using polynomial interpolation, we derive a continuous function $(\overline{X}(t), t \in [0, T])$ such that $\overline{X}(0) = y$ and $\overline{X}(T) = y^+$. Hence,

$$dU = Z^{-1}(d\overline{X} - Y(\overline{X})dt),$$

and the control function U satisfies (5.5) with $\overline{X}(0) = y$, $\overline{X}(T) = y^+$ and $U(0) = 0$. We subtract the resulting Eqs. (5.4) and (5.5), and achieve

$$X(t) - \overline{X}(t) = \int_0^t Y(X(s)) - Y(\overline{X}(s))ds + Z(\beta(t) - U(t)), \quad t \in [0, T].$$

Note that Y is locally Lipschitz continuous because of its continuous differentiability, and the ranges of $X(t)$ and $\overline{X}(t)$ ($t \in [0, T]$) are both compact sets. According to the following property of Brownian motion

$$\mathbb{P}\left(\sup_{0 \le t \le T} \left| \beta(t) - U(t) \right| \le \varepsilon \right) > 0, \quad \forall \, \varepsilon > 0,$$

the Gronwall inequality in a small time interval and its continuation yield that

$$\sup_{0 \le t \le T} \left| X(t) - \overline{X}(t) \right| \to 0 \quad \text{as} \quad \varepsilon \to 0$$

holds with positive probability (see also [139]). For any $\delta > 0$, by choosing $\varepsilon > 0$ small enough, we finally obtain

$$\mathbb{P}\left(|X(T) - y^+| < \delta \right) > 0.$$

In other words, $X(T)$ hits $B(y^+, \delta)$ with positive probability. The irreducibility has been proved.

The above two conditions ensure the uniqueness of the invariant measure of $X(t)$. It suffices to show the existence of invariant measures in the following.

Part 3. Lyapunov condition. A useful tool for proving the existence of invariant measures is provided by Lyapunov functionals, which is introduced in Theorem 1.3. Itô's formula applied to $\|u_M(t)\|_0^2$ implies that

$$d\|u_M(t)\|_0^2 = -2\alpha\|u_M(t)\|_0^2 dt + 2\Re \int_0^1 \overline{u}_M(t)\pi_M dW(t)dx + 2\sum_{k=1}^M \eta_k dt, \quad (5.6)$$

where we have used the fact that

$$\Re\left[i\lambda \int_0^1 \pi_M(|u_M|^2 u_M)\overline{u}_M dx \right] = \Re\left[i\lambda \int_0^1 \left(|u_M|^4 - (Id - \pi_M)(|u_M|^2 u_M)\overline{u}_M \right) dx \right]$$

$$= -\lambda\Im\left((Id - \pi_M)(|u_M|^2 u_M), u_M \right) = 0$$

with Id being the identity operator. Taking the expectation on both sides of (5.6), we obtain

$$\frac{d}{dt}\mathbb{E}\|u_M(t)\|_0^2 = -2\alpha\mathbb{E}\|u_M(t)\|_0^2 + C_M,$$

where $C_M = 2\sum_{k=1}^M \eta_k \leq 2\eta$. It is solved as

$$\mathbb{E}\|u_M(t)\|_0^2 = e^{-2\alpha t}\left(\int_0^t C_M e^{2\alpha s}ds + \mathbb{E}\|u_M(0)\|_0^2\right) \leq e^{-2\alpha t}\mathbb{E}\|u_M(0)\|_0^2 + C, \ \forall t > 0.$$

On the other hand,

$$\|u_M(t)\|_0^2 = \int_0^1 \left|\sum_{k=1}^M a_k(t)e_k(x)\right|^2 dx = \|X(t)\|_{l^2(\mathbb{R}^{2M})}^2.$$

Define $V = \|\cdot\|_{l^2(\mathbb{R}^{2M})} : \mathbb{R}^{2M} \to [0, +\infty]$. The level sets of V are tight by the Heine–Borel theorem. Therefore, $X(t)$ is ergodic according to Theorem 1.5.

We mention that the ergodicity of $X(t)$ is equivalent to the existence of a random variable $\xi = (\xi_1^1, \xi_1^2, \cdots, \xi_M^1, \xi_M^2)$ such that the following convergence holds in distribution

$$\lim_{t\to\infty} X(t) = \xi, \ \text{i.e.,} \ \lim_{t\to\infty} a_k^i(t) = \xi_k^i, \ \forall k = 1, \cdots, M, \ i = 1, 2.$$

It then leads to

$$\lim_{t\to\infty} u_M(t) = \sum_{k=1}^M \left(\xi_k^1 + i\xi_k^2\right)e_k,$$

which shows the ergodicity of $u_M(t)$. □

Remark 5.1 The above proof can be easily generalized for the degenerate case as mentioned in Sect. 3.6: there exists some $N_* \in \mathbb{N}_+$ such that $\eta_k > 0$ for any $k \leq N_*$, in which case we need in addition that $M \geq N_*$. That is, the covariance operator Q may have zero eigenvalues.

According to the proof of the Lyapunov condition, we have the following uniform boundedness for the 0-norm. Moreover, the 1-norm and 2-norm are also uniformly bounded, which is stated in the following proposition. In sequel, all the constants C are independent of the end point T of the time interval and may be different from line to line.

Proposition 5.1 *Assume that $u_0 \in \dot{\mathbb{H}}^1$, $\|Q^{\frac{1}{2}}\|_{\mathscr{L}_2^1} < \infty$ and $p \geq 1$. There exist positive constants c_0 and $C = C(\alpha, p, u_0, c_0, Q)$, such that for any $t > 0$,*

$$(i) \ \mathbb{E}\|u_M(t)\|_0^{2p} \leq e^{-2\alpha pt}\mathbb{E}\|u_M(0)\|_0^{2p} + C \leq C,$$
$$(ii) \ \mathbb{E}[H(u_M(t))^p] \leq e^{-\alpha pt}\mathbb{E}[H(u_M(0))^p] + C \leq C,$$

where $H(u_M(t)) = \frac{1}{2}\|\nabla u_M(t)\|_0^2 - \frac{\lambda}{4}\|u_M(t)\|_{\mathbf{L}^4}^4 + c_0\|u_M(t)\|_0^6$. In addition, if assume further $u_0 \in \dot{\mathbf{H}}^2$ and $\|Q^{\frac{1}{2}}\|_{\mathscr{L}_2^2} < \infty$, we also have

$$(iii) \quad \mathbb{E}\|u_M(t)\|_2^2 \le C.$$

The proof of above proposition is given in Appendix C.1 for the convenience of readers.

Remark 5.2 The uniform boundedness of the original solution u can also be obtained by the same procedure as Proposition 5.1 or [70]. As the $\dot{\mathbf{H}}^2$-regularity for both the original solution and numerical solutions is essential to obtain the time-independent weak error, we need the assumption $u_0 \in \dot{\mathbf{H}}^2$ and $\|Q^{\frac{1}{2}}\|_{\mathscr{L}_2^2} < \infty$ in the error analysis.

5.1.2 Ergodic Full Discretization

In this subsection, a modified implicit Euler scheme is applied in the temporal direction to obtain a full discretization. The well-posedness and uniform boundedness in $\|\cdot\|_0$-norm of the numerical solution is first given such that the Lyapunov condition is satisfied, which implies the existence of invariant measures. Furthermore, the uniqueness of the invariant measure is proved utilizing the uniqueness of the numerical solution, which shows immediately that the numerical solution is ergodic. In addition, the higher regularity of the numerical solution, that is, the uniform boundedness of the numerical solution in $\|\cdot\|_1$- and $\|\cdot\|_2$-norms is also obtained, which is essential to get the order of convergence in weak sense.

We apply a modified implicit Euler scheme to approximate (5.2), and obtain the following scheme

$$\begin{cases} u_M^n - e^{-\alpha\tau}u_M^{n-1} = \left(i\Delta u_M^n + i\lambda\pi_M\left(\frac{|u_M^n|^2 + |e^{-\alpha\tau}u_M^{n-1}|^2}{2}u_M^n\right)\right)\tau + \pi_M\delta_n W \\ u_M^0 = \pi_M u_0(x), \end{cases}$$

(5.7)

where u_M^n is an approximation of $u_M(t_n)$, τ represents the uniform time step-size, $t_n = n\tau$, and $\delta_n W = W(t_n) - W(t_{n-1})$.

The well-posedness of scheme (5.7), together with the uniform boundedness of the numerical solution, is stated in the following proposition.

Proposition 5.2 *Assume that $u_0 \in \dot{\mathbf{H}}^0$. For sufficiently small τ, there exists a unique V_M-valued and $\{\mathscr{F}_{t_n}\}_{n\in\mathbb{N}}$-adapted solution $\{u_M^n\}_{n\in\mathbb{N}}$ of (5.7), which satisfies that for any integer $p \ge 2$, there exists a constant $C = C(p, \alpha, u_M^0) > 0$, such that*

$$\mathbb{E}\|u_M^n\|_0^p \le C, \quad \forall n \in \mathbb{N}.$$

Proof **Step 1**. Existence and uniqueness of the solution.

Similar to [65], we fix a family of deterministic functions $g_n \in V_M$, $n \in \mathbb{N}$. We also fix $\tilde{u}_M^{n-1} \in V_M$, the existence of solution $\tilde{u}_M^n \in V_M$ of

$$\tilde{u}_M^n - e^{-\alpha\tau}\tilde{u}_M^{n-1} = \mathrm{i}\tau\Delta\tilde{u}_M^n + \mathrm{i}\lambda\tau\pi_M\left(\frac{|\tilde{u}_M^n|^2 + |e^{-\alpha\tau}\tilde{u}_M^{n-1}|^2}{2}\tilde{u}_M^n\right) + \sqrt{\tau}g_n \quad (5.8)$$

can be proved by using Brouwer's fixed point theorem. Indeed, multiplying (5.8) by $\overline{\tilde{u}}_M^n$, integrating with respect to x and taking the real part, we obtain

$$\|\tilde{u}_M^n\|_0^2 + \|\tilde{u}_M^n - e^{-\alpha\tau}\tilde{u}_M^{n-1}\|_0^2 - e^{-2\alpha\tau}\|\tilde{u}_M^{n-1}\|_0^2$$

$$= 2\sqrt{\tau}\Re\left[\int_0^1(\overline{\tilde{u}}_M^n - e^{-\alpha\tau}\overline{\tilde{u}}_M^{n-1})g_n dx + \int_0^1(e^{-\alpha\tau}\overline{\tilde{u}}_M^{n-1})g_n dx\right]$$

$$\leq \|\tilde{u}_M^n - e^{-\alpha\tau}\tilde{u}_M^{n-1}\|_0^2 + e^{-2\alpha\tau}\|\tilde{u}_M^{n-1}\|_0^2 + 2\tau\|g_n\|_0^2,$$

i.e.,

$$\|\tilde{u}_M^n\|_0^2 \leq 2e^{-2\alpha\tau}\|\tilde{u}_M^{n-1}\|_0^2 + 2\tau\|g_n\|_0^2. \quad (5.9)$$

Define

$$\Lambda : V_M \times V_M \to \mathscr{P}(\mathbf{L}^2),$$

$$(\tilde{u}_M^{n-1}, g_n) \mapsto \{\tilde{u}_M^n | \tilde{u}_M^n \text{ are solutions of } (5.8)\},$$

where $\mathscr{P}(\mathbf{L}^2)$ is the power set of $\mathbf{L}^2$. The inequality (5.9) implies that Λ is bounded, and its graph is closed by the closed graph theorem. When the spaces are endowed with their Borel σ-algebras, there is a measurable continuous function $\kappa : V_M \times V_M \to \mathbf{L}^2$ such that

$$\kappa(u, g) \in \Lambda(u, g), \quad \forall (u, g) \in V_M \times V_M.$$

Assume that $u_M^{n-1} \in V_M$ is a $\mathscr{F}_{t_{n-1}}$-measurable random variable, then $u_M^n = \kappa(u_M^{n-1}, \frac{\pi_M\delta_n W}{\sqrt{\tau}})$ is an $\mathbf{L}^2$-valued solution of (5.7). Moreover,

$$(1 - \mathrm{i}\Delta\tau)u_M^n = e^{-\alpha\tau}u_M^{n-1} + \mathrm{i}\lambda\tau\pi_M\left(\frac{|u_M^n|^2 + |e^{-\alpha\tau}u_M^{n-1}|^2}{2}u_M^n\right) + \pi_M\delta_n W \in V_M.$$

Hence, u_M^n is actually a V_M-valued solution of (5.7).

For any given u_M^{n-1} and sufficiently small time step-size τ, the solution u_M^n is uniquely defined, which can be proved in a similar procedure as [4]. This fact will be used in proving the ergodicity of the numerical solution $\{u_M^n\}_{n\in\mathbb{N}}$.

In fact, suppose that U and W are two solutions of the scheme, then it follows

$$U - W = \mathbf{i}\tau \Delta(U - W) + \mathbf{i}\lambda \frac{\tau}{2}\pi_M\left[\left(|U|^2 U - |W|^2 W\right) + |e^{-\alpha\tau}u_M^{n-1}|^2(U - W)\right].$$

Multiplying the equation above by $\overline{U} - \overline{W}$, integrating in space and taking the real and imaginary part respectively, we have

$$\|U - W\|_0^2 \le \frac{\tau}{2}\|g(U) - g(W)\|_{\mathbf{L}^{\frac{4}{3}}}\|U - W\|_{\mathbf{L}^4},$$

$$\|\nabla(U - W)\|_0^2 \le \frac{1}{2}\|g(U) - g(W)\|_{\mathbf{L}^{\frac{4}{3}}}\|U - W\|_{\mathbf{L}^4} + \frac{\lambda}{2}\|e^{-\alpha\tau}u_M^{n-1}\|_{\mathbf{L}^4}^2\|U - W\|_{\mathbf{L}^4}^2,$$

where $g(U) := |U|^2 U$ and

$$\|g(U) - g(W)\|_{\mathbf{L}^{\frac{4}{3}}} = \left(\int_0^1 \left||U|^2 U - |W|^2 W\right|^{\frac{4}{3}} dx\right)^{\frac{3}{4}}$$

$$= \left(\int_0^1 \left||U|^2(U - W) + |W|^2(U - W) + UW(\overline{U} - \overline{W})\right|^{\frac{4}{3}} dx\right)^{\frac{3}{4}}$$

$$\le \left(\int_0^1 \left||U|^2 + |W|^2 + |UW|\right|^2 dx\right)^{\frac{1}{2}}\left(\int_0^1 |U - W|^4 dx\right)^{\frac{1}{4}}$$

$$\le \||U| + |W|\|_{\mathbf{L}^4}^2\|U - W\|_{\mathbf{L}^4}.$$

Since

$$\|U - W\|_{\mathbf{L}^4}^4 \le \|U - W\|_0^3\|\nabla(U - W)\|_0$$

$$\le \left(\frac{\tau}{2}\|g(U) - g(W)\|_{\mathbf{L}^{\frac{4}{3}}}\|U - W\|_{\mathbf{L}^4}\right)^{\frac{3}{2}}\left(\frac{1}{2}\|g(U) - g(W)\|_{\mathbf{L}^{\frac{4}{3}}}\|U - W\|_{\mathbf{L}^4}\right)$$

$$+ \frac{|\lambda|}{2}\|e^{-\alpha\tau}u_M^{n-1}\|_{\mathbf{L}^4}^2\|U - W\|_{\mathbf{L}^4}^2\bigg)^{\frac{1}{2}}$$

$$\le \frac{1}{4}\tau^{\frac{3}{2}}\||U| + |W|\|_{\mathbf{L}^4}^3\left(\||U| + |W|\|_{\mathbf{L}^4}^2 + |\lambda|\|u_M^{n-1}\|_{\mathbf{L}^4}^2\right)^{\frac{1}{2}}\|U - W\|_{\mathbf{L}^4}^4$$

$$\le \frac{1}{4}\tau^{\frac{3}{2}}\left(\||U| + |W|\|_{\mathbf{L}^4}^4 + |\lambda|\||U| + |W|\|_{\mathbf{L}^4}^3\|u_M^{n-1}\|_{\mathbf{L}^4}\right)\|U - W\|_{\mathbf{L}^4}^4,$$

if $U \ne W$, then

$$1 \le \frac{1}{4}\tau^{\frac{3}{2}}\left(\||U| + |W|\|_{\mathbf{L}^4}^4 + |\lambda|\||U| + |W|\|_{\mathbf{L}^4}^3\|u_M^{n-1}\|_{\mathbf{L}^4}\right)$$

$$\le C_0\tau^{\frac{3}{2}}\left(\||U| + |W|\|_{\mathbf{L}^4}^4 + |\lambda|\||U| + |W|\|_{\mathbf{L}^4}^6 + |\lambda|\|u_M^{n-1}\|_{\mathbf{L}^4}^2\right).$$

For cases $\lambda = 0$ or -1, the $\mathbf{L}^4$-norm of the solution is uniformly bounded, which leads to a contradiction when τ is sufficiently small. For case $\lambda = 1$, according to the fact that

$$\big\||U| + |W|\big\|_{\mathbf{L}^4}^6 \leq \big\||U| + |W|\big\|_0^{\frac{3}{2}} \big\|\nabla(|U| + |W|)\big\|_0^{\frac{9}{2}} \leq M^{\frac{9}{2}}\big\||U| + |W|\big\|_0^6,$$

we have $C_0 M^{\frac{9}{2}}\tau^{\frac{3}{2}} > 1$, which is also a contradiction when τ is sufficiently small. Thus, the numerical solution for (5.7) is unique.

Step 2. Boundedness of the p-moments.

We use the notation C to denote a generic constant, which does not depend on time and may be different from line to line.

(i) $p = 2$. To show the boundedness, we multiply (5.7) by $\overline{u}_M^n$, integrate in $[0,1]$ with respect to the space variable, take expectation and take the real part,

$$\mathbb{E}\|u_M^n\|_0^2 + \mathbb{E}\|u_M^n - e^{-\alpha\tau}u_M^{n-1}\|_0^2 - e^{-2\alpha\tau}\mathbb{E}\|u_M^{n-1}\|_0^2 = 2\Re\mathbb{E}\int_0^1 \overline{u}_M^n \pi_M \delta_n W dx$$

$$=2\Re\mathbb{E}\int_0^1 \big(\overline{u}_M^n - e^{-\alpha\tau}\overline{u}_M^{n-1}\big)\pi_M \delta_n W dx \leq \mathbb{E}\|u_M^n - e^{-\alpha\tau}u_M^{n-1}\|_0^2 + \mathbb{E}\|\pi_M \delta_n W\|_0^2.$$

We can derive

$$\mathbb{E}\|u_M^n\|_0^2 \leq e^{-2\alpha\tau}\mathbb{E}\|u_M^{n-1}\|_0^2 + C\tau \leq e^{-2\alpha\tau n}\mathbb{E}\|u_M^0\|_0^2$$
$$+ C\tau\big(1 + e^{-2\alpha\tau} + \cdots + e^{-2\alpha\tau(n-1)}\big)$$
$$\leq e^{-2\alpha t_n}\mathbb{E}\|u_M^0\|_0^2 + \frac{C\tau}{1 - e^{-2\alpha\tau}} \leq \mathbb{E}\|u_M^0\|_0^2 + C,$$

where we have used $e^{-2\alpha\tau} < 1 - 2\alpha\tau e^{-2\alpha}$ for $\tau \in (0, 1)$.

(ii) $p = 4$. In the case $p = 2$, without taking the expectation, we have

$$\|u_M^n\|_0^2 - e^{-2\alpha\tau}\|u_M^{n-1}\|_0^2 + \|u_M^n - e^{-\alpha\tau}u_M^{n-1}\|_0^2 = 2\Re\int_0^1 \overline{u}_M^n \pi_M \delta_n W dx.$$

Multiplying both sides by $\|u_M^n\|_0^2$, taking the expectation and taking the real part, we obtain

$$(LHS) = \mathbb{E}\|u_M^n\|_0^4 - e^{-2\alpha\tau}\mathbb{E}\|u_M^{n-1}\|_0^2\|u_M^n\|_0^2 + \mathbb{E}\Big[\|u_M^n - e^{-\alpha\tau}u_M^{n-1}\|_0^2\|u_M^n\|_0^2\Big]$$
$$= \frac{1}{2}\Big(\mathbb{E}\|u_M^n\|_0^4 - e^{-4\alpha\tau}\mathbb{E}\|u_M^{n-1}\|_0^4\Big) + \frac{1}{2}\mathbb{E}\Big(\|u_M^n\|_0^2 - e^{-2\alpha\tau}\|u_M^{n-1}\|_0^2\Big)^2$$
$$+ \mathbb{E}\Big[\|u_M^n - e^{-2\alpha\tau}u_M^{n-1}\|_0^2\|u_M^n\|_0^2\Big]$$

and

$$(RHS) = 2\Re E \int_0^1 \|u_M^n\|_0^2 \overline{u}_M^n \pi_M \delta_n W dx$$

$$= 2\Re E \int_0^1 \left(\|u_M^n\|_0^2 (\overline{u}_M^n - e^{-\alpha\tau} \overline{u}_M^{n-1}) \right) \pi_M \delta_n W dx$$

$$+ 2\Re E \int_0^1 \left((\|u_M^n\|_0^2 - e^{-2\alpha\tau} \|u_M^{n-1}\|_0^2) e^{-\alpha\tau} \overline{u}_M^{n-1} \right) \pi_M \delta_n W dx$$

$$\leq E \left[\|u_M^n - e^{-\alpha\tau} u_M^{n-1}\|_0^2 \|u_M^n\|_0^2 \right] + E \left(\|u_M^n\|_0^2 \|\pi_M \delta_n W\|_0^2 \right)$$

$$+ \frac{1}{4} E \left(\|u_M^n\|_0^2 - e^{-2\alpha\tau} \|u_M^{n-1}\|_0^2 \right)^2 + 4 e^{-2\alpha\tau} E \|\overline{u}_M^{n-1} \pi_M \delta_n W\|_0^2$$

$$\leq E \left[\|u_M^n - e^{-\alpha\tau} u_M^{n-1}\|_0^2 \|u_M^n\|_0^2 \right] + \frac{1}{2} E \left(\|u_M^n\|_0^2 - e^{-2\alpha\tau} \|u_M^{n-1}\|_0^2 \right)^2 + C\tau.$$

Comparing (LHS) with (RHS), we obtain

$$E \|u_M^n\|_0^4 \leq e^{-4\alpha\tau} E \|u_M^{n-1}\|_0^4 + C\tau \leq C.$$

(iii) $p = 3$. Using (i) and (ii), it is easy to check that the following holds true

$$E \|u_M^n\|_0^3 \leq E \frac{\|u_M^n\|_0^2 + \|u_M^n\|_0^4}{2} \leq C.$$

(iv) $p > 4$. By repeating the above procedure, we complete the proof. $\square$

Before showing the weak error between $u_M(t)$ and u_M^n, we need some a priori estimates on $\|u_M^n\|_1$ and $\|u_M^n\|_2$ given in Propositions 5.3 and 5.4, whose proofs are given in Appendices C.2 and C.3.

Proposition 5.3 *Assume that $\lambda = 0$ or -1, $u_0 \in \dot{H}^1$, $u_M^0 = \pi_M u_0$ and $\|Q^{\frac{1}{2}}\|_{\mathscr{L}_2^1} < \infty$. Then for any $p \geq 1$, there exists a constant $C = C(\alpha, u_0, p)$ independent of M and t_n, such that*

$$E[H_n^p] \leq C, \quad \forall n \in \mathbb{N},$$

where $H_n := \|\nabla u_M^n\|_0^2 - \frac{\lambda}{2} \|u_M^n\|_{\mathbf{L}^4}^4$.

Corollary 5.1 *Under the assumptions in Proposition 5.3,*

$$E \|u_M^n - e^{-\alpha\tau} u_M^{n-1}\|_0^{2p} \leq C\tau^p,$$

where constant C is independent of M and t_n.

Proof It is easy to check this by multiplying both sides of (C.9) by $\overline{u}_M^n - e^{-\alpha\tau} \overline{u}_M^{n-1}$, integrating with respect to x and taking the expectation,

$$\mathbb{E}\|u_M^n - e^{-\alpha\tau}u_M^{n-1}\|_0^{2p}$$

$$=\mathbb{E}\Big[\tau\Im\int_0^1\nabla u_M^n\nabla(\overline{u}_M^n - e^{-\alpha\tau}\overline{u}_M^{n-1})dx + \Re\int_0^1\pi_M\delta_n W\left(\overline{u}_M^n - e^{-\alpha\tau}\overline{u}_M^{n-1}\right)dx$$

$$+\frac{\tau}{4}\Im\int_0^1\left(|u_M^n|^2 + |e^{-\alpha\tau}u_M^{n-1}|^2\right)\left(u_M^n + e^{-\alpha\tau}u_M^{n-1}\right)\left(\overline{u}_M^n - e^{-\alpha\tau}\overline{u}_M^{n-1}\right)dx\Big]^p$$

$$\leq C\tau^p\mathbb{E}\left[\|\nabla u_M^n\|_0^p\|\nabla\left(u_M^n - e^{-\alpha\tau}u_M^{n-1}\right)\|_0^p\right]$$

$$+C\tau^p\mathbb{E}\left[\left(\|u_M^n\|_1^{2p} + \|u_M^{n-1}\|_1^{2p}\right)\left(\|u_M^n\|_0^{2p} + \|u_M^{n-1}\|_0^{2p}\right)\right]$$

$$+C\mathbb{E}\|\pi_M\delta_n W\|_0^{2p} + \frac{1}{2}\mathbb{E}\|u_M^n - e^{-\alpha\tau}u_M^{n-1}\|_0^{2p}$$

$$\leq\frac{1}{2}\mathbb{E}\|u_M^n - e^{-\alpha\tau}u_M^{n-1}\|_0^{2p} + C\tau^p.$$

Then we complete the proof by Proposition 5.3. $\qquad\qquad\square$

Proposition 5.4 *Under the assumptions $\lambda = 0$ or -1, $u_0 \in \dot{\mathbf{H}}^2$ and $\|Q^{\frac{1}{2}}\|_{\mathscr{L}_2^2} < \infty$, the uniform boundedness of the 2-norm holds, i.e.,*

$$\mathbb{E}\|u_M^n\|_2^2 \leq C, \quad \forall\, n \in \mathbb{N},$$

where C is also independent of M and t_n.

The numerical solution $\{u_M^n\}_{n\in\mathbb{N}}$ is ergodic possessing a unique invariant measure utilizing the uniform boundedness given above.

Theorem 5.2 *For all τ sufficiently small, the solution $\{u_M^n\}_{n\in\mathbb{N}}$ of scheme (5.7) has a unique invariant measure μ_M^τ. Thus, it is ergodic.*

Proof (i) Lyapunov condition. Based on Proposition 5.2, we can take essentially quadratic function $V(\cdot) = 1 + \|\cdot\|_0^2$ as the Lyapunov functional, and the Lyapunov condition holds.

(ii) Minorization condition (see Assumption 2.1 of [139]). In scheme (5.7), it gives

$$P_M^n = e^{-\alpha\tau}P_M^{n-1} - \tau\Delta Q_M^n$$

$$-\frac{\tau\lambda}{2}\pi_M\left(\left(|P_M^n|^2 + |Q_M^n|^2 + |e^{-\alpha\tau}P_M^{n-1}|^2 + |e^{-\alpha\tau}Q_M^{n-1}|^2\right)Q_M^n\right)$$

$$+\sum_{k=1}^M\sqrt{\eta_k}e_k\delta_n\beta_k^1, \qquad\qquad(5.10)$$

$$Q_M^n = e^{-\alpha\tau}Q_M^{n-1} + \tau\Delta P_M^n$$

$$+\frac{\tau\lambda}{2}\pi_M\left(\left(|P_M^n|^2 + |Q_M^n|^2 + |e^{-\alpha\tau}P_M^{n-1}|^2 + |e^{-\alpha\tau}Q_M^{n-1}|^2\right)P_M^n\right)$$

$$+ \sum_{k=1}^{M} \sqrt{\eta_k} e_k \delta_n \beta_k^2, \tag{5.11}$$

where P_M^n and Q_M^n denote the real and imaginary part of u_M^n respectively, that is $u_M^n = P_M^n + \mathbf{i} Q_M^n$. Also, $\pi_M \delta_n W = \sum_{k=1}^{M} \sqrt{\eta_k} e_k \left(\delta_n \beta_k^1 + \mathbf{i} \delta_n \beta_k^2 \right)$, where $\delta_n \beta_k^1$ and $\delta_n \beta_k^2$ are the real and imaginary part of $\delta_n W$ respectively.

For any $y_1 = a_1 + \mathbf{i} b_1$, $y_2 = a_2 + \mathbf{i} b_2 \in V_M$ with a_i and b_i denoting the real and imaginary part of y_i $(i = 1, 2)$ respectively, as $\{e_k\}_{k=1}^{M}$ is a basis of V_M, $\{\delta_n \beta_k^1, \delta_n \beta_k^2\}_{k=1}^{M}$ can be uniquely determined to ensure that $(P_M^{n-1}, Q_M^{n-1}) = (a_1, b_1)$ and $(P_M^n, Q_M^n) = (a_2, b_2)$, which implies the irreducibility of u_M^n.

As stated in Proposition 5.2, the $\mathscr{F}_{t_n}$-measurable solution $\{u_M^n\}_{n \in \mathbb{N}}$ is uniquely defined by $u_M^n = \kappa(u_M^{n-1}, \frac{\pi_M \delta_n W}{\sqrt{\tau}})$, where $\delta_n W$ has a $\mathbf{C}^\infty$ density. Thus, the transition kernel $P_1(x, G)$ with $G \in \mathscr{B}(V_M)$ possesses a jointly continuous density $p_1(x, y)$. Furthermore, densities $p_n(x, y)$ are achieved by the time-homogeneous property of Markov chain $\{u_M^n\}_{n \in \mathbb{N}}$.

In conclusion, the minorization condition together with the Lyapunov condition leads to the existence and uniqueness of the invariant measure according to Theorem 2.5 in [139]. $\qquad\qquad\qquad\qquad\qquad\qquad\qquad\qquad\qquad\qquad\qquad\quad\square$

5.1.3 Weak Error and Error of Invariant Measures

In this section, the error analysis in weak sense is studied for the considered model (5.1) in the linear sense (i.e., $\lambda = 0$). This result can be extended to the globally Lipschitz case as is investigated in [67]. For non-globally Lipschitz case, the exponential integrability of both the exact and numerical solutions are need in addition to get the order of convergence in both mean-square and weak sense, see for instance [55, 57] for the study of stochastic NLSEs and [112–115, 118] for other kinds of SDEs or SPDEs.

We first establish the weak convergence error for the spatial semi-discretization (5.2) utilizing a transformation of $u_M(t)$ and the corresponding Kolmogorov equation.

Theorem 5.3 *Let* $\lambda = 0$. *Assume that* $u_0 \in \dot{\mathbf{H}}^2$ *and* $\|Q^{\frac{1}{2}}\|_{\mathscr{L}_2^2} < \infty$. *For any* $f \in \mathbf{C}_b^2(\mathbf{L}^2)$, *there exists a constant* $C = C(u_0, f, Q)$ *independent of* T *such that*

$$\left| \mathbb{E}\left[f\left(u_M(T) \right) \right] - \mathbb{E}\left[f\left(u(T) \right) \right] \right| \leq C M^{-2}, \quad \forall\, T > 0.$$

Before proving Theorem 5.3, the error between semigroups of the original system and the spatial semi-discretization is given in the following lemma.

Lemma 5.1 *Assume that $S(t)$ and π_M are defined as before. We have the following estimation*

$$\|S(t) - S(t)\pi_M\|_{\mathscr{L}(\dot{\mathbf{H}}^s, \mathbf{L}^2)} \leq Ce^{-\alpha t} M^{-s}.$$

Proof For any $u \in \dot{\mathbf{H}}^s$, we have

$$\|S(t)u - S(t)\pi_M u\|_0 = e^{-\alpha t}\|u - \pi_M u\|_0 = e^{-\alpha t}\left(\sum_{k=M+1}^{\infty} |(u, e_k)|^2\right)^{\frac{1}{2}}$$

$$\leq e^{-\alpha t}|\lambda_M|^{-\frac{s}{2}}\left(\sum_{k=M+1}^{\infty} |\lambda_k^{\alpha}|^s |(u, e_k)|^2\right)^{\frac{1}{2}} \leq Ce^{-\alpha t} M^{-s}\|u\|_s.$$

$\square$

Proof (of Theorem 5.3) We split the proof into three steps.
Step 1. Calculation of $\mathbb{E}[f(u(T))]$.

For the case $\lambda = 0$, the mild solution of the considered Eq. (5.1) is

$$u(T) = S(T)u_0 + \int_0^T S(T - t)dW(t),$$

which leads to

$$\mathbb{E}[f(u(T))] = \mathbb{E}\left[f\left(S(T)u_0 + \int_0^T S(T - t)dW(t)\right)\right]. \tag{5.12}$$

Denoting $v(T - t, y) := \mathbb{E}[f(u(T))|S(T - t)u(t) = y]$, it follows

$$\frac{\partial v(T - t, y)}{\partial t} = -\frac{1}{2}Tr\left[(S(T - t)Q^{\frac{1}{2}})^* D^2 v(T - t, y)S(T - t)Q^{\frac{1}{2}}\right].$$

In addition, according to the mild solution of (5.1) given above, we have

$$v(T - t, y) = \mathbb{E}\left[f\left(y + \int_t^T S(T - s)dW(s)\right)\right].$$

For any $h \in \mathbf{L}^2$, similar to [67, Lemma 5.13], we have

$$(Dv(T - t, y), h) = \mathbb{E}\left[\left(Df\left(y + \int_t^T S(T - s)dW\right), h\right)\right],$$

which satisfies

$$|(Dv(T - t, y), h)| \leq C\|f\|_{\mathbf{C}_b^1}\|h\|_0. \tag{5.13}$$

Similarly, we have

$$\left| \left((D^2 v(T-t, y), h), h \right) \right| \leq C \|f\|_{C_b^2} \|h\|_0^2. \tag{5.14}$$

Step 2. Calculation of $\mathbb{E}\left[f(u_M(T))\right]$.

The mild solution of the spatial semi-discretization (5.2) is

$$u_M(t) = S(t)\pi_M u_0 + \int_0^T S(t-s)\pi_M dW(s).$$

We consider an auxiliary stochastic process:

$$Y_M(t) = S(T-t)u_M(t),$$

which satisfies

$$dY_M(t) = S(T-t)\pi_M dW(t).$$

Itô's formula applied to $t \mapsto v(T-t, Y_M(t))$ yields

$$
\begin{aligned}
dv(T-t, Y_M(t)) = &\frac{\partial v}{\partial t}(T-t, Y_M(t))dt \\
&+ (Dv(T-t, Y_M(t)), S(T-t)\pi_M dW(t)) \\
&+ \frac{1}{2}Tr\left[(S(T-t)\pi_M Q^{\frac{1}{2}})^* D^2 v(T-t, Y_M(t)) S(T-t)\pi_M Q^{\frac{1}{2}}\right]dt \\
= &(Dv(T-t, Y_M(t)), S(T-t)\pi_M dW(t)) \\
&+ \frac{1}{2}Tr\left[(S(T-t)\pi_M Q^{\frac{1}{2}})^* D^2 v(T-t, Y_M(t)) S(T-t)\pi_M Q^{\frac{1}{2}}\right]dt \\
&- \frac{1}{2}Tr\left[(S(T-t)Q^{\frac{1}{2}})^* D^2 v(T-t, Y_M(t)) S(T-t)Q^{\frac{1}{2}}\right]dt.
\end{aligned}
$$

Therefore,

$$
\begin{aligned}
v(0, Y_M(T)) = &v(T, Y_M(0)) + \int_0^T (Dv(T-s, Y_M(s)), S(T-s)\pi_M dW(s)) \\
&+ \frac{1}{2}\int_0^T Tr\left[(S(T-t)\pi_M Q^{\frac{1}{2}})^* D^2 v(T-t, Y_M(t)) S(T-t)\pi_M Q^{\frac{1}{2}}\right]dt \\
&- \frac{1}{2}\int_0^T Tr\left[(S(T-t)Q^{\frac{1}{2}})^* D^2 v(T-t, Y_M(t)) S(T-t)Q^{\frac{1}{2}}\right]dt.
\end{aligned}
$$

$$\tag{5.15}$$

According to the definition of Y_M, we have

$$Y_M(0) = S(T)\pi_M u_0 \quad \text{and} \quad Y_M(T) = u_M(T).$$

According to the representation of v, we have

$$v(0, Y_M(T)) = \mathbb{E}[f(u(T))|u(T) = Y_M(T)] = \mathbb{E}[f(u_M(T))|u(T) = u_M(T)]$$

and

$$v(T, Y_M(0)) = \mathbb{E}[f(u(T))|S(T)u(0) = Y_M(0)]$$
$$= \mathbb{E}\left[f\left(S(T)\pi_M u_0 + \int_0^T S(T-t)dW(t)\right)\Big|S(T)u(0) = S(T)\pi_M u_0\right].$$

Taking the expectation of the two sides of (5.15) and we obtian

$$\mathbb{E}[f(u_M(T))] = \mathbb{E}\left[f\left(S(T)\pi_M u_0 + \int_0^T S(T-t)dW(t)\right)\right]$$
$$+ \frac{1}{2}\mathbb{E}\int_0^T \left\{Tr\left[(S(T-t)\pi_M Q^{\frac{1}{2}})^* D^2 v(T-t, Y_M(t))S(T-t)\pi_M Q^{\frac{1}{2}}\right]\right.$$
$$\left. - Tr\left[(S(T-t)Q^{\frac{1}{2}})^* D^2 v(T-t, Y_M(t))S(T-t)Q^{\frac{1}{2}}\right]\right\}dt. \qquad (5.16)$$

Step 3. Weak error of the solutions.
Subtracting the resulting Eqs. (5.12) and (5.16) leads to

$$\mathbb{E}[f(u_M(T))] - \mathbb{E}[f(u(T))]$$
$$= \mathbb{E}\left[f\left(S(T)\pi_M u_0 + \int_0^T S(T-t)Q^{\frac{1}{2}}dW(t)\right) - f\left(S(T)u_0 + \int_0^T S(T-t)Q^{\frac{1}{2}}dW(t)\right)\right]$$
$$+ \frac{1}{2}\mathbb{E}\int_0^T \left\{Tr\left[(S(T-t)\pi_M Q^{\frac{1}{2}})^* D^2 v(T-t, Y_M(t))S(T-t)\pi_M Q^{\frac{1}{2}}\right]\right.$$
$$\left. - Tr\left[(S(T-t)Q^{\frac{1}{2}})^* D^2 v(T-t, Y_M(t))S(T-t)Q^{\frac{1}{2}}\right]\right\}dt$$
$$=: I + II. \qquad (5.17)$$

Due to Lemma 5.1, term I is estimated as

$$|I| \leq C\|f\|_{C_b^1}\mathbb{E}\|S(T)u_0 - S(T)\pi_M u_0\|_0$$
$$\leq Ce^{-\alpha T}\|f\|_{C_b^1}\mathbb{E}\|u_0\|_2 M^{-2}$$
$$\leq Ce^{-\alpha T}M^{-2}. \qquad (5.18)$$

Let us now estimate term II. As $(S(T-t)\pi_M - S(T-t))Q^{\frac{1}{2}}$ is a bounded linear operator and so is $D^2 v$ as shown in (5.14), we have

$$\left|Tr\left[(S(T-t)\pi_M Q^{\frac{1}{2}})^* D^2 v(T-t, Y_M(t))S(T-t)\pi_M Q^{\frac{1}{2}}\right]\right.$$

$$- Tr\left[(S(T-t)Q^{\frac{1}{2}})^* D^2 v(T-t, Y_M(t)) S(T-t) Q^{\frac{1}{2}} \right] \bigg|$$

$$= \left| Tr\left[((S(T-t)\pi_M - S(T-t))Q^{\frac{1}{2}})^* D^2 v(T-t, Y_M(t))(S(T-t)\pi_M + S(T-t))Q^{\frac{1}{2}} \right] \right|$$

$$\leq C \|S(T-t)\pi_M - S(T-t)\|_{\mathscr{L}(\dot{H}^2, L^2)} \|Q^{\frac{1}{2}}\|_{\mathscr{L}_2} \|f\|_{C_b^2} \|S(T-t)\|_{\mathscr{L}(L^2, L^2)} \|Q^{\frac{1}{2}}\|_{\mathscr{L}_2^0}$$

$$\leq C e^{-\alpha(T-t)} M^{-2}.$$

Hence, integrating the above equation leads to

$$|II| \leq \frac{C}{\alpha} M^{-2}. \tag{5.19}$$

Plugging (5.18) and (5.19) into (5.17), we conclude

$$\left| \mathbb{E}\left[f\left(u_M(T) \right) \right] - \mathbb{E}\left[f\left(u(T) \right) \right] \right| \leq C\left(e^{-\alpha T} + \frac{1}{\alpha} \right) M^{-2} \leq C M^{-2}, \tag{5.20}$$

in which C is independent of time T. $\square$

We still use modified processes to calculate the weak error of the fully discrete scheme in temporal direction. Denote $S_\tau = (Id - i\tau\Delta)^{-1} e^{-\alpha\tau}$, then scheme (5.7) is rewritten as

$$u_M^n = S_\tau u_M^{n-1} + i\lambda\tau e^{\alpha\tau} S_\tau \pi_M \left(\frac{|u_M^n|^2 + |e^{-\alpha\tau} u_M^{n-1}|^2}{2} u_M^n \right) + e^{\alpha\tau} S_\tau \pi_M \delta_n W$$

$$= S_\tau^n u_M^0 + i\lambda\tau e^{\alpha\tau} \sum_{l=1}^n S_\tau^{n+1-l} \pi_M \left(\frac{|u_M^l|^2 + |e^{-\alpha\tau} u_M^{l-1}|^2}{2} u_M^l \right)$$

$$+ e^{\alpha\tau} \sum_{l=1}^n S_\tau^{n+1-l} \pi_M \delta_l W. \tag{5.21}$$

Lemma 5.2 *For any $k \in \mathbb{N}$ and sufficiently small τ, we have the following estimates,*

$$(i) \ \|S_\tau^n - S(t)\|_{\mathscr{L}(\dot{H}^2, L^2)} \leq C(t+\tau)^{\frac{1}{2}} e^{-\alpha t} \tau^{\frac{1}{2}}, \quad t \in [t_{n-1}, t_{n+1}],$$

$$(ii) \ \|S_\tau^n - S(t)\|_{\mathscr{L}(\dot{H}^1, \dot{H}^1)} \leq C e^{-\alpha t}, \quad t \in [t_{n-1}, t_{n+1}],$$

where the constant $C = C(\alpha)$ is independent of n and τ.

Proof **Step 1.** Let $t = t_n$. As $S(t)$ is the operator semigroup generated by the linear operator of the following equation

$$du(t) = (i\Delta - \alpha Id)u(t)dt, \ u(0) = u^0 \in \dot{H}^2, \tag{5.22}$$

and $S_\tau = (Id - i\tau\Delta)^{-1} e^{-\alpha\tau}$ is the discrete operator semigroup, we have

$$S_\tau^n u(0) = u^n = e^{-\alpha\tau} u^{n-1} + i\tau \Delta u^n, \tag{5.23}$$

$$S(t_n) u(0) = u(t_n) = e^{-\alpha\tau} u(t_{n-1}) + \int_{t_{n-1}}^{t_n} i e^{-\alpha(t_n-s)} \Delta u(s) ds, \tag{5.24}$$

where u^n is the numerical solution of the proposed scheme applied to (5.22) and $u(t_n)$ is the exact solution of (5.22) at time t_n. Denote $E_n = u^n - u(t_n) = (S_\tau^n - S(t_n)) u(0)$ with $E_0 = 0$, then

$$E_n = e^{-\alpha\tau} E_{n-1} + i\tau \Delta E_n + i \int_{t_{n-1}}^{t_n} \left[\Delta u(t_n) - e^{-\alpha(t_n-s)} \Delta u(s) \right] ds.$$

Multiplying the above formula by $\overline{E}_n$, integrating with respect to x, and taking the real part, we derive

$$\frac{1}{2} \left[\|E_n\|_0^2 + \|E_n - e^{-\alpha\tau} E_{n-1}\|_0^2 - e^{-2\alpha\tau} \|E_{n-1}\|_0^2 \right]$$

$$= \Re \left[i \int_0^1 \int_{t_{n-1}}^{t_n} \Delta \overline{E}_n \int_s^{t_n} i e^{-\alpha(t_n-r)} \Delta u(r) dr ds dx \right]$$

$$\leq C \int_{t_{n-1}}^{t_n} \int_s^{t_n} \|\Delta u^n - \Delta u(t_k)\|_0 \|\Delta u(r)\|_0 dr ds$$

$$\leq C e^{-2\alpha t_n} \|\Delta u(0)\|_0^2 \tau^2,$$

where we have used the fact that

$$\|\Delta u^n\|_0^2 \leq e^{-2\alpha t_n} \|\Delta u^0\|_0^2, \quad \|\Delta u(t)\|_0 \leq C e^{-\alpha t} \|\Delta u(0)\|_0.$$

In fact, multiplying (5.23) by $\Delta \overline{u}^n - e^{-\alpha\tau} \Delta \overline{u}^{n-1}$, integrating in space and taking the imaginary part, we obtain

$$\|\Delta u^n\|_0^2 \leq e^{-2\alpha\tau} \|\Delta u^{n-1}\|_0^2 \leq e^{-2\alpha t_n} \|\Delta u^0\|_0^2.$$

Then it is easy to check that the fact

$$\|E_n\|_0^2 \leq e^{-2\alpha\tau} \|E_{n-1}\|_0^2 + C e^{-2\alpha t_n} \|\Delta u(0)\|_0^2 \tau^2$$

leads to

$$\|E_n\|_0^2 \leq C t_n e^{-2\alpha t_n} \|\Delta u(0)\|_0^2 \tau, \tag{5.25}$$

which finally yields $\|S_\tau^n - S(t_n)\|_{\mathscr{L}(\dot{H}^2, L^2)} \leq C t_n^{\frac{1}{2}} e^{-\alpha t_n} \tau^{\frac{1}{2}}$ in (i).

For (ii), by noticing that $\Delta e_k = -k^2\pi^2 e_k$, we have

$$\left\|\left(S_\tau^n - S(t_n)\right)u(0)\right\|_1^2 = \sum_{k=1}^\infty \left|e^{-\alpha t_n}\left((1 + i\tau k^2\pi^2)^{-n} - e^{-ik^2\pi^2 t_n}\right)(u(0), e_k)\right|^2 |\lambda_k^\alpha|$$

$$\leq 4e^{-2\alpha t_n} \sum_{k=1}^\infty |(u(0), e_k)|^2 |\lambda_k^\alpha| = 4e^{-2\alpha t_n}\|u(0)\|_1^2.$$

In the following two steps, we only give the proof of (i), and (ii) can be proved in a same procedure.

Step 2. If $t \in [t_{n-1}, t_n]$, then

$$\|S_\tau^n - S(t)\|_{\mathscr{L}(\dot{H}^2, L^2)} \leq \|S_\tau^n - S(t_n)\|_{\mathscr{L}(\dot{H}^2, L^2)} + \|S(t_n) - S(t)\|_{\mathscr{L}(\dot{H}^2, L^2)}$$

$$\leq Ct_n^{\frac{1}{2}} e^{-\alpha t_n}\tau^{\frac{1}{2}} + e^{-\alpha t}|e^{-\alpha(t_n - t)} - 1|$$

$$\leq Ct_n^{\frac{1}{2}} e^{-\alpha t_n}\tau^{\frac{1}{2}} + e^{-\alpha t}\sum_{n=1}^\infty \frac{1}{n!}(\alpha\tau)^n$$

$$\leq Ct_n^{\frac{1}{2}} e^{-\alpha t_n}\tau^{\frac{1}{2}} + e^{-\alpha t}\alpha\tau\frac{e^{\alpha\tau} - 1}{\alpha\tau}$$

$$\leq C(t + \tau)^{\frac{1}{2}} e^{-\alpha t}\tau^{\frac{1}{2}}.$$

We have used the fact that $\frac{e^{\alpha\tau} - 1}{\alpha\tau}$ is uniformly bounded for $\alpha\tau \in [0, 1]$.

Step 3. If $t \in [t_n, t_{n+1}]$, we have

$$\|S_\tau^n - S(t)\|_{\mathscr{L}(\dot{H}^2, L^2)} \leq \|S_\tau^n - S(t_n)\|_{\mathscr{L}(\dot{H}^2, L^2)} + \|S(t_n) - S(t)\|_{\mathscr{L}(\dot{H}^2, L^2)}$$

$$\leq Ct_n^{\frac{1}{2}} e^{-\alpha t_n}\tau^{\frac{1}{2}} + e^{-\alpha t}|e^{-\alpha(t_n - t)} - 1|$$

$$\leq Ct_n^{\frac{1}{2}} e^{-\alpha t} e^{\alpha(t - t_n)}\tau^{\frac{1}{2}} + e^{-\alpha t}\alpha\tau\frac{e^{\alpha\tau} - 1}{\alpha\tau}$$

$$\leq C(t + \tau)^{\frac{1}{2}} e^{-\alpha t}\tau^{\frac{1}{2}},$$

where the fact $e^{\alpha(t - t_n)} \leq e^{\alpha\tau} \leq e$ has been used.   □

Remark 5.3 From (5.23), we can also prove that

$$\|S_\tau^n\|_{\mathscr{L}(L^2, L^2)} \leq Ce^{-\alpha t},$$

where n and t satisfy $t \in [t_{n-1}, t_{n+1}]$.

The next theorem gives the time-independent weak error of the solutions for different cases.

Theorem 5.4 *Let* $\lambda = 0$. *Assume that* $u_0 \in \dot{\mathbf{H}}^2$, $u_M^0 = u_M(0) = \pi_M u_0$ *and* $\|Q^{\frac{1}{2}}\|_{\mathscr{L}_2^2}^2 < \infty$. *The weak error between* (5.2) *and* (5.7) *is independent of time and of order* $\frac{1}{2}$, *i.e., for any* $f \in \mathbf{C}_b^2(\mathbf{L}^2)$, *there exists a constant* $C = C(u_0, f)$ *independent of* M, T *and* N *such that*

$$\left| \mathbb{E}[f(u_M(T))] - \mathbb{E}[f(u_M^N)] \right| \leq C\tau^{\frac{1}{2}}, \quad \forall T = N\tau.$$

Corollary 5.2 *Under above assumptions,*

$$\left| \mathbb{E}[f(u_M(t))] - \mathbb{E}[f(u_M^N)] \right| \leq C\tau^{\frac{1}{2}}$$

holds for any $t \in [(N-1)\tau, (N+1)\tau]$.

Proof As

$$\left| \mathbb{E}[f(u_M(t))] - \mathbb{E}[f(u_M^N)] \right|$$
$$= \left| \mathbb{E}[f(u_M(T))] - \mathbb{E}[f(u_M(t))] \right| + \left| \mathbb{E}[f(u_M(T))] - \mathbb{E}[f(u_M^N)] \right|$$

and

$$\left| \mathbb{E}[f(u_M(T))] - \mathbb{E}[f(u_M(t))] \right| \leq \|f\|_{\mathbf{C}_b^1} \mathbb{E}\|u_M(T) - u_M(t)\|_0$$
$$\leq \|f\|_{\mathbf{C}_b^1}(T-t) \sup_{t \geq 0} \left[\mathbb{E}\|u_M(t)\|_2 + \mathbb{E}\|u_M(t)\|_0 + \mathbb{E}\|u_M(t)\|_1^2 \|u_M(t)\|_0 \right]$$
$$+ \|f\|_{\mathbf{C}_b^1} \mathbb{E}\|\pi_M(W(T) - W(t))\|_0 \leq C\tau^{\frac{1}{2}},$$

we then complete the proof according to Theorem 5.4.   □

Proof (of Theorem 5.4) We split it into several steps.
Step 1. Calculation of $\mathbb{E}[f(u_M(T))]$.

Recall the process we constructed in the proof of Theorem 5.3,

$$dY_M(t) = S(T-t)\pi_M dW(t).$$

Denote $v_M(T-t, y) = \mathbb{E}[f(Y_M(T))|Y_M(t) = y]$. Then

$$v_M(0, Y_M(T))$$
$$= v_M(T, Y_M(0)) + \int_0^T \Big(Dv_M(T-t, Y_M(t)), S(T-t)\pi_M dW(t) \Big), \quad (5.26)$$

where

$$v_M(0, Y_M(T)) = \mathbb{E}[f(u_M(T))|Y_M(T) = u_M(T)]$$

and

$$v_M(T, Y_M(0)) = \mathbb{E}[f(Y_M(T))|Y_M(0) = S(T)u_M(0)]$$
$$= \mathbb{E}\left[f\left(S(T)u_M(0) + \int_0^T S(T-s)\pi_M dW\right)\bigg| Y_M(0) = S(T)u_M(0)\right].$$

The expectation of (5.26) implies,

$$\mathbb{E}[f(u_M(T))] = \mathbb{E}\left[f\left(S(T)u_M(0) + \int_0^T S(T-s)\pi_M dW\right)\right]. \qquad (5.27)$$

Step 2. Calculation of $\mathbb{E}[f(u_M^N)]$.

Similar to [71], we define a discrete modified process

$$Y_M^n := S_\tau^{N-n} u_M^n$$
$$= S_\tau^N u_M^0 + e^{\alpha\tau} \sum_{l=1}^n S_\tau^{N+1-l} \pi_M \delta_l W.$$

Consider the following time continuous interpolation of Y_M^n, which is also V_M-valued and $\{\mathscr{F}_t\}_{t\geq 0}$-adapted,

$$\tilde{Y}_M(t) := S_\tau^N u_M^0 + e^{\alpha\tau} \int_0^t \sum_{l=1}^N S_\tau^{N+1-l} \pi_M 1_l(s) dW(s),$$

where $1_l(s) = 1$ if $s \in [t_{l-1}, t_l]$, otherwise, $1_l(s) = 0$. In particular for $t \in [t_{l-1}, t_l]$,

$$\tilde{Y}_M(t) = Y_M^{l-1} + e^{\alpha\tau} S_\tau^{N+1-l} \pi_M\left(W(t) - W(t_{l-1})\right), \qquad (5.28)$$

or equivalently,

$$\tilde{Y}_M(t) = Y_M^l + e^{\alpha\tau} S_\tau^{N+1-l} \pi_M\left(W(t) - W(t_l)\right). \qquad (5.29)$$

Apply Itô's formula to $t \mapsto v_M(T - t, \tilde{Y}_M(t))$,

$$dv_M(T - t, \tilde{Y}_M(t))$$
$$= \frac{\partial v_M}{\partial t}(T - t, \tilde{Y}_M(t))dt + \left(Dv_M, e^{\alpha\tau} \sum_{l=1}^N S_\tau^{N+1-l} \pi_M 1_l(t) dW(t)\right)$$

$$+ \frac{1}{2} Tr \left[\left(e^{\alpha \tau} \sum_{l=1}^{N} S_{\tau}^{N+1-l} \pi_M Q^{\frac{1}{2}} 1_l(t) \right)^{*} D^2 v_M \left(e^{\alpha \tau} \sum_{l=1}^{N} S_{\tau}^{N+1-l} \pi_M Q^{\frac{1}{2}} 1_l(t) \right) \right] dt$$

$$= \left(D v_M, e^{\alpha \tau} \sum_{l=1}^{N} S_{\tau}^{N+1-l} \pi_M 1_l(t) dW(t) \right)$$

$$+ \frac{1}{2} \sum_{l=1}^{N} Tr \left[\left(e^{\alpha \tau} S_{\tau}^{N+1-l} \pi_M Q^{\frac{1}{2}} \right)^{*} D^2 v_M \left(e^{\alpha \tau} S_{\tau}^{N+1-l} \pi_M Q^{\frac{1}{2}} \right) \right] 1_l(t) dt$$

$$- \frac{1}{2} \sum_{l=1}^{N} Tr \left[\left(S(T-t) \pi_M Q^{\frac{1}{2}} \right)^{*} D^2 v_M \left(S(T-t) \pi_M Q^{\frac{1}{2}} \right) \right] 1_l(t) dt,$$

where $D v_M$ and $D^2 v_M$ are evaluated at $(T - t, \tilde{Y}_M(t))$.

The same as before, integrating the formula above from 0 to T, and taking the expectation based on the fact that

$$v_M(0, \tilde{Y}_M(T)) = \mathbb{E}[f(Y_M(T)) | Y_M(T) = \tilde{Y}_M(T)] = \mathbb{E}[f(u_M^N) | Y_M(T) = u_M^N]$$

and

$$v_M(T, \tilde{Y}_M(0))$$
$$= \mathbb{E}[f(Y_M(T)) | Y_M(0) = \tilde{Y}_M(0)]$$
$$= \mathbb{E}\left[f\left(S_{\tau}^N u_M(0) + \int_0^T S(T-s) \pi_M dW \right) \Big| Y_M(0) = S_{\tau}^N u_M(0) \right],$$

we obtain

$$\mathbb{E}[f(u_M^N)] = \mathbb{E}\left[f\left(S_{\tau}^N u_M(0) + \int_0^T S(T-s) \pi_M dW \right) \right]$$
$$+ \frac{1}{2} \sum_{l=1}^{N} \mathbb{E} \int_0^T Tr \left[\left(e^{\alpha \tau} S_{\tau}^{N+1-l} \pi_M Q^{\frac{1}{2}} \right)^{*} D^2 v_M \left(e^{\alpha \tau} S_{\tau}^{N+1-l} \pi_M Q^{\frac{1}{2}} \right) \right.$$
$$\left. - \left(S(T-t) \pi_M Q^{\frac{1}{2}} \right)^{*} D^2 v_M \left(S(T-t) \pi_M Q^{\frac{1}{2}} \right) \right] 1_l(t) dt. \qquad (5.30)$$

Step 3. Weak convergence order.

Subtracting (5.27) from (5.30), we derive

$$\mathbb{E}[f(u_M^N)] - \mathbb{E}[f(u_M(T))]$$
$$= \mathbb{E}\left[f\left(S_{\tau}^N u_M(0) + \int_0^T S(T-s) \pi_M dW \right) - f\left(S(T) u_M(0) + \int_0^T S(T-s) \pi_M dW \right) \right]$$
$$+ \frac{1}{2} \sum_{l=1}^{N} \mathbb{E} \int_0^T Tr \left[\left(e^{\alpha \tau} S_{\tau}^{N+1-l} \pi_M Q^{\frac{1}{2}} \right)^{*} D^2 v_M \left(e^{\alpha \tau} S_{\tau}^{N+1-l} \pi_M Q^{\frac{1}{2}} \right) \right.$$

$$- \left(S(T-t)\pi_M Q^{\frac{1}{2}} \right)^* D^2 v_M \left(S(T-t)\pi_M Q^{\frac{1}{2}} \right) \Big] 1_l(t) dt.$$

$$=: I + II.$$

Now we estimate terms I and II separately. The constants C below may be different and are all independent of T and τ.

$$|I| = \left| \mathbb{E} \left[f \left(S_\tau^N u_M(0) + \int_0^T S(T-s)\pi_M dW \right) \right] \right.$$
$$\left. - \mathbb{E} \left[f \left(S(T) u_M(0) + \int_0^T S(T-s)\pi_M dW \right) \right] \right|$$
$$\leq C \|f\|_{C_b^1} \|S_\tau^N u_M(0) - S(T) u_M(0)\|_0$$
$$\leq C \|f\|_{C_b^1} \|S_\tau^N - S(T)\|_{\mathscr{L}(\dot{\mathbf{H}}^2, \mathbf{L}^2)} \|u_M(0)\|_2$$
$$\leq C(T+\tau)^{\frac{1}{2}} e^{-\alpha T} \tau^{\frac{1}{2}}, \tag{5.31}$$

where we have used Lemma 5.2 and $u_M(0) = \pi_M u_0 \in \dot{\mathbf{H}}^2$.

For term II, similar to the same part in the proof of Theorem 5.3, we have

$$II = \frac{1}{2} \sum_{l=1}^N \mathbb{E} \int_0^T Tr \Big[\left(e^{\alpha\tau} S_\tau^{N+1-l} \pi_M Q^{\frac{1}{2}} \right)^* D^2 v_M \left(e^{\alpha\tau} S_\tau^{N+1-l} \pi_M Q^{\frac{1}{2}} \right)$$
$$- \left(S(T-t)\pi_M Q^{\frac{1}{2}} \right)^* D^2 v_M \left(S(T-t)\pi_M Q^{\frac{1}{2}} \right) \Big] 1_l(t) dt$$
$$= \frac{1}{2} \sum_{l=1}^N \mathbb{E} \int_0^T Tr \Big[\left(\left(e^{\alpha\tau} S_\tau^{N+1-l} - S(T-t) \right) \pi_M Q^{\frac{1}{2}} \right)^* D^2 v_M$$
$$\cdot \left(\left(e^{\alpha\tau} S_\tau^{N+1-l} - S(T-t) \right) \pi_M Q^{\frac{1}{2}} \right) \Big]$$
$$+ 2 Tr \Big[\left(\left(e^{\alpha\tau} S_\tau^{N+1-l} - S(T-t) \right) \pi_M Q^{\frac{1}{2}} \right)^* D^2 v_M \left(S(T-t)\pi_M Q^{\frac{1}{2}} \right) \Big] 1_l(t) dt$$
$$= \frac{1}{2} \sum_{l=1}^N \mathbb{E} \int_0^T Tr \Big[e^{2\alpha\tau} \left(\left(S_\tau^{N+1-l} - S(T-t) \right) \pi_M Q^{\frac{1}{2}} \right)^* D^2 v_M$$
$$\cdot \left(\left(S_\tau^{N+1-l} - S(T-t) \right) \pi_M Q^{\frac{1}{2}} \right)$$
$$+ 2 e^{2\alpha\tau} \left(\left(S_\tau^{N+1-l} - S(T-t) \right) \pi_M Q^{\frac{1}{2}} \right)^* D^2 v_M \left(S(T-t)\pi_M Q^{\frac{1}{2}} \right)$$
$$+ (e^{2\alpha\tau} - 1) \left(S(T-t)\pi_M Q^{\frac{1}{2}} \right)^* D^2 v_M \left(S(T-t)\pi_M Q^{\frac{1}{2}} \right) \Big] 1_l(t) dt$$
$$=: \frac{1}{2} \sum_{l=1}^N \mathbb{E} \int_0^T (A_l + 2B_l + C_l) 1_l(t) dt,$$

where A_l, B_l and C_l satisfy

$$\mathbb{E}|A_l| \leq C\|S_\tau^{N+1-l} - S(T-t))\|_{\mathscr{L}(\dot{\mathbf{H}}^2,\mathbf{L}^2)}^2 \|\pi_M Q^{\frac{1}{2}}\|_{\mathscr{L}(\mathbf{L}^2,\dot{\mathbf{H}}^2)}^2 \|f\|_{\mathbf{C}_b^2}$$

$$\leq C(T-t+\tau)e^{-2\alpha(T-t)}\tau,$$

$$\mathbb{E}|B_l| \leq C\|S_\tau^{N+1-l} - S(T-t))\|_{\mathscr{L}(\dot{\mathbf{H}}^2,\mathbf{L}^2)}\|\pi_M Q^{\frac{1}{2}}\|_{\mathscr{L}(\mathbf{L}^2,\dot{\mathbf{H}}^2)}^2 \|f\|_{\mathbf{C}_b^2} \|S(T-t)\|_{\mathscr{L}(\mathbf{L}^2,\mathbf{L}^2)}$$

$$\leq C(T-t+\tau)^{\frac{1}{2}}e^{-2\alpha(T-t)}\tau^{\frac{1}{2}}$$

and

$$\mathbb{E}|C_l| \leq C\tau\|\pi_M Q^{\frac{1}{2}}\|_{\mathscr{L}(\mathbf{L}^2,\mathbf{L}^2)}^2 \|f\|_{\mathbf{C}_b^2} \|S(T-t)\|_{\mathscr{L}(\mathbf{L}^2,\mathbf{L}^2)}^2 \leq Ce^{-2\alpha(T-t)}\tau.$$

It follows

$$|II| \leq C\tau^{\frac{1}{2}}. \tag{5.32}$$

We then conclude from (5.31) and (5.32) that

$$\left|\mathbb{E}[f(u_M(T))] - \mathbb{E}[f(u_M^N)]\right| \leq C\tau^{\frac{1}{2}},$$

where C is independent of M, T and N. $\qquad\square$

Remark 5.4 For the linear case ($\lambda = 0$), the weak convergence order depends heavily on the regularity of the solution, which depends only on the regularity of the initial value and noise. We can achieve higher order by increasing the regularity of the initial value and the noise. For example, the weak order turns out to be 1 if we assume $u_0 \in \dot{\mathbf{H}}^4$ and $\|Q^{\frac{1}{2}}\|_{\mathscr{L}_2^4} < \infty$.

Based on the ergodicity of stochastic processes u and u_M, for any deterministic $u_0 \in \dot{\mathbf{H}}^2$, we have the following two equations

$$\lim_{T\to\infty} \frac{1}{T}\int_0^T \mathbb{E}f(u(t))dt = \int_{\mathbf{L}^2} f(y)d\mu(y),$$

$$\lim_{T\to\infty} \frac{1}{T}\int_0^T \mathbb{E}f(u_M(t))dt = \int_{V_M} f(y)d\mu_M(y)$$

for any $f \in \mathbf{C}_b^2(\mathbf{L}^2)$. Due to the time-independence of the weak error in Theorem 5.3, it turns out for any fixed α and M,

$$\left|\int_{\mathbf{L}^2} f(y)d\mu(y) - \int_{V_M} f(y)d\mu_M(y)\right| = \left|\lim_{T\to\infty} \frac{1}{T}\int_0^T \mathbb{E}f(u(t)) - \mathbb{E}f(u_M(t))dt\right|$$

$$\leq \lim_{T\to\infty} \frac{1}{T}\int_0^T \left|\mathbb{E}f(u(t)) - \mathbb{E}f(u_M(t))\right| dt$$

$$\leq \lim_{T \to \infty} \frac{1}{T} \int_0^T C \left(e^{-\alpha t} + \frac{1}{\alpha} \right) M^{-2} dt$$

$$\leq \frac{C}{\alpha} M^{-2},$$

which implies that μ_M is a proper approximation of μ. Thus, we give the following theorem.

Theorem 5.5 *Let* $\lambda = 0$. *Assume that* $u_0 \in \dot{\mathbf{H}}^2$, $\|Q^{\frac{1}{2}}\|_{\mathscr{L}_2^3} < \infty$ *and* $f \in \mathbf{C}_b^2(\mathbf{L}^2)$. *The error between the invariant measures* μ *and* μ_M *is of order* 2, *i.e.,*

$$\left| \int_{\mathbf{L}^2} f(y) d\mu(y) - \int_{V_M} f(y) d\mu_M(y) \right| < \frac{C}{\alpha} M^{-2}.$$

Remark 5.5 Although the time-independent weak error between u and u_M is obtained under the assumption $\|Q^{\frac{1}{2}}\|_{\mathscr{L}_2^2} < \infty$, it is necessary to assume in addition that $\|Q^{\frac{1}{2}}\|_{\mathscr{L}_2^3} < \infty$ in order to get the unique ergodicity of u given in Theorem 3.10.

The error between the invariant measures μ_M and μ_M^τ is derived in the same way.

Theorem 5.6 *Let* $\lambda = 0$. *Assume that* $u_0 \in \dot{\mathbf{H}}^2$, $\|Q^{\frac{1}{2}}\|_{\mathscr{L}_2^2} < \infty$ *and* $f \in \mathbf{C}_b^2(\mathbf{L}^2)$, *the error between invariant measures* μ_M *and* μ_M^τ *is of order* $\frac{1}{2}$, *i.e.,*

$$\left| \int_{V_M} f(y) d\mu_M(y) - \int_{V_M} f(y) d\mu_M^\tau(y) \right| < C\tau^{\frac{1}{2}}.$$

5.1.4 Numerical Experiments

This section provides numerical experiments to test the longtime behavior of scheme (5.7) for the case $\lambda = 0$. Based on the spatial semi-discretization (5.3) with $\lambda_k^\alpha = \mathbf{i}(k\pi)^2 + \alpha$:

$$da_k(t) = -\mathbf{i}(k\pi)^2 a_k(t) dt - \alpha a_k(t) dt + \sqrt{\eta_k} d\beta_k(t), \quad 1 \leq k \leq M,$$

we derive an equivalent form of the full discretization (5.7) as

$$\mathbf{a}^n - e^{-\alpha\tau} \mathbf{a}^{n-1} = -\mathbf{i}\tau\pi^2 \begin{pmatrix} 1 & & \\ & \ddots & \\ & & M^2 \end{pmatrix} \mathbf{a}^n + \begin{pmatrix} \sqrt{\eta_1}\delta_n\beta_1 \\ \vdots \\ \sqrt{\eta_M}\delta_n\beta_M \end{pmatrix},$$

where $\mathbf{a}^n := (a_1^n, \cdots, a_M^n)^\top$ is an approximation of $\mathbf{a}(t) := (a_1(t), \cdots, a_M(t))^\top$ and $\delta_n\beta_k = \beta_k(t_n) - \beta_k(t_{n-1})$ for $1 \leq k \leq M$. In the sequel, we take $\alpha = 1$, $M = 100$. Furthermore, we choose 1000 realizations to approximate the expectations.

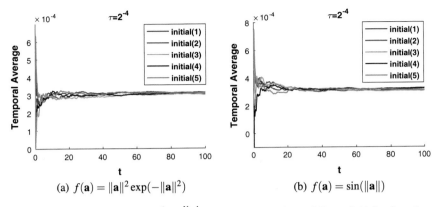

Fig. 5.1 The temporal averages $\frac{1}{N}\sum_{n=0}^{N-1}\mathbb{E}[f(\mathbf{a}^n)]$ started from different initial values ($\tau = 2^{-4}$, $T = 100$)

Ergodic limit. Figure 5.1 shows the temporal averages $\frac{1}{N}\sum_{n=0}^{N-1}\mathbb{E}[f(\mathbf{a}^n)]$ of the fully discrete scheme starting from five different initial values

$$\text{initial}(1) = (1, 0, \cdots, 0)^\top,$$

$$\text{initial}(2) = (0.0003\mathbf{i}, 0, \cdots, 0)^\top,$$

$$\text{initial}(3) = \left(\sin\left(\frac{1}{101}\pi\right), \sin\left(\frac{2}{101}\pi\right), \cdots, \sin\left(\frac{100}{101}\pi\right) \right)^\top,$$

$$\text{initial}(4) = \left(\frac{2+\mathbf{i}}{20}\right)(1, 2, \cdots, 100)^\top,$$

$$\text{initial}(5) = \left(\exp\left(-\frac{\mathbf{i}}{50}\right), \exp\left(-\frac{2\mathbf{i}}{50}\right), \cdots, \exp\left(-\frac{100\mathbf{i}}{50}\right) \right)^\top.$$

They converge to the same value with error $\tau^{\frac{1}{2}}$ before time T, where $\tau = 2^{-4}$ and $T = 100$. We choose $\eta_k = k^{-3}$ in this experiment. This result verifies that the temporal averages converge to the spatial average, which is a constant, for almost every initial values in the whole space.

Weak error. In Figs. 5.2 and 5.3, we fix the initial value $u_0(x)$ as $\sqrt{2}\sin(\pi x)$, such that $a_k(0) = (u_0, e_k)$ and $\mathbf{a}^0 = \mathbf{a}(0) = (1, 0, \cdots, 0)^\top$. Figure 5.2 displays the weak error $\mathbb{E}[f(\mathbf{a}(t_n)) - f(\mathbf{a}^n)]$ over long time $T = 10^3$ for different time step-sizes and test functions:

$$\text{(a) } \tau = 2^{-2}, f(\mathbf{a}) = \|\mathbf{a}\|^2 \exp(-\|\mathbf{a}\|^2),$$

$$\text{(b) } \tau = 2^{-4}, f(\mathbf{a}) = \|\mathbf{a}\|^2 \exp(-\|\mathbf{a}\|^2),$$

$$\text{(c) } \tau = 2^{-2}, f(\mathbf{a}) = \sin(\|\mathbf{a}\|),$$

$$\text{(d) } \tau = 2^{-4}, f(\mathbf{a}) = \sin(\|\mathbf{a}\|).$$

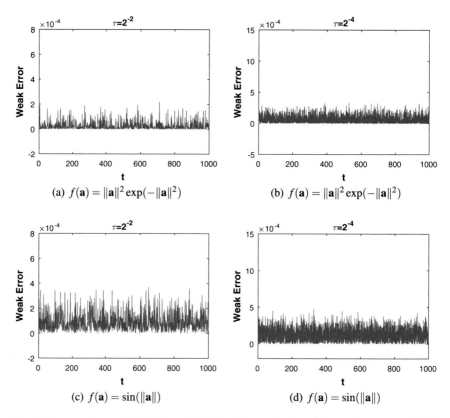

Fig. 5.2 The weak error $\mathbb{E}[f(\mathbf{a}(t_n)) - f(\mathbf{a}^n)]$ for different f and step-size τ with $t_n = n\tau \in [0, T]$ and $T = 10^3$

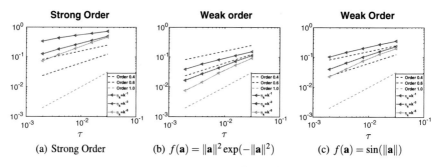

Fig. 5.3 The strong and weak orders for noise in $\mathbf{L}^2$, $\dot{\mathbf{H}}^2$ and $\dot{\mathbf{H}}^4$, i.e., $\eta_k = k^{-1}, k^{-3}, k^{-5}$ ($T = 1$, $\tau \in \{2^{-i}, 5 \le i \le 9\}$)

The reference values are generated for the time step-size $\tau = 2^{-8}$, and the noise is chosen in $\dot{\mathbf{H}}^2$, i.e., $\eta_k = k^{-3}$. Figure 5.2 shows that the weak error is independent of time interval and can be controlled by $C\tau^{\frac{1}{2}}$.

Convergence order. Figure 5.3 displays

    (a) the strong convergence order,

    (b) the rate of weak convergence for $f(\mathbf{a}) = \|\mathbf{a}\|^2 \exp(-\|\mathbf{a}\|^2)$,

    (c) the rate of weak convergence for $f(\mathbf{a}) = \sin(\|\mathbf{a}\|)$.

The reference values are generated for the time step-size $\tau = 2^{-12}$. As the initial value $u_0(x) = \sqrt{2}\sin(\pi x)$ is regular enough, both the strong and weak convergence order depend heavily on the regularity of the noise for the linear case. It shows in Fig. 5.3 that the orders slightly increase as the noise from $\mathbf{L}^2$ via $\dot{\mathbf{H}}^2$ to $\dot{\mathbf{H}}^4$ (i.e., η_k from k^{-1} via k^{-3} to k^{-5}), which verifies Remark 5.4. Note that the orders are a little bit better than the theoretical results, because the truncation of the noise makes the noise more regular than it should be. Numerical tests also show that the weak convergence order is almost the same as the strong convergence order, which is similar to the statement in Remark 5.11, [67].

5.2 Ergodic and Conformal Multi-symplectic Full Approximation

In this section, we further take the geometric structure into consideration, and aim to propose a fully discrete scheme to inherit both the ergodicity and the conformal multi-symplecticity of the original system (5.1).

5.2.1 Numerical Schemes

We apply the central difference scheme to (5.1) in the spatial direction and obtain

$$\begin{cases} du_j = \left(i\dfrac{u_{j+1} - 2u_j + u_{j-1}}{h^2} - \alpha u_j + i\lambda|u_j|^2 u_j\right)dt + \varepsilon\sum_{k=1}^{K}\sqrt{\eta_k}e_k(x_j)d\beta_k(t), \\ u_0(t) = u_{M+1}(t) = 0, \\ u_j(0) = u_0(x_j), \end{cases}$$

$$(5.33)$$

where h is the uniform spatial step-size and $u_j := u_j(t)$ is an approximation of $u(x_j, t)$ with $x_j = jh$, $j = 1, 2, \cdots, M$ and $(M+1)h = 1$. With the notation

$$U = (u_1, \cdots, u_M)^\top \in \mathbb{C}^M, \quad \beta = (\beta_1, \cdots, \beta_K)^\top \in \mathbb{C}^K,$$
$$F(U) = \text{diag}\{|u_1|^2, \cdots, |u_M|^2\}, \quad \Lambda = \text{diag}\{\sqrt{\eta_1}, \cdots, \sqrt{\eta_K}\},$$

$$A = \begin{pmatrix} -2 & 1 & & \\ 1 & -2 & 1 & \\ & \ddots & \ddots & \ddots \\ & & 1 & -2 \end{pmatrix} \in \mathbb{R}^{M \times M} \quad \text{and} \quad \sigma = \begin{pmatrix} e_1(x_1) & \cdots & e_K(x_1) \\ \vdots & & \vdots \\ e_1(x_M) & \cdots & e_K(x_M) \end{pmatrix} \in \mathbb{R}^{M \times K},$$

we rewrite (5.33) into a finite dimensional SDE

$$\begin{cases} dU = \left(\mathrm{i} \dfrac{1}{h^2} AU - \alpha U + \mathrm{i} \lambda F(U)U \right) dt + \varepsilon \sigma \Lambda d\beta, \\ U(0) = (u_0(x_1), \cdots, u_0(x_M))^\top. \end{cases} \tag{5.34}$$

In the sequel, we still denote by $\| \cdot \|$ the Euclidean norm for vectors or matrices, i.e., $\|v\| = \left(\sum_{j=1}^M |v_j|^2 \right)^{1/2}$ for a vector $v = (v_1, \cdots, v_M)^\top \in \mathbb{C}^M$ and $\|A\| = \max\{\sqrt{\mu} : \mu \text{ is an eigenvalue of } A^\top A\}$.

Lemma 5.3 *Matrix A is uniformly bounded for any dimension M, more precisely,* $\|A\| \le 4$.

Proof Based on the definition of $\|A\|$, we only need to show that the maximum absolute value $\lambda_* := \max\{|\lambda| : \det(\lambda Id - A) = 0\}$ of eigenvalues of $A \in \mathbb{R}^{M \times M}$ is uniformly bounded with respect to dimension M, and no larger than 4, then the lemma holds.

In fact, according to the Gerschgorin theorem and the properties of A, all the eigenvalues λ_i, $i = 1, \cdots, M$, of A lie in the Gerschgorin area $G^M = \bigcup_{i=1}^M G_i(A)$ with

$$G_i(A) := \{z \in \mathbb{R} : |z + 2| \le R_i\}, \quad i = 1, \cdots, M.$$

Here, $R_i := \sum_{j=1, j \ne i}^M |a_{ij}| = 2$ for any $i = 1, \cdots, M$ with a_{ij} being the components of matrix $A = (a_{ij})_{M \times M}$. As a result, we have

$$\lambda_i \in G^M = \{z \in \mathbb{R} : -4 \le z \le 0\}$$

for all $i = 1, \cdots, M$, which completes the proof. $\qquad\square$

The solution of (5.34) is uniformly bounded, which is stated in the following proposition.

Proposition 5.5 *Assume that $\mathbb{E}\|u_0\|_{\mathbf{L}^2}^2 < \infty$, then the solution U of (5.34) is uniformly bounded with*

$$h\mathbb{E}\|U(t)\|^2 \le e^{-2\alpha t} h\mathbb{E}\|U(0)\|^2 + \frac{2\varepsilon^2 \eta^{(K)}}{\alpha} (1 - e^{-2\alpha t}), \tag{5.35}$$

where $\eta^{(K)} := \sum_{k=1}^K \eta_k$.

Proof Similar to the proof of Proposition 4.1, we apply Itô's formula to $\|U(s)\|^2$ and obtain

$$
\begin{aligned}
d\|U(s)\|^2 &= 2\Re[\overline{U}^\top dU] + (\varepsilon\sigma\Lambda d\overline{\beta})^\top(\varepsilon\sigma\Lambda d\beta) \\
&= -2\alpha\|U(s)\|^2 ds + 2\Re[\overline{U}^\top \varepsilon\sigma\Lambda d\beta] \\
&\quad + \varepsilon^2 \sum_{j=1}^{M}\left[\left(\sum_{k=1}^{K}\sqrt{\eta_k}e_k(x_j)d\overline{\beta}_k\right)^\top\left(\sum_{k=1}^{K}\sqrt{\eta_k}e_k(x_j)d\beta_k\right)\right].
\end{aligned} \tag{5.36}
$$

Taking the expectation on both sides of (5.36) leads to

$$
d\mathbb{E}\|U(s)\|^2 = -2\alpha\mathbb{E}\|U(s)\|^2 ds + 2\varepsilon^2\sum_{j=1}^{M}\sum_{k=1}^{K}\eta_k e_k^2(x_j)ds.
$$

Thus, multiplying the above equation by $he^{2\alpha s}$ and integrating from 0 to t leads to

$$
\int_0^t he^{2\alpha s}d\mathbb{E}\|U(s)\|^2 + \int_0^t 2\alpha he^{2\alpha s}\mathbb{E}\|U(s)\|^2 ds = 2\varepsilon^2 h\sum_{j=1}^{M}\sum_{k=1}^{K}\eta_k e_k^2(x_j)\int_0^t e^{2\alpha s}ds.
$$

Based on the fact that $\sum_{j=1}^{M}e_n^2(x_j) \le 2M \le 2h^{-1}$, we have

$$
\begin{aligned}
e^{2\alpha t}h\mathbb{E}\|U(t)\|^2 - h\mathbb{E}\|U(0)\|^2 &= \frac{\varepsilon^2 h}{\alpha}(e^{2\alpha t}-1)\sum_{j=1}^{M}\sum_{k=1}^{K}\eta_k e_k^2(x_j) \\
&\le \frac{2\varepsilon^2\eta^{(K)}}{\alpha}(e^{2\alpha t}-1)
\end{aligned} \tag{5.37}
$$

which completes the proof.

In addition, noting that for $h = 1/(M+1)$ and $x_j = jh$, $j = 1, \cdots, M$,

$$
\begin{aligned}
\mathbb{E}\|u_0\|_{L^2}^2 &= \mathbb{E}\sum_{j=1}^{M}\int_{x_{j-1}}^{x_j}|u_0(x)|^2 dx \\
&= \mathbb{E}\sum_{j=1}^{M}|u_0(x_j)|^2 h + O(h) = h\mathbb{E}\|U(0)\|^2 + O(h),
\end{aligned}
$$

the uniform boundedness under the assumption $\mathbb{E}\|u_0\|_{L^2}^2 < \infty$ is obtained. □

Remark 5.6 Scheme (5.33) is equivalent to the symplectic Euler scheme applied to (4.21), i.e.,

$$\begin{cases} p_{j+1} - p_j = hv_{j+1}, \\ q_{j+1} - q_j = hw_{j+1}, \\ v_{j+1} - v_j = h(q_j)_t + \alpha h q_j - h((p_j)^2 + (q_j)^2)p_j - \varepsilon \sum_{k=1}^{K} \sqrt{\eta_k} e_k(x_j) \frac{d\beta_k^2(t)}{dt}, \\ w_{j+1} - w_j = -h(p_j)_t - \alpha h p_j - h((p_j)^2 + (q_j)^2)q_j + \varepsilon \sum_{k=1}^{K} \sqrt{\eta_k} e_k(x_j) \frac{d\beta_k^1(t)}{dt}. \end{cases}$$

For the construction of fully discrete schemes, the splitting technique is applied such that the proposed scheme could inherit the properties of (5.1). We drop the linear terms and the stochastic term for the moment and consider the following equation

$$dU(t) - i\lambda F(U(t))U(t)dt = 0 \tag{5.38}$$

first. Multiplying $\overline{F(U(t))}$ to both sides of (5.38) and taking the imaginary part, we obtain $\|U(t)\|^2 = \|U(0)\|^2$, which implies that $F(U(t)) = F(U(0))$. Thus, (5.38) is shown to possess a unique solution $U(t) = e^{i\lambda F(U(0))t}U(0)$.

For the linear equation

$$dU(t) - i\left(\frac{1}{h^2}AU(t) + i\alpha U(t)\right)dt = \varepsilon\sigma\Lambda d\beta,$$

a modified midpoint scheme is applied to obtain its full discretization. Now we define the following splitting scheme initialized with $U^0 = U(0)$:

$$U^{n+1} = e^{-\frac{1}{2}\alpha\tau}\tilde{U}^n + \frac{i\tau}{h^2}A\frac{U^{n+1} + e^{-\frac{1}{2}\alpha\tau}\tilde{U}^n}{2} - \alpha\tau\frac{U^{n+1} + e^{-\frac{1}{2}\alpha\tau}\tilde{U}^n}{4} + \varepsilon\sigma\Lambda\delta_{n+1}\beta, \tag{5.39a}$$

$$\tilde{U}^n = e^{i\lambda F(U^n)\tau}U^n, \tag{5.39b}$$

where $U^n = (u_1^n, \cdots, u_M^n)^\top \in \mathbb{C}^M$, τ denotes the uniform time step-size, $\delta_{n+1}\beta = \beta(t_{n+1}) - \beta(t_n)$ and $t_n = n\tau$, $n \in \mathbb{N}$. Note that scheme (5.39) can be rewritten as

$$\begin{aligned} U^{n+1} - e^{f(U^n)}U^n = &i\frac{\tau}{2h^2}A\left(U^{n+1} + e^{f(U^n)}U^n\right) \\ &- \frac{1}{4}\alpha\tau\left(U^{n+1} + e^{f(U^n)}U^n\right) + \varepsilon\sigma\Lambda\delta_{n+1}\beta, \end{aligned} \tag{5.40}$$

which can also be expressed in the following explicit form

$$U^{n+1} = \left(Id - \frac{i\tau}{2h^2}A + \frac{1}{4}\alpha\tau Id\right)^{-1}\left(Id + \frac{i\tau}{2h^2}A - \frac{1}{4}\alpha\tau Id\right)e^{f(U^n)}U^n$$

$$+ \left(Id - \frac{i\tau}{2h^2}A + \frac{1}{4}\alpha\tau Id\right)^{-1}\varepsilon\sigma\Lambda\delta_{n+1}\beta \tag{5.41}$$

with $f(U^n) = \left(-\frac{1}{2}\alpha Id + i\lambda F(U^n)\right)\tau$. Thus, there uniquely exists an adapted solution $\{U^n\}_{n\in\mathbb{N}_+}$ of (5.40) for sufficiently small τ.

As for the proposed splitting scheme (5.39), the solution of the one-step approximation (5.39b) coincides with the exact solution of the Hamiltonian system $dU(t) - i\lambda F(U(t))U(t)dt = 0$. That is, the phase flow $U^n \mapsto \tilde{U}^n$ preserves the symplectic structure. It then suffices to show that (5.39a) possesses the conformal multi-symplectic conservation law, which is stated in the following theorem.

Theorem 5.7 *The one-step approximation defined through (5.39a) possesses the discrete conformal multi-symplectic conservation law*

$$e^{-\alpha\tau}\frac{dz_j^{n+1}\wedge M_4 dz_j^{n+1} - dz_j^n\wedge M_4 dz_j^n}{\tau} + \frac{dz_j^{n+\frac{1}{2}}\wedge(K_4^1 dz_{j+1}^{n+\frac{1}{2}} - K_4^2 dz_{j-1}^{n+\frac{1}{2}})}{h}$$

$$= -\frac{1}{2}\alpha dz_j^{n+\frac{1}{2}}\wedge M_4 dz_j^{n+\frac{1}{2}},$$

where $z_j^n = (p_j^n, q_j^n, v_j^n, w_j^n)^\top$, $z_j^{n+\frac{1}{2}} = \frac{1}{2}(z_j^{n+1} + e^{-\frac{1}{2}\alpha\tau}z_j^n)$, $v_{j+1}^n := (p_{j+1}^n - p_j^n)h^{-1}$ *and* $w_{j+1}^n := (q_{j+1}^n - q_j^n)h^{-1}$ *with* p_j^n *and* q_j^n *being the real and imaginary parts of* u_j^n, *respectively. Moreover,*

$$K_4^1 = \begin{pmatrix} 0 & 0 & -1 & 0 \\ 0 & 0 & 0 & -1 \\ 0 & 0 & 0 & 0 \\ 0 & 0 & 0 & 0 \end{pmatrix}, \quad K_4^2 = \begin{pmatrix} 0 & 0 & 0 & 0 \\ 0 & 0 & 0 & 0 \\ 1 & 0 & 0 & 0 \\ 0 & 1 & 0 & 0 \end{pmatrix}$$

such that $K_4 := K_4^1 + K_4^2$ *and* M_4 *are the same as those in Sect. 4.1.*

Proof We denote $\tilde{U}^n$ still by U^n for convenience in this proof, and we have

$$U^{n+1} = e^{-\frac{1}{2}\alpha\tau}U^n + \frac{i\tau}{h^2}A\frac{U^{n+1} + e^{-\frac{1}{2}\alpha\tau}U^n}{2} - \alpha\tau\frac{U^{n+1} + e^{-\frac{1}{2}\alpha\tau}U^n}{4} + \varepsilon\sigma\Lambda\delta_{n+1}\beta$$

with $U^n = (u_1^n, \cdots, u_M^n)^\top \in \mathbb{C}^M$. Denote by $\delta_{n+1}\beta^1$ and $\delta_{n+1}\beta^2$ the real and imaginary parts of $\delta_{n+1}\beta$, respectively. Noticing that the jth component of $h^{-2}AU^n$ can be expressed as $h^{-1}(v_{j+1}^n - v_j^n) + ih^{-1}(w_{j+1}^n - w_j^n)$, we decompose (5.39a) with its real and imaginary parts respectively and derive

$$\begin{cases} \dfrac{p_j^{n+1} - e^{-\frac{1}{2}\alpha\tau}p_j^n}{\tau} + \dfrac{w_{j+1}^{n+1} - w_j^{n+1}}{2h} + e^{-\frac{1}{2}\alpha\tau}\dfrac{w_{j+1}^n - w_j^n}{2h} \\ \qquad\qquad = -\dfrac{1}{4}\alpha(p_j^{n+1} + e^{-\frac{1}{2}\alpha\tau}p_j^n) + \varepsilon\sigma\Lambda\delta_{n+1}\beta^1, \\[2ex] \dfrac{q_j^{n+1} - e^{-\frac{1}{2}\alpha\tau}q_j^n}{\tau} - \dfrac{v_{j+1}^{n+1} - v_j^{n+1}}{2h} - e^{-\frac{1}{2}\alpha\tau}\dfrac{v_{j+1}^n - v_j^n}{2h} \\ \qquad\qquad = -\dfrac{1}{4}\alpha(q_j^{n+1} + e^{-\frac{1}{2}\alpha\tau}q_j^n) + \varepsilon\sigma\Lambda\delta_{n+1}\beta^2. \end{cases}$$

Combining formula $v_{j+1}^n = (p_{j+1}^n - p_j^n)h^{-1}$, $w_{j+1}^n = (q_{j+1}^n - q_j^n)h^{-1}$ with above equations, we have

$$M_4\frac{z_j^{n+1} - e^{-\frac{1}{2}\alpha\tau}z_j^n}{\tau} + K_4^1\frac{z_{j+1}^{n+\frac{1}{2}} - z_j^{n+\frac{1}{2}}}{h} + K_4^2\frac{z_j^{k+\frac{1}{2}} - z_{j-1}^{n+\frac{1}{2}}}{h} = -\frac{1}{2}\alpha M_4 z_j^{n+\frac{1}{2}} + \xi_j^{n+\frac{1}{2}},$$

where $\xi_j^{n+\frac{1}{2}} := (\varepsilon\sigma\Lambda\delta_{n+1}\beta^2, -\varepsilon\sigma\Lambda\delta_{n+1}\beta^1, v_j^{n+\frac{1}{2}}, w_j^{n+\frac{1}{2}})^\top$. Taking differential in phase space on both sides of the above equation, and performing the wedge product with $dz_j^{n+\frac{1}{2}}$ respectively, we show the discrete conformal multi-symplectic conservation law based on the symmetry of matrix $-K_4^1 + K_4^2$ and the fact $dz_j^{n+\frac{1}{2}} \wedge (-K_4^1 + K_4^2)dz_j^{n+\frac{1}{2}} = 0$, $dz_j^{n+\frac{1}{2}} \wedge d\xi_j^{n+\frac{1}{2}} = 0$. $\qquad\square$

Remark 5.7 It is also feasible to show that scheme (5.39) are conformal symplectic in time, which together with Remark 5.6, yields the conformal multi-symplecticity of the fully discrete scheme (5.40).

Proposition 5.6 *Assume that $\mathbb{E}\|u_0\|_{\mathbf{L}^2}^2 < \infty$, $Q^{\frac{1}{2}} \in \mathscr{L}_2^2$ and $K \leq C_*(M+1)$ for some constant $C_* \geq 1$, then the solution $\{U^n\}_{n\in\mathbb{N}_+}$ of (5.40) is uniformly bounded, i.e.,*

$$h\mathbb{E}\|U^n\|^2 \leq e^{-\alpha t_n}h\mathbb{E}\|U^0\|^2 + C \tag{5.42}$$

with $t_n = n\tau$ and the constant C depending on α, ε, Q and C_*.

Proof We multiply $\overline{(U^{n+1} + e^{f(U^n)}U^n)}^\top$ to (5.40), take the real part and expectation, and obtain

$$\mathbb{E}\|U^{n+1}\|^2 - e^{-\alpha\tau}\mathbb{E}\|U^n\|^2$$
$$= -\frac{1}{4}\alpha\tau\mathbb{E}\|U^{n+1} + e^{f(U^n)}U^n\|^2 + \mathbb{E}\left[\Re\left[\overline{(U^{n+1} - e^{f(U^n)}U^n)}^\top \varepsilon\sigma\Lambda\delta_{n+1}\beta\right]\right]$$
$$= -\frac{1}{4}\alpha\tau\mathbb{E}\|U^{n+1} + e^{f(U^n)}U^n\|^2 + \mathbb{E}\left[\Re\left[\left(-\mathrm{i}\frac{\tau}{2h^2}A\overline{(U^{n+1} + e^{f(U^n)}U^n)}\right.\right.\right.$$

$$- \frac{1}{4} \alpha \tau \overline{\left(U^{n+1} + e^{f(U^n)} U^n \right)} + \overline{\varepsilon \sigma \Lambda \delta_{n+1} \beta} \right)^\top \varepsilon \sigma \Lambda \delta_{n+1} \beta \Bigg] \Bigg]$$

$$\leq - \frac{1}{4} \alpha \tau \mathbb{E} \| U^{n+1} + e^{f(U^n)} U^n \|^2 + \frac{1}{8} \alpha \tau \mathbb{E} \| U^{n+1} + e^{f(U^n)} U^n \|^2$$

$$+ C \tau \mathbb{E} \| h^{-2} A \varepsilon \sigma \Lambda \delta_{n+1} \beta \|^2 + \frac{1}{8} \alpha \tau \mathbb{E} \| U^{n+1} + e^{f(U^n)} U^n \|^2$$

$$+ C \tau \mathbb{E} \| \varepsilon \sigma \Lambda \delta_{n+1} \beta \|^2 + \mathbb{E} \| \varepsilon \sigma \Lambda \delta_{n+1} \beta \|^2. \tag{5.43}$$

For the smooth functions $e_k(x), k = 1, \cdots, K$, we have

$$\left| \Delta e_k(x_j) - \frac{e_k(x_{j+1}) - 2e_k(x_j) + e_k(x_{j-1})}{h^2} \right| \leq C k^4 h^2 \leq C k^2, \quad k \geq 1$$

based on the fact $kh \leq K(M+1)^{-1} \leq C_*$. Thus,

$$\mathbb{E} \| h^{-2} A \varepsilon \sigma \Lambda \delta_{n+1} \beta \|^2 = \varepsilon^2 \sum_{j=1}^{M} \mathbb{E} \left| \sum_{k=1}^{K} \sqrt{\eta_k} \frac{e_k(x_{j+1}) - 2e_k(x_j) + e_k(x_{j-1})}{h^2} \delta_{n+1} \beta_k \right|^2$$

$$\leq 2 \varepsilon^2 \sum_{j=1}^{M} \sum_{k=1}^{K} \eta_k \left(| \Delta e_k(x_j) | + C k^2 \right)^2 \tau \leq C M \tau \sum_{k=1}^{K} k^4 \eta_k \leq C h^{-1} \tau. \tag{5.44}$$

In the last step, we have used the fact $\sum_{k=1}^{K} k^4 \eta_k \leq C \| Q^{\frac{1}{2}} \|_{\mathscr{L}_2^2} \leq C$. Similarly,

$$\mathbb{E} \| \varepsilon \sigma \Lambda \delta_{n+1} \beta \|^2 = \varepsilon^2 \sum_{j=1}^{M} \mathbb{E} \left| \sum_{k=1}^{K} \sqrt{\eta_k} e_k(x_j) \delta_{n+1} \beta_k \right|^2 \leq C M \eta \tau \leq C h^{-1} \tau. \tag{5.45}$$

Substituting (5.44) and (5.45) into (5.43), we obtain

$$h \mathbb{E} \| U^{n+1} \|^2 \leq e^{-\alpha \tau} h \mathbb{E} \| U^n \|^2 + C \tau \leq e^{-\alpha t_{n+1}} h \mathbb{E} \| U^0 \|^2 + C \tau \frac{1 - e^{-\alpha t_n}}{1 - e^{-\alpha \tau}},$$

which, together with the fact $\frac{1 - e^{-\alpha t_n}}{1 - e^{-\alpha \tau}} \leq \frac{1}{\alpha \tau}$, completes the proof.   □

Theorem 5.8 *Under the assumptions in Proposition 5.6 and $\eta_k > 0$ for $k = 1, \cdots,$ K, the solution $\{U^n\}_{n \in \mathbb{N}_+}$ of (5.40) is uniquely ergodic with a unique invariant measure, denoted by μ_h^τ, satisfying*

$$\lim_{N \to \infty} \frac{1}{N} \sum_{n=0}^{N-1} \mathbb{E} f(U^n) = \int_{\mathbb{C}^M} f d\mu_h^\tau, \quad \forall f \in C_b(\mathbb{C}^M). \tag{5.46}$$

Proof For any fixed $h > 0$, we choose $V(\cdot) := h\| \cdot \|^2$ as the Lyapunov functional, which satisfies that the level sets $K_c := \{u \in \mathbb{C}^M : V(u) \leq c\}$ are compact for any $c > 0$ and $\mathbb{E}[V(U^n)] \leq V(U^0) + C$ for any $n \in \mathbb{N}$. Thus, the Markov chain $\{U^n\}_{n \in \mathbb{N}}$ possesses an invariant measure according to Theorem 1.3 (see also [58, Proposition 7.10]).

Next we show that $\{U^n\}_{n \in \mathbb{N}}$ is irreducible and strong Feller (also known as the minorization condition in Assumption 2.1 of [139]), which yields the uniqueness of the invariant measure. In fact, for any $u, v \in \mathbb{C}^M$, we can derive from (5.40) that $\delta_1 \beta$ can be chosen as

$$\varepsilon \sigma \Lambda \delta_1 \beta = v - e^{f(u)} u - \mathbf{i} \frac{\tau}{2h^2} A \left(v + e^{f(u)} u \right) + \frac{1}{4} \alpha \tau \left(v + e^{f(u)} u \right)$$

such that $U^0 = u$, $U^1 = v$, where we have used the fact that σ is full rank and Λ is invertible. Thus, we can conclude based on the homogenous property of the Markov chain $\{U^n\}_{n \in \mathbb{N}}$ that the transition kernel $P_n(u, A) := \mathbb{P}(U^n \in A | U^0 = u) > 0$, which implies the irreducibility of the chain. On the other hand, as $\delta_1 \beta$ has a $\mathbf{C}^\infty$ density, it follows from (5.41) that U^1 also has a $\mathbf{C}^\infty$ density for any deterministic initial value $U^0 = u$. Then explicit construction shows that $\{U^n\}_{n \in \mathbb{N}}$ possesses a family of $\mathbf{C}^\infty$ densities and is strong Feller.   □

The conformal multi-symplecticity, uniform boundedness of the charge and ergodicity for scheme (5.40) are clearly consistent with the continuous results (4.22), (4.27) and (3.12), respectively. The next result concerns the error estimation of the proposed scheme, where the truncation technique will be used to deal with the non-globally Lipschitz nonlinearity.

5.2.2 Convergence in Probability

In this section, we focus on the approximate error for the proposed scheme in temporal direction. As the nonlinear term is not global Lipschitz, we consider the following truncated function first

$$dU_R = \left(\mathbf{i} \frac{1}{h^2} A U_R - \alpha U_R + \mathbf{i} \lambda F_R(U_R) U_R \right) dt + \varepsilon \sigma \Lambda d\beta, \qquad (5.47)$$

with $U_R := U_R(t) = (u_{R,1}(t), \cdots, u_{R,M}(t))^\top$ and the initial value $U_R(0) = U(0)$. Here $F_R(v) := \theta \left(\frac{\|v\|}{R} \right) F(v)$ for any vector $v \in \mathbb{C}^M$ and a cut-off function $\theta \in \mathbf{C}^\infty(\mathbb{R})$ satisfying $\theta(x) = 1$ for $x \in [0, 1]$ and $\theta(x) = 0$ for $x \geq 2$ (see also [67, 136]). In addition, we have

$$\|F_R(U_R)\| = \theta \left(\frac{\|U_R\|}{R} \right) \max_{1 \leq j \leq M} |u_{R,j}|^2 \leq \theta \left(\frac{\|U_R\|}{R} \right) \|U_R\|^2 \leq 4R^2.$$

As a result, the nonlinear term $F_R(U_R)U_R$ is globally Lipschitz with respect to the norm $\|\cdot\|$. The proposed scheme (5.41) applied to the truncated Eq. (5.47) yields the following scheme

$$
\begin{aligned}
U_R^{n+1} &= \left(Id - \frac{i\tau}{2h^2}A + \frac{1}{4}\alpha\tau Id\right)^{-1}\left(Id + \frac{i\tau}{2h^2}A - \frac{1}{4}\alpha\tau Id\right)e^{f_R(U_R^n)}U_R^n \\
&+ \left(Id - \frac{i\tau}{2h^2}A + \frac{1}{4}\alpha\tau Id\right)^{-1}\varepsilon\sigma\Lambda\delta_{n+1}\beta,
\end{aligned}
\tag{5.48}
$$

where $f_R(U_R^n) := \left(-\frac{1}{2}\alpha Id + i\lambda F_R(U_R^n)\right)\tau$ and $U_R^n = (u_{R,1}^n, \cdots, u_{R,M}^n)^\top$.

Theorem 5.9 *For Eq. (5.47) and scheme (5.48), assume that $\mathbb{E}\|u_0\|_{\mathbf{L}^2}^2 < \infty$, $Q^{\frac{1}{2}} \in \mathscr{L}_2^2$, $\alpha \geq \frac{1}{2}$ and $\tau = O(h^4)$. For $T = N\tau$, there exists a constant C_R which depends on $\alpha, \varepsilon, R, Q, u_0$ and is independent of T and N such that*

$$
h\mathbb{E}\|U_R(T) - U_R^N\|^2 \leq C_R\tau^2.
$$

Proof Denote the semigroup operator $S(t) := e^{Et}$ which is generated by the linear operator $E := i\frac{1}{h^2}A - \frac{\alpha}{2}Id$, then the mild solution of (5.47) is

$$
\begin{aligned}
U_R(t_{n+1}) &= S(\tau)U_R(t_n) + \int_{t_n}^{t_{n+1}} S(t_{n+1} - s)i\lambda F_R(U_R(s))U_R(s)ds \\
&- \int_{t_n}^{t_{n+1}} S(t_{n+1} - s)\frac{\alpha}{2}U_R(s)ds + \int_{t_n}^{t_{n+1}} S(t_{n+1} - s)\varepsilon\sigma\Lambda d\beta(s).
\end{aligned}
\tag{5.49}
$$

Subtracting (5.48) from (5.49), we obtain

$$
\begin{aligned}
&U_R(t_{n+1}) - U_R^{n+1} \\
={}&S(\tau)U_R(t_n) - \left(Id - \frac{1}{2}E\tau\right)^{-1}\left(Id + \frac{1}{2}E\tau\right)e^{f_R(U_R^n)}U_R^n \\
&+ \int_{t_n}^{t_{n+1}} S(t_{n+1} - s)i\lambda F_R(U_R(s))U_R(s)ds - \int_{t_n}^{t_{n+1}} S(t_{n+1} - s)\frac{\alpha}{2}U_R(s)ds \\
&+ \int_{t_n}^{t_{n+1}} \left(S(t_{n+1} - s) - \left(Id - \frac{1}{2}E\tau\right)^{-1}\right)\varepsilon\sigma\Lambda d\beta(s) \\
={}&S(\tau)\left(U_R(t_n) - U_R^n\right) + \left[S(\tau) - \left(Id - \frac{1}{2}E\tau\right)^{-1}\left(Id + \frac{1}{2}E\tau\right)\right]U_R^n \\
&+ \left(Id - \frac{1}{2}E\tau\right)^{-1}\left(Id + \frac{1}{2}E\tau\right)\left(U_R^n - e^{f_R(U_R^n)}U_R^n\right) \\
&+ \int_{t_n}^{t_{n+1}} S(t_{n+1} - s)i\lambda F_R(U_R(s))U_R(s)ds - \int_{t_n}^{t_{n+1}} S(t_{n+1} - s)\frac{1}{2}\alpha U_R(s)ds
\end{aligned}
$$

$$+ \int_{t_n}^{t_{n+1}} \left(S(t_{n+1} - s) - \left(Id - \frac{1}{2} E\tau \right)^{-1} \right) \varepsilon \sigma \Lambda d\beta(s)$$

$$= : I + II + III + IV + V + VI.$$

To show the strong convergence order of (5.48), we give the estimates of above terms, respectively. For terms I and II, we have

$$\mathbb{E}\|I\|^2 = \mathbb{E} \left\| e^{(i \frac{1}{h^2} A - \frac{\alpha}{2} Id)\tau} (U_R(t_n) - U_R^n) \right\|^2 = e^{-\alpha\tau} \mathbb{E}\|U_R(t_n) - U_R^n\|^2 \qquad (5.50)$$

and

$$\mathbb{E}\|II\|^2 \leq C\mathbb{E}\|(E\tau)^3 U_R^n\|^2 \leq C\tau^6 \|E^3\|^2 \mathbb{E}\|U_R^n\|^2$$
$$\leq Ch^{-13}\tau^6 \|A\|^6 \leq Ch^{-13}\tau^6 \qquad (5.51)$$

based on $\left| e^x - (1 - \frac{x}{2})^{-1}(1 + \frac{x}{2}) \right| = O(x^3)$ as $x \to 0$, Lemma 5.3 and Proposition 5.6.

For term VI, the Taylor expansion yields that

$$\mathbb{E}\|VI\|^2 \leq 2\mathbb{E} \left\| \int_{t_n}^{t_{n+1}} (S(t_{n+1} - s) - S(\tau)) \varepsilon \sigma \Lambda d\beta(s) \right\|^2$$
$$+ 2\mathbb{E} \left\| \left(S(\tau) - \left(Id - \frac{1}{2} E\tau \right)^{-1} \right) \varepsilon \sigma \Lambda \delta_{n+1} \beta \right\|^2 \qquad (5.52)$$
$$\leq C\tau^2 \mathbb{E}\|E\varepsilon\sigma \Lambda \delta_{n+1}\beta\|^2 \leq Ch^{-1}\tau^3.$$

It then remains to estimate terms III, IV and V. We obtain the following equation in the same way as that of (5.43)

$$\|U_R^{n+1}\|^2 - e^{-\alpha\tau}\|U_R^n\|^2 \leq C\tau \|h^{-2} A\varepsilon\sigma \Lambda \delta_{n+1}\beta\|^2 + C\|\varepsilon\sigma \Lambda \delta_{n+1}\beta\|^2.$$

Multiplying the above equation by $\|U_R^{n+1}\|^2$, we derive

$$\|U_R^{n+1}\|^4 + \left(\|U_R^{n+1}\|^2 - e^{-\alpha\tau}\|U_R^n\|^2 \right)^2 - e^{-2\alpha\tau}\|U_R^n\|^4$$
$$\leq C\tau \left(\|U_R^{n+1}\|^2 - e^{-\alpha\tau}\|U_R^n\|^2 \right) \|h^{-2} A\varepsilon\sigma \Lambda \delta_{n+1}\beta\|^2$$
$$+ C\tau e^{-\alpha\tau}\|U_R^n\|^2 \|h^{-2} A\varepsilon\sigma \Lambda \delta_{n+1}\beta\|^2$$
$$+ C \left(\|U_R^{n+1}\|^2 - e^{-\alpha\tau}\|U_R^n\|^2 \right) \|\varepsilon\sigma \Lambda \delta_{n+1}\beta\|^2 + Ce^{-\alpha\tau}\|U_R^n\|^2 \|\varepsilon\sigma \Lambda \delta_{n+1}\beta\|^2$$
$$\leq \left(\|U_R^{n+1}\|^2 - e^{-\alpha\tau}\|U_R^n\|^2 \right)^2 + \tau e^{-2\alpha\tau}\|U_R^n\|^4 + C\tau \|h^{-2} A\varepsilon\sigma \Lambda \delta_{n+1}\beta\|^4$$
$$+ \frac{C}{\tau}\|\varepsilon\sigma \Lambda \delta_{n+1}\beta\|^4.$$

Based on (5.44) and (5.45), we take the expectation of the above equation and derive

$$\mathbb{E}\|U_R^{n+1}\|^4 \leq (1+\tau)e^{-2\alpha\tau}\mathbb{E}\|U_R^n\|^4 + Ch^{-2}\tau$$
$$\leq (1+\tau)^{n+1}e^{-2\alpha\tau(n+1)}\mathbb{E}\|U_R^0\|^4 + Ch^{-2} \leq Ch^{-2},$$

where we have used the fact that $\mathbb{E}\|U_R^0\|^4 \leq (\mathbb{E}\|U_R^0\|^2)^2 \leq Ch^{-2}$ and $(1+\tau)e^{-2\alpha\tau} < 1$ for $\alpha \geq \frac{1}{2}$. Similarly, we derive $\mathbb{E}\|U_R^n\|^8 \leq Ch^{-4}$, which implies that

$$\mathbb{E}\|F_R(U_R^n)\|^4 = \mathbb{E}\left(\sum_{j=1}^M |u_{R,j}^n|^4\right)^2 \leq \mathbb{E}\|U_R^n\|^8 \leq Ch^{-4}, \quad \forall n \in \mathbb{N}.$$

Thus, by the Taylor expansion, we have

$$III + IV + V \tag{5.53}$$

$$= \left(Id - \frac{1}{2}E\tau\right)^{-1}\left(Id + \frac{1}{2}E\tau\right)\left(-f_R(U_R^n) + O(f(U_R^n)^2)\right)U_R^n$$

$$+ \int_{t_n}^{t_{n+1}} S(t_{n+1}-s)i\lambda F_R(U_R(s))U_R(s)ds - \int_{t_n}^{t_{n+1}} S(t_{n+1}-s)\frac{1}{2}\alpha U_R(s)ds$$

$$= i\lambda \int_{t_n}^{t_{n+1}}\left[S(t_{n+1}-s)F_R(U_R(s))U_R(s) - \left(Id - \frac{1}{2}E\tau\right)^{-1}\left(Id + \frac{1}{2}E\tau\right)F_R(U_R^n)U_R^n\right]ds$$

$$- \frac{1}{2}\alpha \int_{t_n}^{t_{n+1}}\left[S(t_{n+1}-s)U_R(s) - \left(Id - \frac{1}{2}E\tau\right)^{-1}\left(Id + \frac{1}{2}E\tau\right)U_R^n\right]ds$$

$$+ \left(Id - \frac{1}{2}E\tau\right)^{-1}\left(Id + \frac{1}{2}E\tau\right)O(f_R(U_R^n)^2)U_R^n$$

$$=: \tilde{III} + \tilde{IV} + \tilde{V}. \tag{5.54}$$

Now we estimate the above terms one by one. For $\tilde{III}$, we have

$$\tilde{III} = i\lambda \int_{t_n}^{t_{n+1}}\left[S(t_{n+1}-s) - \left(Id - \frac{1}{2}E\tau\right)^{-1}\left(Id + \frac{1}{2}E\tau\right)\right]F_R(U_R^n)U_R^n ds$$

$$+ i\lambda \int_{t_n}^{t_{n+1}} S(t_{n+1}-s)\left[F_R(U_R(s))F_R(U_R(s)) - F_R(U_R^n)U_R^n\right]ds$$

$$=: \tilde{III}_1 + \tilde{III}_2,$$

which satisfies

$$\mathbb{E}\|\tilde{III}_1\|^2 \leq \tau \int_{t_n}^{t_{n+1}} \mathbb{E}\left\|\left[S(t_{n+1}-s) - \left(Id - \frac{1}{2}E\tau\right)^{-1}\left(Id + \frac{1}{2}E\tau\right)\right]F_R(U_R^n)U_R^n\right\|^2 ds$$

$$\leq Ch^{-12}\tau^8\mathbb{E}\|F_R(U_R^n)U_R^n\|^2 \leq Ch^{-12}\tau^8\mathbb{E}\left[\sum_{j=1}^M |u_{R,j}^n|^6\right] \leq Ch^{-15}\tau^8$$

and

$$\mathbb{E}\|\tilde{I}\tilde{I}_2\|^2 \leq \tau \int_{t_n}^{t_{n+1}} \mathbb{E}\left\| S(t_{n+1}-s)\left[F_R(U_R(s))F_R(U_R(s)) - F_R(U_R^n)U_R^n \right]\right\|^2 ds$$

$$\leq C\tau \int_{t_n}^{t_{n+1}} \mathbb{E}\|U_R(s) - U_R(t_n)\|^2 ds + C\tau^2 e^{-\alpha\tau}\mathbb{E}\|U_R(t_n) - U_R^n\|^2.$$

Note that

$$\mathbb{E}\|U_R(s) - U_R(t_n)\|^2 = \mathbb{E}\left\| \int_{t_n}^{s} S(s-r)\mathbf{i}\lambda F_R(U_R(r))U_R(r)dr - \int_{t_n}^{s} S(s-r)\frac{\alpha}{2}U_R(r)dr \right.$$

$$\left. + \int_{t_n}^{s} S(s-r)\varepsilon\sigma\, \Lambda d\beta(r) \right\|^2 \leq h^{-3}\tau^2.$$

We then get

$$\mathbb{E}\|\tilde{I}\tilde{I}\|^2 \leq Ch^{-15}\tau^8 + Ch^{-3}\tau^4 + C\tau^2 e^{-\alpha\tau}\mathbb{E}\|U_R(t_n) - U_R^n\|^2. \qquad (5.55)$$

Term $\tilde{IV}$ can be estimated in the same way as the estimation of $\tilde{III}$. Term $\tilde{V}$ turns to be

$$\mathbb{E}\|\tilde{V}\|^2 \leq C\mathbb{E}\left\| f_R(U_R^n)^2 U_R^n \right\|^2 \leq C\tau^4\mathbb{E}\left[\sup_{1\leq j\leq M}\left| -\frac{1}{2}\alpha + \mathbf{i}\lambda|u_{R,j}^n|^2\right|^4 \|U_R^n\|^2 \right]$$

$$\leq C\tau^4 \left(\mathbb{E}\left(\sum_{j=1}^{M}|u_{R,j}^n|^2\right)^8\right)^{\frac{1}{2}} \left(\mathbb{E}\|U_R^n\|^4\right)^{\frac{1}{2}} \leq Ch^{-5}\tau^4. \qquad (5.56)$$

From (5.50)–(5.56), we conclude

$$h\mathbb{E}\|U_R(t_{n+1}) - U_R^{n+1}\|^2$$

$$\leq h(1 + C\tau^2)e^{-\alpha\tau}\mathbb{E}\|U_R(t_n) - U_R^n\|^2 + C\tau^3 + Ch^{-4}\tau^4 + Ch^{-12}\tau^6 + Ch^{-14}\tau^8$$

$$\leq C\tau^2 + Ch^{-4}\tau^3 + Ch^{-12}\tau^5 + Ch^{-14}\tau^7 \leq C\tau^2,$$

where in the last two steps we have used the facts that $\tau = O(h^4)$ and $(1 + C\tau^2)e^{-\alpha\tau} < 1$ for sufficiently small τ.   □

Based on the estimates on the truncated equation and its numerical scheme, we are now in the position to give the approximate error between $U(t)$ and U^n. The proof of the following theorem is motivated by [67, 136] and holds for any fixed $T > 0$ without other restrictions.

Theorem 5.10 For Eq. (5.34) and scheme (5.41), assume that $\mathbb{E}\|u_0\|_{\mathbf{L}^2}^2 < \infty$, $Q^{\frac{1}{2}} \in \mathscr{L}_2^2$, $\alpha \geq \frac{1}{2}$ and $\tau = O(h^4)$. For any $T > 0$, scheme (5.41) converges with order one in probability, i.e.,

$$\lim_{C \to \infty} \mathbb{P}\left(\sup_{1 \le n \le [T/\tau]} \sqrt{h} \|U(t_n) - U^n\| \ge C\tau \right) = 0. \tag{5.57}$$

Proof For any $\gamma \in (0, 1)$, we define $n_\gamma := \inf\{1 \le n \le [T/\tau] : \|U(t_n) - U^n\| \ge \gamma\}$ and then deduce that

$$\left\{ \sup_{1 \le n \le [T/\tau]} \|U(t_n) - U^n\| \ge \gamma \right\}$$

$$\subset \left[\left(\left\{ \sup_{0 \le n \le n_\gamma} \|U(t_n)\| \ge R - 1 \right\} \cap \left\{ \sup_{1 \le n \le [T/\tau]} \|U(t_n) - U^n\| \ge \gamma \right\} \right) \right.$$

$$\left. \cup \left(\left\{ \sup_{0 \le n \le n_\gamma} \|U(t_n)\| < R - 1 \right\} \cap \left\{ \sup_{1 \le n \le [T/\tau]} \|U(t_n) - U^n\| \ge \gamma \right\} \right) \right]$$

$$\subset \left[\left\{ \sup_{0 \le n \le n_\gamma} \|U(t_n)\| \ge R - 1 \right\} \right.$$

$$\left. \cup \left(\left\{ \sup_{0 \le n \le n_\gamma} \|U(t_n)\| < R - 1 \right\} \cap \left\{ \sup_{1 \le n \le [T/\tau]} \|U(t_n) - U^n\| \ge \gamma \right\} \right) \right].$$

If $\left\{ \sup_{0 \le n \le n_\gamma} \|U(t_n)\| < R - 1 \right\}$ happens, it is easy to show that $\|U^n\| \le \|U(t_n) - U^n\| + \|U(t_n)\| < R - 1 + \gamma < R$, $F_R(U_R^k) = F(U_R^k)$, $U_R^k = U^n$ for $k = 0, 1, \cdots, n_\gamma - 1$ and $U_R(t_n) = U(t_n)$ for $0 \le n \le n_\gamma$. Furthermore, comparing scheme (5.48) with (5.41) and noting that

$$f_R(U_R^{n_\gamma - 1}) = \left(-\frac{1}{2} \alpha Id + i\lambda F_R(U_R^{n_\gamma - 1}) \right) \tau$$

$$= \left(-\frac{1}{2} \alpha Id + i\lambda F(U^{n_\gamma - 1}) \right) \tau = f(U^{n_\gamma - 1}),$$

we have $U_R^{n_\gamma} = U^{n_\gamma}$, which implies

$$\|U_R(t_{n_\gamma}) - U_R^{n_\gamma}\| = \|U(t_{n_\gamma}) - U^{n_\gamma}\| \ge \gamma.$$

We conclude that for any $\gamma \in (0, 1)$, there exists $n_\gamma \in \mathbb{N}$ such that

$$\left\{ \sup_{0 \le n \le n_\gamma} \|U(t_n)\| < R - 1 \right\} \cap \left\{ \sup_{1 \le n \le [T/\tau]} \|U(t_n) - U^n\| \ge \gamma \right\}$$

$$\subset \{\|U_R(t_{n_\gamma}) - U_R^{n_\gamma}\| \ge \gamma\}.$$

Thus, for some constants $C, C_1 > 0$, choosing $\gamma = \sqrt{h^{-1}}C\tau$ and $R = \sqrt{h^{-1}}C_1$, we deduce

$$
\mathbb{P}\left(\sup_{1 \le n \le [T/\tau]} \sqrt{h}\|U(t_n) - U^n\| \ge C\tau\right)
$$

$$
\le \mathbb{P}\left(\sup_{0 \le n \le n_\gamma} \sqrt{h}\|U(t_n)\| \ge C_1\right) + \mathbb{P}\left(\sqrt{h}\|U_R(t_{n_\gamma}) - U_R^{n_\gamma}\| \ge C\tau\right)
$$

$$
\le \frac{h\mathbb{E}\left[\sup_{0 \le n \le n_\gamma} \|U(t_n)\|^2\right]}{C_1^2} + \frac{h\mathbb{E}\|U_R(t_{n_\gamma}) - U_R^{n_\gamma}\|^2}{C^2\tau^2}. \tag{5.58}
$$

We claim that $e^{2\alpha t}\|U(t)\|^2$ is a submartingale, which ensures that

$$
h\mathbb{E}\left[\sup_{0 \le n \le n_\gamma} \|U(t_n)\|^2\right] \le h\mathbb{E}\left[\sup_{0 \le n \le n_\gamma} e^{2\alpha t_n}\|U(t_n)\|^2\right]
$$

$$
\le e^{2\alpha T}h\mathbb{E}\left[\|U(t_{n_\gamma})\|^2\right] \le Ce^{2\alpha T}
$$

based on a martingale inequality and Proposition 5.5. In fact, denoting $C_{M,K} := \sum_{j=1}^{M} \sum_{k=1}^{K} \eta_k e_k^2(x_j)$ and applying Itô's formula to $e^{2\alpha t}\|U(t)\|^2$ similar to (5.36), we derive

$$
e^{2\alpha t}\|U(t)\|^2 = \|U(0)\|^2 + 2\int_0^T e^{2\alpha s}\Re\left[\overline{U}(s)\varepsilon\sigma\Lambda d\beta(s)\right] + \frac{C_{M,K}}{\alpha}\left(e^{2\alpha t} - 1\right),
$$

where $2\int_0^T e^{2\alpha s}\Re\left[\overline{U}(s)\varepsilon\sigma\Lambda d\beta(s)\right]$ is a martingale. Apparently, we have

$$
\mathbb{E}\left[e^{2\alpha t}\|U(t)\|^2 \big| \mathscr{F}_r\right] = \|U(0)\|^2 + 2\int_0^r e^{2\alpha s}\Re\left[\overline{U}(s)\varepsilon\sigma\Lambda d\beta(s)\right] + \frac{C_{M,K}}{\alpha}\left(e^{2\alpha t} - 1\right)
$$

$$
\ge \|U(0)\|^2 + 2\int_0^r e^{2\alpha s}\Re\left[\overline{U}(s)\varepsilon\sigma\Lambda d\beta(s)\right] + \frac{C_{M,K}}{\alpha}\left(e^{2\alpha r} - 1\right)
$$

$$
= e^{2\alpha r}\|U(r)\|^2
$$

for $r \le t$, which completes the claim. Hence, based on the above claim and Theorem 5.9, inequality (5.58) turns to be

$$
\mathbb{P}\left(\sup_{1 \le n \le [T/\tau]} \sqrt{h}\|U(t_n) - U^n\| \ge C\tau\right) \le \frac{Ce^{2\alpha T}}{C_1^2} + \frac{C_R}{C^2},
$$

which approaches to 0 as $C_1, C \to +\infty$ for any $T > 0$. $\qquad\qquad\square$

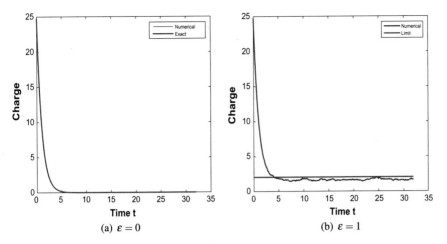

Fig. 5.4 Evolution of the discrete charge $h\mathbb{E}\|U^n\|^2$ with $t = n\tau$ for **a** $\varepsilon = 0$ and **b** $\varepsilon = 1$ ($h = 0.1$, $\tau = 2^{-6}$, $T = 32$)

5.2.3 Numerical Experiments

In this section, we provide several numerical experiments to illustrate the accuracy and capability of the fully discrete scheme (5.40), which can be calculated explicitly. We investigate the good performance in longtime simulation of the proposed scheme and check the temporal accuracy by fixing the space step-size. In the sequel, we take $\lambda = 1$, $\alpha = 0.5$, truncate the infinite series of Wiener process to $K = 100$ terms and choose 500 realizations to approximate the expectation.

Charge evolution. For the semi-discretization, the charge of the solution satisfies the evolution formula (5.37). To investigate the recurrence relation for the discrete charge of the fully discrete scheme, Fig. 5.4 plots the discrete charge for different values of ε with initial value $u_0(x) = \sin(\pi x)$, $\eta_k = k^{-6}$, $h = 1/(M + 1) = 0.1$, $\tau = 2^{-6}$ and $T = 32$. We can observe that the discrete charge inherits the charge dissipation law without the noise term, i.e., $\varepsilon = 0$, and preserves the charge dissipation law approximately with a limit $\frac{\varepsilon^2 h}{\alpha} \sum_{j=1}^{M} \sum_{k=1}^{K} \eta_k e_k^2(x_j)$ calculated through (5.37) for $\varepsilon = 1$.

Ergodic limit. Based on the definition of ergodicity, if numerical solution U^n is ergodic, its temporal averages $\frac{1}{N} \sum_{n=1}^{N-1} \mathbb{E}[f(U^n)]$ starting from different initial values will converge to the spatial average $\int_{\mathbb{C}^M} f d\mu_h^\tau$. To verify this property, Fig. 5.5 shows the temporal averages of the fully discrete scheme starting from five different initial values

$$\text{initial}(1) = (1, \, 0, \, \cdots, \, 0)^\top,$$
$$\text{initial}(2) = (0.0003\mathbf{i}, \, 0, \, \cdots, \, 0)^\top,$$

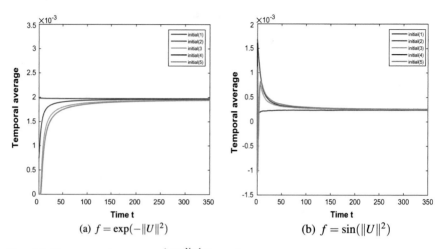

Fig. 5.5 The temporal averages $\frac{1}{N}\sum_{n=1}^{N-1}\mathbb{E}[f(U^n)]$ starting from different initial values for bounded functions **a** $f = \exp(-\|U\|^2)$ and **b** $f = \sin(\|U\|^2)$ ($h = 0.1$, $\varepsilon = 1$, $\tau = 2^{-6}$, $T = 350$)

$$\text{initial}(3) = (\sin\left(\frac{1}{101}\pi\right), \sin\left(\frac{2}{101}\pi\right), \cdots, \sin\left(\frac{100}{101}\pi\right))^\top,$$

$$\text{initial}(4) = \frac{2+\mathbf{i}}{20}(1, 2, \cdots, 100)^\top,$$

$$\text{initial}(5) = (\exp(-\frac{\mathbf{i}}{50}), \exp(-\frac{2\mathbf{i}}{50}), \cdots, \exp(-\frac{100\mathbf{i}}{50}))^\top.$$

These temporal averages tend to the same value with test functions: (a) $f(U) = \exp(-\|U\|^2)$, (b) $f(U) = \sin(\|U\|^2)$.

Weak error. As stated in Theorems 5.9 and 5.10, the mean-square convergence error $\left(h\mathbb{E}\|U_R(T) - U_R^N\|^2\right)^{\frac{1}{2}}$ with respect to the truncated Eq. (5.47) is independent of time T, and convergence in probability sense with respect to the original equation is also independent of time T. To clarify this property, we define the mean-square convergence error as

$$\mathscr{E}_{h,\tau} := \left(h\mathbb{E}\|U(T) - U^N\|^2\right)^{\frac{1}{2}}, \quad T = N\tau.$$

Figure 5.6 displays the error $\mathscr{E}_{h,\tau}$ till time $T = 10^3$ for different time step-sizes: (a) $\tau = 2^{-8}$ and (b) $\tau = 2^{-10}$ with $h = 0.25$, and shows that the mean-square convergence error is independent of time interval, which coincides with our theoretical results.

Convergence order. We investigate the mean-square convergence order in temporal direction of the proposed method (5.40) in this experiment. Let $h = 0.1$, $T = 1$ and the initial value $u_0(x) = \sin(\pi x)$. We plot $\mathscr{E}_{h,\tau}$ against τ on a log-log scale with

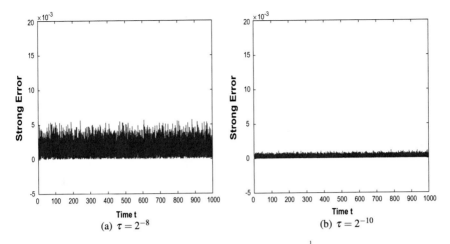

Fig. 5.6 The mean-square convergence error $\left(h\mathbb{E}\|U(T) - U^N\|^2\right)^{\frac{1}{2}}$ for step-sizes **a** $\tau = 2^{-8}$ and **b** $\tau = 2^{-10}$ ($h = 0.25$, $\varepsilon = 1$, $T = 10^3$)

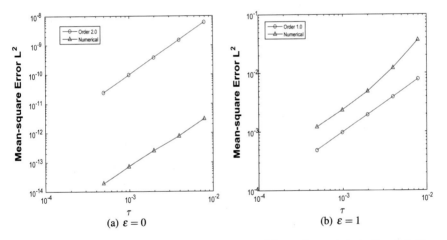

Fig. 5.7 Rates of convergence of (5.40) for **a** $\varepsilon = 0$ and **b** $\varepsilon = 1$, respectively ($h = 0.1$, $T = 1$, $\tau = 2^{-l}$, $11 \leq l \leq 14$)

various combinations of (α, ε) and take method (5.40) with small time step-size $\tau = 2^{-16}$ as the reference solution. We then compare it with method (5.40) evaluated with time step-sizes $(2^2\tau, 2^3\tau, 2^4\tau, 2^5\tau)$ in order to show the rate of convergence. Figure 5.7 presents the mean-square convergence order for the error $\mathscr{E}_{h,\tau}$ with various sizes of ε. Figure 5.7 shows that the proposed scheme (5.40) is of order 2 for the deterministic case, i.e., $\varepsilon = 0$, and of order 1 for the stochastic case with $\varepsilon = 1$, which coincides with the theoretical analysis.

Summary

This chapter focuses on the numerical approximations of the invariant measure for the damped stochastic NLSE. An ergodic full discretization combining the spectral Galerkin method in space and modified implicit Euler scheme in time is proposed in Sect. 5.1. The error for invariant measures is given based on the weak error between the exact solution and the numerical one. As for the damped stochastic NLSE, it possesses the conformal symplectic conservation law, an ergodic scheme with the discrete conformal symplectic structure is then constructed in Sect. 5.2 to inherit the internal properties of the original system as more as possible.

We would like to mention that the error between invariant measures, for ergodic systems and schemes, can also be derived directly utilizing the expansion of the solution of the Kolmogorov equation without requiring that the weak error is independent of time interval. We refer to [178, 180] for the analysis of the Euler and Milstein schemes. This idea could be modified to analyze the approximation error of the ergodic limit in Chap. 6.

Except for the approximation of invariant measures, there is also a lot of work related to the density for solutions of SDEs and error analysis between the density for exact solution and that for the numerical one. We refer to [12, 24, 42, 151, 153] for the study of densities for both finite and infinite dimensional stochastic systems, and to [13, 14] for the convergence of the density for the Euler scheme applied to SDEs under the Hörmander condition. The existence, regularity and convergence of the density for numerical schemes of SPDEs are also interesting and important topics, and are needed further investigations.

Chapter 6
Approximation of Ergodic Limit for Conservative Stochastic Nonlinear Schrödinger Equations

This chapter is mainly developed to consider stochastic NLSEs possessing the stochastic multi-symplectic conservation law and the charge conservation law (see e.g. [64, 119]). For this kind of conservative equations, it then suffices to consider its dynamic behavior on the unit sphere without loss of generality. We show in Sect. 6.1 that the finite dimensional approximation (FDA) based on the midpoint scheme is ergodic with a unique invariant measure. Also, the ergodic limit of this FDA can be approximated via the temporal average of an ergodic fully discrete scheme (FDS), see Sect. 6.3. We also prove in Sect. 6.2 that the FDS could inherit the internal properties, i.e., charge conservation law and multi-symplecticity, of the original system.

6.1 Finite Dimensional Ergodic Approximation

In Sects. 6.1 and 6.2, we concentrate on the stochastic NLSE on a bounded interval $\mathscr{O} = (0, 1)$ with $\sigma = 1$:

$$\begin{cases} du = \mathbf{i}\big(\Delta u + \lambda|u|^2u\big)dt + \mathbf{i}u \circ dW, \\ u(t, 0) = u(t, 1) = 0, \ t > 0, \\ u(0, x) = u_0(x), \ x \in (0, 1). \end{cases} \tag{6.1}$$

Here, $\lambda = \pm 1$, and W is a Q-Wiener process on $\mathbf{L}^2(\mathscr{O}; \mathbb{R})$ under the Dirichlet boundary condition. The covariance operator Q is symmetric and positive definite such that the Karhunen–Loève expansion of W is as follows

$$W = \sum_{k=1}^{\infty} Q^{\frac{1}{2}}e_k\beta_k = \sum_{k=1}^{\infty} \sqrt{\eta_k}e_k\beta_k, \quad \eta_k > 0, \ k \in \mathbb{N},$$

© Springer Nature Singapore Pte Ltd. 2019
J. Hong and X. Wang, *Invariant Measures for Stochastic Nonlinear Schrödinger Equations*, Lecture Notes in Mathematics 2251,
https://doi.org/10.1007/978-981-32-9069-3_6

where $\{e_k := \sqrt{2}\sin(k\pi \cdot)\}_{k\in\mathbb{N}}$ is an orthogonal basis for $\mathbf{L}^2(\mathcal{O}; \mathbb{R})$ with homogeneous boundary condition and $\{\beta_k\}_{k\in\mathbb{N}}$ is a sequence of $\mathbb{R}$-valued mutually independent and identically distributed Brownian motions. In addition, Q is assumed to commute with the Laplacian and hence $\{e_k\}_{k\in\mathbb{N}}$ is also an eigenbasis of the Dirichlet Laplacian Δ in $\mathbf{L}^2(\mathcal{O}; \mathbb{R})$.

The well-posedness, as well as the corresponding assumptions on u_0 and Q, is given in Sect. 3.3.2. Now we turn our attention to the numerical approximation of (6.1).

To investigate longtime behaviors of (6.1) numerically, we first apply the central finite difference scheme to (6.1) in the spatial variable to obtain an FDA, which is also a Hamiltonian system, in a finite dimensional space $\mathbb{R}^M$ for some $M \in \mathbb{N}_+$. It is worth noticing that the charge conservation law for the FDA also holds, which allows us to consider properties of the FDA restricted onto the unit sphere $\mathbb{S} \subset \mathbb{R}^M$. When showing the ergodicity of this conservative system, we construct an invariant control set $\mathbb{M}_0 \subset \mathbb{S}$ with respect to a control function introduced in Sect. 6.1.2. The FDA is proved to be ergodic in $\mathbb{M}_0$ based on the Krylov–Bogoliubov theorem and the Hörmander condition.

6.1.1 Finite Dimensional Approximation

Based on the central finite difference scheme and the notation $u_j := u_j(t)$, $j = 1, \cdots, M$, we consider the following spatial semi-discretization

$$du_j = \mathbf{i}\left[\frac{u_{j+1} - 2u_j + u_{j-1}}{h^2} + \lambda|u_j|^2 u_j\right]dt + \mathbf{i}u_j\sum_{k=1}^{K}\sqrt{\eta_k}e_k(x_j) \circ d\beta_k(t)$$

with a truncated noise $\sum_{k=1}^{K}\sqrt{\eta_k}e_k(x)\beta_k(t)$, $K \in \mathbb{N}_+$, a given uniform spatial stepsize $h = \frac{1}{M+1}$ for some $M \leq K$ and $x_j = jh$, $j = 1, \cdots, M$. The condition $M \leq K$ here ensures the existence of the solution for the control function. Denoting vectors $U := U(t) = (u_1, \cdots, u_M)^\top \in \mathbb{C}^M$, $\beta(t) = (\beta_1(t), \cdots, \beta_K(t))^\top \in \mathbb{R}^K$, and matrices $F(U) = \mathrm{diag}\{|u_1|^2, \cdots, |u_M|^2\}$, $E_k = \mathrm{diag}\{e_k(x_1), \cdots, e_k(x_M)\}$, $\Lambda = \mathrm{diag}\{\sqrt{\eta_1}, \cdots, \sqrt{\eta_K}\}$, $Z(U) = \mathrm{diag}\{u_1, \cdots, u_M\}E_{MK}\Lambda$,

$$A = \begin{pmatrix} -2 & 1 & & \\ 1 & -2 & 1 & \\ & \ddots & \ddots & \ddots \\ & & 1 & -2 \end{pmatrix} \in \mathbb{R}^{M\times M}, \quad E_{MK} = \begin{pmatrix} e_1(x_1) & \cdots & e_K(x_1) \\ \vdots & & \vdots \\ e_1(x_M) & \cdots & e_K(x_M) \end{pmatrix}_{M\times K},$$

then the FDA is in the following form

$$\begin{cases} dU = \mathrm{i}\left[\dfrac{1}{h^2}AU + \lambda F(U)U\right]dt + \mathrm{i}Z(U)\circ d\beta(t), \\ U(0) = c_*\,(u_0(x_1),\cdots,u_0(x_M))^{\top}, \end{cases} \tag{6.2}$$

where c_* is a normalized constant such that $U(0) \in \mathbb{S}$. The noise term in (6.2) has an equivalent Itô form

$$\mathrm{i}Z(U)\circ d\beta(t) = \mathrm{i}\sum_{k=1}^{K}\sqrt{\eta_k}\,E_k U\circ d\beta_k(t) = -\frac{1}{2}\sum_{k=1}^{K}\eta_k E_k^2 U dt + \mathrm{i}\sum_{k=1}^{K}\sqrt{\eta_k}\,E_k U d\beta_k(t)$$

$$=: -\hat{E}U dt + \mathrm{i}\sum_{k=1}^{K}\sqrt{\eta_k}\,E_k U d\beta_k(t) \tag{6.3}$$

with $\hat{E} = \frac{1}{2}\sum_{k=1}^{K}\eta_k E_k^2$.

In the sequel, denote by $\|\cdot\|$ the Euclidean norm for both matrices and vectors, which satisfies $\|BV\| \le \|B\|\|V\|$ for any matrices $B \in \mathbb{C}^{m\times n}$ and vectors $V \in \mathbb{C}^n$, $m, n \in \mathbb{N}$, and denote by $\|\cdot\|_F$ the Frobenius norm for matrices. Matrix A is bounded in $\|\cdot\|$-norm independent of M as stated in Lemma 5.3.

Proposition 6.1 *FDA (6.2) possesses the charge conservation law, i.e.,*

$$\|U(t)\|^2 = \|U(0)\|^2, \quad \forall\, t \ge 0, \quad \mathbb{P}\text{-}a.s.,$$

where $\|U(t)\| = (\|P(t)\|^2 + \|Q(t)\|^2)^{\frac{1}{2}} = \left(\sum_{m=1}^{M}(|p_m(t)|^2 + |q_m(t)|^2)\right)^{\frac{1}{2}}$, $P(t) = (p_1(t),\cdots,p_M(t))^{\top}$ *and* $Q(t) = (q_1(t),\cdots,q_M(t))^{\top}$ *are the real and imaginary parts of* $U(t)$ *respectively.*

Proof Noticing that matrices A and $F(U)$ are symmetric and the linear function $Z(U)$ satisfies

$$\overline{U}^{\top} Z(U) = (\overline{u_1},\cdots,\overline{u_M})\begin{pmatrix} u_1 & & \\ & \ddots & \\ & & u_M \end{pmatrix} E_{MK}\begin{pmatrix} \sqrt{\eta_1} & & \\ & \ddots & \\ & & \sqrt{\eta_K} \end{pmatrix}$$

$$= (|u_1|^2,\cdots,|u_M|^2)E_{MK}\begin{pmatrix} \sqrt{\eta_1} & & \\ & \ddots & \\ & & \sqrt{\eta_K} \end{pmatrix} \in \mathbb{R}^{K}, \tag{6.4}$$

where $\overline{U}$ denotes the conjugate of U, we multiply (6.2) by $\overline{U}^{\top}$, take the real part, and then get the charge conservation law for U. $\qquad\square$

In the sequel, without pointing it out explicitly, all equations hold in the sense $\mathbb{P}$-a.s.

Remark 6.1 Equation (6.1) can be rewritten into an infinite dimensional Hamiltonian system (see [119]). It is easy to verify that the central finite difference scheme (6.2) applied to (6.1) is equivalent to the symplectic Euler scheme applied to the infinite dimensional Hamiltonian form of (6.1), which implies the symplecticity of (6.2).

6.1.2   Unique Ergodicity

As the charge of (6.2) is conserved shown in Proposition 6.1, for any fixed initial value $U(0)$, the solution $U(t)$ of (6.2) is trapped in the equipotential surface, and there will not be a unique invariant measure on the whole domain. Without loss of generality, we assume that $U(0) \in \mathbb{S}$ and only investigate the unique ergodicity of (6.2) on $\mathbb{S}$. The most challenging thing is that the diffusion term $Z(U)$ depends linearly on the solution U. As a result, the Hörmander condition, which is frequently used to show the ergodicity of SDEs with multiplicative noises, does not uniformly hold for $U \in \mathbb{S}$. It depends on the number of zero entries of the vector U. Instead, we need to construct an invariant control set, which is the union of several disjoint subsets, such that the Hörmander condition holds in each subset.

Definition 6.1 *(see e.g. [9])* A subset $\mathbb{M} \neq \varnothing$ of $\mathbb{S}$ is called an invariant control set for the control system

$$d\phi = \mathbf{i}\left[\frac{1}{h^2}A\phi + \lambda F(\phi)\phi\right]dt + \mathbf{i}Z(\phi)d\Psi(t) \tag{6.5}$$

of (6.2) with a differentiable deterministic function Ψ, if $\overline{\mathcal{O}^+(x)} = \overline{\mathbb{M}}$ for any $x \in \mathbb{M}$, and $\mathbb{M}$ is maximal with respect to inclusion, where $\mathcal{O}^+(x)$ denotes the set of points reachable from x (i.e., connected with x) in any finite time and $\overline{\mathbb{M}}$ denotes the closure of $\mathbb{M}$.

The uniqueness of the invariant measure on an invariant set is given through the following theorem, which will be used to show the uniqueness of the invariant measure for FDA (6.2).

Theorem 6.1 *(Theorem 5.1, [9]) Let $\mathbb{M}$ be an invariant control set. Assume that there exists a point $x_0 \in \mathbb{M}$ such that the Hörmander condition given in Theorem 2.2 holds for x_0. Then there is at most one invariant probability measure μ with $\mathrm{supp}\mu = \overline{\mathbb{M}}$ and $\mu(\mathbb{M}) = 1$.*

We state one of our main results in the following theorem.

Theorem 6.2 *FDA (6.2) possesses a unique invariant probability measure μ_h on an invariant control set $\mathbb{M}_0$ with*

$$\mathrm{supp}(\mu_h) = \mathbb{S} \quad and \quad \mu_h(\mathbb{M}_0) = 1,$$

which implies the ergodicity of (6.2).

Proof **Step 1. Existence of Invariant Measures**
From Proposition 6.1, we find $\pi_t(U(0), \mathbb{S}) = 1$ for all $t \geq 0$, where $\pi_t(U(0), \cdot)$ denotes the transition probability (probability kernel) of $U(t)$. As the finite dimensional unit sphere $\mathbb{S}$ is tight, the family of measures $\{\pi_t(U(0), \cdot)\}_{t \geq 0}$ is tight, which implies the existence of invariant measures by the Krylov–Bogoliubov theorem [58].

Step 2. Invariant Control Set
Denote $U = P + \mathbf{i}Q$ with $P = (P_1, \cdots, P_M)^\top$, $Q = (Q_1, \cdots, Q_M)^\top \in \mathbb{R}^M$ being the real and imaginary parts of U, respectively. Note that the subset

$$\mathbb{S}_0 := \{U \in \mathbb{S} : \exists\, 1 \leq i \leq M \text{ s.t. } P_i = 0 \text{ or } Q_i = 0\}$$

is a union of finite number of lower dimensional unit spheres. We have $m(\mathbb{S}_0) = 0$ with $m(\cdot)$ being the Lebesgue measure in $\mathbb{R}^M$. We denote further $X_i := P_i$ and $X_{M+i} = Q_i$ for $1 \leq i \leq M$ for convenience.

We first consider the following subset of $\mathbb{S}$ as an example:

$$\mathbb{S}_{(1, \cdots, M)} := \{U \in \mathbb{S}\backslash\mathbb{S}_0 : X_i > 0,\ 1 \leq i \leq M\}.$$

For any $t > 0$, $y, z \in \mathbb{S}_{(1, \cdots, M)}$, there exists a differentiable function ϕ satisfying $\phi(s) = (\phi_1(s), \cdots, \phi_M(s))^\top \in \mathbb{S}_{(1, \cdots, M)}, s \in [0, t], \phi(0) = y$ and $\phi(t) = z$ by polynomial interpolation argument. As $\text{rank}(Z(\phi(s))) = M$ for $\phi(s) \in \mathbb{S}_{(1, \cdots, M)}$ and $M \leq K$, the linear system

$$Z(\phi(s))X = -\mathbf{i}\phi'(s) - \left[\frac{1}{h^2}A\phi(s) + \lambda F(\phi(s))\phi(s)\right]$$

possesses a solution $X \in \mathbb{C}^M$. As in addition $Z(\phi(s)) = \text{diag}\{\phi_1(s), \cdots, \phi_M(s)\}$ $E_{MK}\Lambda$, where $\text{diag}\{\phi_1(s), \cdots, \phi_M(s)\}$ is invertible for $\phi(s) \in \mathbb{S}_{(1, \cdots, M)}$, the solution X depends continuously on s and is denoted by $X(s)$. Thus, there exists a differentiable function $\Psi(\cdot) := \int_0^\cdot X(s)ds$ which, together with ϕ defined above, satisfies the control system (6.5) with the initial datum $\Psi(0) = 0$. That is, for any $y, z \in \mathbb{S}_{(1, \cdots, M)}$, y and z are connected, denoted by $y \leftrightarrow z$. We also call $\mathbb{S}_{(1, \cdots, M)}$ a self-connected set. The above argument also holds for the following subsets

$$\mathbb{S}_\sigma^+ := \{U \in \mathbb{S}\backslash\mathbb{S}_0 : \sigma = (\sigma_1, \cdots, \sigma_M),\ X_{\sigma_k} > 0,\ k = 1, \cdots, M\},$$
$$\mathbb{S}_\sigma^- := \{U \in \mathbb{S}\backslash\mathbb{S}_0 : \sigma = (\sigma_1, \cdots, \sigma_M),\ X_{\sigma_k} < 0,\ k = 1, \cdots, M\},$$

where

$$\sigma \in \Sigma := \{v = (v_1, \cdots, v_M) : 1 \leq v_1 < v_2 < \cdots < v_M \leq 2M\}.$$

That is, each element in the collection $\{\mathbb{S}_\sigma^+, \mathbb{S}_\sigma^-\}_{\sigma \in \Sigma}$ is self-connected.

Moreover, for any two different elements $\mathbb{S}_1, \mathbb{S}_2 \in \{\mathbb{S}_\sigma^+, \mathbb{S}_\sigma^-\}_{\sigma \in \Sigma}$ and any $y \in \mathbb{S}_1$, $z \in \mathbb{S}_2$, there exist $\mathbb{S}_3 \in \{\mathbb{S}_\sigma^+, \mathbb{S}_\sigma^-\}_{\sigma \in \Sigma}$, r_1 and r_2, satisfying $r_1 \in \mathbb{S}_1 \cap \mathbb{S}_3 \neq \varnothing$, $r_2 \in \mathbb{S}_2 \cap \mathbb{S}_3 \neq \varnothing$ and $y \leftrightarrow r_1 \leftrightarrow r_2 \leftrightarrow z$. Thus,

$$\mathbb{M}_0 := \bigcup_{\sigma \in \Sigma} (\mathbb{S}_\sigma^+ \cup \mathbb{S}_\sigma^-) = \mathbb{S} \backslash \mathbb{S}_0$$

is an invariant control set for (6.5) with $\overline{\mathbb{M}_0} = \mathbb{S}$ since it is the maximal set in which any two points are connected.

Step 3. Uniqueness of the Invariant Measure

We rewrite (6.2) with P and Q according to its equivalent form in the Itô sense and obtain

$$d\begin{pmatrix} P \\ Q \end{pmatrix} = \begin{pmatrix} -\hat{E} & -\frac{1}{h^2}A - \lambda F(P, Q) \\ \frac{1}{h^2}A + \lambda F(P, Q) & -\hat{E} \end{pmatrix} \begin{pmatrix} P \\ Q \end{pmatrix} dt$$

$$+ \sum_{k=1}^{K} \sqrt{\eta_k} \begin{pmatrix} 0 & -E_k \\ E_k & 0 \end{pmatrix} \begin{pmatrix} P \\ Q \end{pmatrix} d\beta_k(t)$$

$$=: X_0(P, Q)dt + \sum_{k=1}^{K} X_k(P, Q)d\beta_k(t). \tag{6.6}$$

To derive the uniqueness of the invariant measure, we consider the Lie algebra generated by the diffusions of (6.6)

$$L(X_0, X_1, \cdots, X_K) = \text{span}\left\{ X_l, [X_i, X_j], \big[X_l, [X_i, X_j]\big], \cdots, 0 \leq l, i, j \leq K \right\}.$$

Choosing $p_* = 0$ and $q_* = \frac{-1}{\sqrt{M}}(1, \cdots, 1)^\top$ such that $z_* := p_* + iq_* \in \mathbb{S}_{(M+1, \cdots, 2M)}^- \subset \mathbb{M}_0$, we derive that the following vectors

$$X_k(p_*, q_*) = \sqrt{\frac{\eta_k}{M}} \begin{pmatrix} e_k(x_1) \\ \vdots \\ e_k(x_M) \\ 0 \\ \vdots \\ 0 \end{pmatrix}, \quad [X_0, X_k](p_*, q_*) = \sqrt{\frac{\eta_k}{M}} \begin{pmatrix} -\hat{E}\begin{pmatrix} e_k(x_1) \\ \vdots \\ e_k(x_M) \end{pmatrix} \\ (\frac{1}{h^2}A + \frac{1}{M}Id)\begin{pmatrix} e_k(x_1) \\ \vdots \\ e_k(x_M) \end{pmatrix} \end{pmatrix}$$

are independent of each other for $k = 1, \cdots, M$, which hence implies the following Hörmander condition

$$\dim L(X_0, X_1, \cdots, X_K)(z_*) = 2M.$$

Then there is at most one invariant measure with $\text{supp}(\mu_h) = \mathbb{S}$ according to Theorem 6.1. Actually, according to the above procedure, we obtain that Hörmander condition holds uniformly for any $z \in \mathbb{M}_0$.

Combining the three steps above, we conclude that there exists a unique invariant measure μ_h on $\mathbb{M}_0$ for the FDA, with $\mu_h(\mathbb{M}_0) = 1$.   □

For some other nonlinearities such that the equation still possesses the charge conservation law, e.g., $iF(x, |u|)u$ with F being some real valued potential function, we can still get the ergodicity of the finite dimensional approximation of the original equation through the procedure used in Theorem 6.2. The procedure could also applied to higher dimensional Schrödinger equations with proper well-posed assumptions, but it may be more technical to verify the Hörmander condition.

Remark 6.2 According to the ergodicity of (6.2) and noticing that $1 = \mu_h(\mathbb{M}_0) \le \mu_h(\mathbb{S}) \le 1$, we have

$$\lim_{T \to \infty} \frac{1}{T} \int_0^T \mathbb{E} f(U(t)) dt = \int_{\mathbb{S}} f d\mu_h, \quad \forall f \in \mathbf{B}_b(\mathbb{S}), \quad \text{in } \mathbf{L}^2(\mathbb{S}, \mu_h),$$

where $\int_{\mathbb{S}} f d\mu_h$ is known as the ergodic limit with respect to the invariant measure μ_h.

6.2 Multi-symplectic Ergodic Fully Discrete Scheme

An FDS with the discrete multi-symplectic structure and the discrete charge conservation law is constructed in this section, which also inherits the unique ergodicity of the FDA. More precisely, we apply the midpoint scheme to (6.2), and obtain the following FDS

$$\begin{cases} U^{n+1} - U^n = i\dfrac{\tau}{h^2} A U^{n+\frac{1}{2}} + i\lambda\tau F(U^{n+\frac{1}{2}})U^{n+\frac{1}{2}} + iZ(U^{n+\frac{1}{2}})\delta_{n+1}\beta, \\ U^0 = U(0) \in \mathbb{S}, \end{cases} \quad (6.7)$$

where τ denotes the uniform time step-size, $t_n = n\tau$, $U^n = (u_1^n, \cdots, u_M^n) \in \mathbb{C}^M$, $U^{n+\frac{1}{2}} = \frac{1}{2}(U^{n+1} + U^n)$ and $\delta_{n+1}\beta = \beta(t_{n+1}) - \beta(t_n)$. For FDS (6.7), which is implicit in both drift and diffusion terms, its well-posedness is stated in the following proposition, and it converges to FDA (6.2) in Stratonovich sense since it is consistent with the Stratonovich rule.

Proposition 6.2 *For any initial value $U^0 = U(0) \in \mathbb{S}$, there exists a unique solution $\{U^n\}_{n\in\mathbb{N}}$ of (6.7), and it possesses the discrete charge conservation law, i.e.,*

$$\|U^{n+1}\|^2 = \|U^n\|^2 = 1, \quad \forall n \in \mathbb{N}.$$

Proof We multiply both sides of (6.7) by $\overline{U^{n+\frac{1}{2}}}$, take the real part, and obtain the existence of the numerical solution by Brouwer's fixed point theorem as well as the discrete charge conservation law.

For the uniqueness, we assume that $X = (X_1, \cdots, X_M)^\top$ and $Y = (Y_1, \cdots, Y_M)^\top$ are two solutions of (6.7) with $U^n = z = (z_1, \cdots, z_M)^\top \in \mathbb{S}$. It follows that $X, Y \in \mathbb{S}$ and

$$X - Y = i\frac{\tau}{h^2} A \frac{X - Y}{2} + \frac{i\lambda\tau}{8} H(X, Y, z) + iZ \left(\frac{X - Y}{2} \right) \delta_{n+1}\beta, \qquad (6.8)$$

where

$$H(X, Y, z) = \begin{pmatrix} |X_1 + z_1|^2 (X_1 + z_1) - |Y_1 + z_1|^2 (Y_1 + z_1) \\ \vdots \\ |X_M + z_M|^2 (X_M + z_M) - |Y_M + z_M|^2 (Y_M + z_M) \end{pmatrix}.$$

Based on the fact that $|a|^2 a - |b|^2 b = |a|^2 (a - b) + |b|^2 (a - b) + ab(\bar{a} - \bar{b})$ for any $a, b \in \mathbb{C}$, we have

$$\Im \left[(\overline{X} - \overline{Y})^\top H(X, Y, z) \right] = \Im \left[\sum_{m=1}^{M} (X_m + z_m)(Y_m + z_m)(\overline{X_m} - \overline{Y_m})^2 \right]$$

with $\Im[V]$ denoting the imaginary part of V. Multiplying (6.8) by $(\overline{X} - \overline{Y})^\top$ and taking the real part, we obtain

$$\|X - Y\|^2 = -\frac{\lambda\tau}{8} \Im \left[(\overline{X} - \overline{Y})^\top H(X, Y, z) \right]$$

$$\leq \frac{\tau}{8} \left(\max_{1 \leq m \leq M} |X_m + z_m||Y_m + z_m| \right) \|X - Y\|^2 \leq \frac{\tau}{2}\|X - Y\|^2,$$

where we have used the fact $X, Y, z \in \mathbb{S}$ and (6.4). For $\tau < 1$, we get $X = Y$ and complete the proof. □

The proposition above shows that (6.7) possesses the discrete charge conservation law. Furthermore, (6.7) also inherits the unique ergodicity of the FDA and the stochastic multi-symplecticity of the original equation, which are stated in the following two theorems.

Theorem 6.3 *FDS (6.7) is also ergodic with a unique invariant measure μ_h^τ on the control set $\mathbb{M}_0$, such that $\mu_h^\tau(\mathbb{M}_0) = 1$. Also,*

$$\lim_{N \to \infty} \frac{1}{N} \sum_{n=0}^{N-1} f(U^n) = \int_{\mathbb{S}} f d\mu_h^\tau, \quad \forall f \in \mathbf{B}_b(\mathbb{S}), \quad in \ \mathbf{L}^2(\mathbb{S}, \mu_h^\tau).$$

Proof Based on the charge conservation law for $\{U^n\}_{n\in\mathbb{N}}$, we obtain the existence of the invariant measure similar to the proof of Theorem 6.2.

To obtain the uniqueness of the invariant measure, we show that the Markov chain $\{U^{3n}\}_{n\in\mathbb{N}}$ satisfies Assumption 2.2. Firstly, Proposition 6.2 implies that for a given $U^n \in \mathbb{S}$, solution U^{n+1} can be defined through a continuous function $U^{n+1} = \kappa(U^n, \delta_{n+1}\beta)$. As $\delta_{n+1}\beta$ has a $\mathbf{C}^\infty$ density, we derive a jointly continuous density for U^{n+1}. Secondly, similar to Theorem 6.2, for any given $y, z \in \mathbb{M}_0$, there must exist $\sigma_1, \sigma_2, \sigma_3 \in \Sigma$ and $r_1, r_2 \in \mathbb{M}_0$, such that $y \in \mathbb{S}_{\sigma_1}, z \in \mathbb{S}_{\sigma_2}, r_1 \in \mathbb{S}_{\sigma_1} \cap \mathbb{S}_{\sigma_3}$ and $r_2 \in \mathbb{S}_{\sigma_2} \cap \mathbb{S}_{\sigma_3}$. As $\frac{y+r_1}{2} \in \mathbb{S}_{\sigma_1}$ and $Z(\frac{y+r_1}{2})$ is invertible, $\delta_{3n+1}\beta$ can be chosen to ensure that

$$r_1 - y = \mathrm{i}\frac{\tau}{h^2}A\frac{y+r_1}{2} + \mathrm{i}\lambda\tau F\left(\frac{y+r_1}{2}\right)\frac{y+r_1}{2} + \mathrm{i}Z\left(\frac{y+r_1}{2}\right)\delta_{3n+1}\beta$$

holds, i.e., $r_1 = \kappa(y, \delta_{3n+1}\beta)$. Similarly, based on the fact $\frac{r_1+r_2}{2} \in \mathbb{S}_{\sigma_k}$ and $\frac{r_2+z}{2} \in \mathbb{S}_{\sigma_k}$, we have $r_2 = \kappa(r_1, \delta_{3n+2}\beta)$ and $z = \kappa(r_2, \delta_{3n+3}\beta)$. That is, for any given $y, z \in \mathbb{M}_0, \delta_{3n+1}\beta, \delta_{3n+2}\beta, \delta_{3n+3}\beta$ can be chosen to ensure that $U^{3n} = y$ and $U^{3(n+1)} = z$. Finally we obtain that, for any $\delta > 0$,

$$\mathbb{P}_3(y, B(z, \delta)) := \mathbb{P}\left(U^3 \in B(z, \delta)\big|U^0 = y\right) > 0,$$

where $B(z, \delta)$ denotes the open ball of radius δ centered at z. $\square$

The infinite dimensional system (6.1) has been shown to preserve the stochastic multi-symplectic conservation law locally (see e.g. [119])

$$d(dp \wedge dq) - \partial_x(dp \wedge dv + dq \wedge dw)dt = 0$$

with p, q denoting the real and imaginary parts of the solution u, respectively, and $v = p_x, w = q_x$ being the derivatives of p and q with respect to x. We now show that this ergodic FDS (6.7) not only possesses the discrete charge conservation law as shown in Proposition 6.2 but also preserves the discrete stochastic multi-symplectic structure.

Theorem 6.4 *The implicit FDS (6.7) preserves the discrete multi-symplectic structure*

$$\frac{1}{\tau}(dp_j^{n+1} \wedge dq_j^{n+1} - dp_j^n \wedge dq_j^n) - \frac{1}{h}(dp_j^{n+\frac{1}{2}} \wedge dv_{j+1}^{n+\frac{1}{2}} - dp_{j-1}^{n+\frac{1}{2}} \wedge dv_j^{n+\frac{1}{2}})$$
$$- \frac{1}{h}(dq_j^{n+\frac{1}{2}} \wedge dw_{j+1}^{n+\frac{1}{2}} - dq_{j-1}^{n+\frac{1}{2}} \wedge dw_j^{n+\frac{1}{2}}) = 0,$$

where p_j^n, q_j^n denote the real and imaginary parts of u_j^n, respectively, $v_j = \frac{1}{h}(p_j^n - p_{j-1}^n)$ and $w_j = \frac{1}{h}(q_j^n - q_{j-1}^n)$ with $j = 1, \cdots, M-1$ and $n \in \mathbb{N}$.

Proof Rewriting (6.7) with the real and imaginary parts of the components u_j^n of U^n, we have

$$
\begin{cases}
\frac{1}{\tau}(q_j^{n+1} - q_j^n) - \frac{1}{h}(v_{j+1}^{n+\frac{1}{2}} - v_j^{n+\frac{1}{2}}) = \left((p_j^{n+\frac{1}{2}})^2 + (q_j^{n+\frac{1}{2}})^2\right)p_j^{n+\frac{1}{2}} + p_j^{n+\frac{1}{2}}\zeta_j^K, \\[2mm]
-\frac{1}{\tau}(p_j^{n+1} - p_j^n) - \frac{1}{h}(w_{j+1}^{n+\frac{1}{2}} - w_j^{n+\frac{1}{2}}) = \left((p_j^{n+\frac{1}{2}})^2 + (q_j^{n+\frac{1}{2}})^2\right)q_j^{n+\frac{1}{2}} + q_j^{n+\frac{1}{2}}\zeta_j^K, \\[2mm]
\frac{1}{h}(p_j^{n+\frac{1}{2}} - p_{j-1}^{n+\frac{1}{2}}) = v_j^{n+\frac{1}{2}}, \\[2mm]
\frac{1}{h}(q_j^{n+\frac{1}{2}} - q_{j-1}^{n+\frac{1}{2}}) = w_j^{n+\frac{1}{2}},
\end{cases}
$$

(6.9)

where $\zeta_j^K = \sum_{k=1}^{K}\sqrt{\eta_k}e_k(x_j)\delta_{n+1}\beta_k(t)$. Denoting $z_j^{n+\frac{1}{2}} = (p_j^{n+\frac{1}{2}}, q_j^{n+\frac{1}{2}}, v_j^{n+\frac{1}{2}}, w_j^{n+\frac{1}{2}})^\top$ and taking differential in the phase space on both sides of (6.9), we obtain

$$
\frac{1}{\tau}d\begin{pmatrix} q_j^{n+1} - q_j^n \\ -(p_j^{n+1} - p_j^n) \\ 0 \\ 0 \end{pmatrix} + \frac{1}{h}d\begin{pmatrix} -(v_{j+1}^{n+\frac{1}{2}} - v_j^{n+\frac{1}{2}}) \\ -(w_{j+1}^{n+\frac{1}{2}} - w_j^{n+\frac{1}{2}}) \\ p_j^{n+\frac{1}{2}} - p_{j-1}^{n+\frac{1}{2}} \\ q_j^{n+\frac{1}{2}} - q_{j-1}^{n+\frac{1}{2}} \end{pmatrix}
$$

$$
= \nabla^2 S_1(z_j^{n+\frac{1}{2}})dz_j^{n+\frac{1}{2}} + \nabla^2 S_2(z_j^{n+\frac{1}{2}})dz_j^{n+\frac{1}{2}}\zeta_j^K,
$$

(6.10)

where

$$
S_1(z_j^{n+\frac{1}{2}}) = \frac{1}{4}\left((p_j^{n+\frac{1}{2}})^2 + (q_j^{n+\frac{1}{2}})^2\right)^2 + \frac{1}{2}\left(v_j^{n+\frac{1}{2}}\right)^2 + \frac{1}{2}\left(w_j^{n+\frac{1}{2}}\right)^2
$$

and

$$
S_2(z_j^{n+\frac{1}{2}}) = \frac{1}{2}\left(p_j^{n+\frac{1}{2}}\right)^2 + \frac{1}{2}\left(q_j^{n+\frac{1}{2}}\right)^2.
$$

Then the wedge product between $dz_j^{n+\frac{1}{2}}$ and (6.10) concludes the proof based on the symmetry of $\nabla^2 S_1$ and $\nabla^2 S_2$.   □

Before giving the approximate error of the ergodic limit, we give some essential a priori estimates about the stability of FDS (6.7) and FDA (6.2). In the following, C denotes a generic constant independent of T, N, τ and h while C_h denotes a generic constant depending on h.

Lemma 6.1 *For any initial value $U^0 \in \mathbb{S}$ and $\gamma \geq 1$, if the covariance operator $Q^{\frac{1}{2}} \in \mathscr{L}_2^{\frac{1}{2}+\varepsilon}$ for some $\varepsilon > 0$, then there exists a constant $C = C(\gamma, \varepsilon)$ such that the solution $\{U^n\}_{n\in\mathbb{N}}$ of (6.7) satisfies*

$$
\mathbb{E}\left\|U^{n+1} - U^n\right\|^{2\gamma} \leq C(\tau^{2\gamma}h^{-4\gamma} + \tau^\gamma), \quad \forall\, n \in \mathbb{N}.
$$

Proof As proved in Proposition 6.2 that $\|U^n\| = 1$ for any $n \in \mathbb{N}$, for the nonlinear term, we have

$$\mathbb{E}\left\|F(U^{n+\frac{1}{2}})U^{n+\frac{1}{2}}\right\|^{2\gamma} = \mathbb{E}\left[\sum_{m=1}^{M}\left|u_m^{n+\frac{1}{2}}\right|^{6\gamma}\right] \leq \mathbb{E}\left[\sum_{m=1}^{M}\left|u_m^{n+\frac{1}{2}}\right|^2 \cdot 1^{6\gamma-2}\right] \leq 1$$

by the convexity of $\mathbb{S}$, i.e., $\|U^{n+\frac{1}{2}}\| \leq 1$ and $|u_m^{n+\frac{1}{2}}| \leq 1$, a.s. The noise term can be estimated as

$$\mathbb{E}\left\|Z(U^{n+\frac{1}{2}})\delta_{n+1}\beta\right\|^{2\gamma} = \mathbb{E}\left(\sum_{m=1}^{M}\left|\sum_{k=1}^{K} u_m^{n+\frac{1}{2}} e_k(x_m)\sqrt{\eta_k}\delta_{n+1}\beta_k\right|^2\right)^{\gamma}$$

$$= \mathbb{E}\left(\sum_{m=1}^{M}\left[\left|u_m^{n+\frac{1}{2}}\right|^2\left(\sum_{k=1}^{K} e_k(x_m)\sqrt{\eta_k}\delta_{n+1}\beta_k\right)^2\right]\right)^{\gamma}$$

$$\leq \mathbb{E}\left(2\sum_{m=1}^{M}\left|u_m^{n+\frac{1}{2}}\right|^2\left(\sum_{k=1}^{K}\sqrt{\eta_k}|\delta_{n+1}\beta_k|\right)^2\right)^{\gamma}$$

$$\leq 2^{\gamma}\mathbb{E}\left(\sum_{k=1}^{K}\eta_k^{\frac{1}{2}}|\delta_{n+1}\beta_k|\right)^{2\gamma}$$

$$\leq C\left[\sum_{k=1}^{K}\left(\eta_k^{\frac{1}{2}}k^{\frac{1+2\varepsilon}{2\gamma}}\right)^{\frac{2\gamma}{2\gamma-1}}\right]^{2\gamma-1}\mathbb{E}\left[\sum_{k=1}^{K}k^{-(1+2\varepsilon)}|\delta_{n+1}\beta_k|^{2\gamma}\right]$$

$$\leq C\tau^{\gamma}, \tag{6.11}$$

where we have used the fact

$$\sum_{k=1}^{K}\left(\eta_k^{\frac{1}{2}}k^{\frac{1+2\varepsilon}{2\gamma}}\right)^{\frac{2\gamma}{2\gamma-1}} = \sum_{k=1}^{K}\left[\left(\eta_k^{\frac{\gamma}{2\gamma-1}}k^{\frac{\gamma}{2\gamma-1}(1+2\varepsilon)}\right)\left(k^{\frac{1-\gamma}{2\gamma-1}(1+2\varepsilon)}\right)\right]$$

$$\leq \left[\sum_{k=1}^{K}\left(\eta_k^{\frac{\gamma}{2\gamma-1}}k^{\frac{\gamma(1+2\varepsilon)}{2\gamma-1}}\right)^{\frac{2\gamma-1}{\gamma}}\right]^{\frac{\gamma}{2\gamma-1}}\left[\sum_{k=1}^{K}\left(k^{\frac{(1-\gamma)(1+2\varepsilon)}{2\gamma-1}}\right)^{\frac{2\gamma-1}{\gamma-1}}\right]^{\frac{\gamma-1}{2\gamma-1}}$$

$$= \left[\sum_{k=1}^{K}\left(\eta_k k^{1+2\varepsilon}\right)\right]^{\frac{\gamma}{2\gamma-1}}\left[\sum_{k=1}^{K}k^{-(1+2\varepsilon)}\right]^{\frac{\gamma-1}{2\gamma-1}} \leq C$$

according to the assumption $Q^{\frac{1}{2}} \in \mathscr{L}_2^{\frac{1}{2}+\varepsilon}$. In conclusion,

$$
\begin{aligned}
&\mathbb{E} \left\| U^{n+1} - U^n \right\|^{2\gamma} \\
&\leq C \left(\mathbb{E} \left\| \frac{\tau}{h^2} A U^{n+\frac{1}{2}} \right\|^{2\gamma} + \mathbb{E} \left\| \lambda \tau F(U^{n+\frac{1}{2}}) U^{n+\frac{1}{2}} \right\|^{2\gamma} + \mathbb{E} \left\| Z(U^{n+\frac{1}{2}}) \delta_{n+1}\beta \right\|^{2\gamma} \right) \\
&\leq \frac{C\tau^{2\gamma}}{h^{4\gamma}} \mathbb{E} \left\| U^{n+\frac{1}{2}} \right\|^{2\gamma} + C\tau^{2\gamma} + C\tau^{\gamma} \leq C \left(\tau^{2\gamma} h^{-4\gamma} + \tau^{\gamma} \right),
\end{aligned}
$$

where we have used the fact that $\|A\| \leq 4$ in Lemma 5.3. $\qquad\square$

Lemma 6.2 *For any initial value $U(0) \in \mathbb{S}$ and $\gamma \geq 1$, there exists a constant C such that the solution $U(t)$ of (6.2) satisfies*

$$
\mathbb{E}\|U(t_{n+1}) - U(t_n)\|^{2\gamma} \leq C(\tau^{2\gamma} h^{-4\gamma} + \tau^{\gamma}), \quad \forall n \in \mathbb{N}.
$$

Proof From (6.2) and (6.3), based on Hölder's inequality, we obtain

$$
\begin{aligned}
&\mathbb{E}\|U(t_{n+1}) - U(t_n)\|^{2\gamma} \\
&= \mathbb{E} \left\| \int_{t_n}^{t_{n+1}} \left[i\frac{1}{h^2} AU + i\lambda F(U)U - \hat{E}U \right] dt + \int_{t_n}^{t_{n+1}} iZ(U)d\beta(t) \right\|^{2\gamma} \\
&\leq C \left(\int_{t_n}^{t_{n+1}} \mathbb{E} \left\| i\frac{1}{h^2} AU + i\lambda F(U)U - \hat{E}U \right\|^{2\gamma} dt \left(\int_{t_n}^{t_{n+1}} 1^{\frac{2\gamma}{2\gamma-1}} dt \right)^{2\gamma-1} \right. \\
&\qquad \left. + \mathbb{E} \left\| \int_{t_n}^{t_{n+1}} iZ(U)d\beta(t) \right\|^{2\gamma} \right) \\
&\leq C\tau^{2\gamma-1} \left\| \frac{1}{h^2} A \right\|^{2\gamma} \int_{t_n}^{t_{n+1}} \mathbb{E}\|U\|^{2\gamma} dt + C\tau^{2\gamma} + C\tau^{\gamma} \\
&\leq C(\tau^{2\gamma} h^{-4\gamma} + \tau^{\gamma}),
\end{aligned}
$$

where we have used the boundedness of $F(U)U$ in $\mathbb{S}$ similar to that in Lemma 6.1. In the third step of the equation above, we also used

$$
\begin{aligned}
\mathbb{E}\|\hat{E}U\|^{2\gamma} &\leq C\mathbb{E} \left(\sum_{m=1}^{M} \left| \sum_{k=1}^{K} \eta_k e_k^2(x_m) u_m \right|^2 \right)^{\gamma} \\
&\leq C\mathbb{E} \left(\sum_{m=1}^{M} |u_m|^2 \left(\sum_{k=1}^{K} \eta_k \right)^2 \right)^{\gamma} \leq C\eta^{2\gamma} \mathbb{E}\|U\|^{2\gamma} \leq C
\end{aligned}
$$

and

$$
\mathbb{E} \left\| \int_{t_n}^{t_{n+1}} iZ(U)d\beta(t) \right\|^{2\gamma} \le C \left(\int_{t_n}^{t_{n+1}} \left(\mathbb{E} \|Z(U)\|_F^{2\gamma} \right)^{\frac{1}{\gamma}} dt \right)^{\gamma}
$$

$$
\le C \left(\int_{t_n}^{t_{n+1}} \left(\mathbb{E} \left(\sum_{m=1}^{M} \sum_{k=1}^{K} |u_m e_k(x_m) \sqrt{\eta_k}|^2 \right)^{\gamma} \right)^{\frac{1}{\gamma}} dt \right)^{\gamma}
$$

$$
\le C \left(\int_{t_n}^{t_{n+1}} \left(\mathbb{E} \left(2\eta \|U\|^2 \right)^{\gamma} \right)^{\frac{1}{\gamma}} dt \right)^{\gamma} \le C\tau^{\gamma}
$$

according to the Burkholder–Davis–Gundy inequality, where $\| \cdot \|_F$ denotes the Frobenius norm. □

6.3 Approximate Error of the Ergodic Limit

To approximate the ergodic limit of (6.2) and get the approximate error, we give an estimate of the local weak convergence error between $U(\tau)$ and U^1 based on the Poisson equation associated to (6.2) (see also [140]). Recall that SDE (6.2) in the Stratonovich sense has an equivalent Itô form

$$
dU = \left[i\frac{1}{h^2} AU + i\lambda F(U)U - \hat{E}U \right] dt + iZ(U)d\beta(t)
$$

$$
= : b(U)dt + \sigma(U)d\beta(t) \tag{6.12}
$$

based on (6.3). For any fixed $f \in \mathbf{W}^{4,\infty}(\mathbb{S})$, let $\hat{f} := \int_{\mathbb{S}} f d\mu_h$ and φ be the unique solution of the Poisson equation $\mathscr{L}\varphi = f - \hat{f}$, where $\mathscr{L}$ denotes the infinitesimal generator defined in (2.2). It is easy to find out that (6.12) satisfies the hypoelliptic setting (see e.g. [140]) according to the Hörmander condition in Theorem 6.2. Thus, $\varphi \in \mathbf{W}^{4,\infty}(\mathbb{S})$ according to Theorem 4.1 in [140]. Note that φ has a unique extension to a function in $\mathbf{W}^{4,\infty}(B(0,1))$ with $B(0,1)$ being the unit open ball centered at 0. In fact, the harmonic equation

$$
\begin{cases} \Delta g = 0 & \text{in } B(0,1) \\ g = \varphi & \text{on } \mathbb{S} \end{cases}
$$

has a smooth solution in $\mathbf{W}^{4,\infty}(B(0,1))$ such that $g|_{\mathbb{S}} = \varphi$. For simplicity, we still use the notation φ to denote the extension on $B(0,1)$.

Based on the existence and uniqueness of the numerical solution $\{U^n\}_{n \in \mathbb{N}}$, (6.7) can be rewritten in the following explicit form

$$U^{n+1} = U^n + \tau \Phi(U^n, \tau, h, \delta_{n+1}\beta) \tag{6.13}$$

for some function Φ. Denoting by $D^k\varphi(u)(\Phi_1, \cdots, \Phi_k)$ the kth order Fréchet derivative evaluated in the directions Φ_j, $j = 1, \cdots, k$ with $D^k\varphi(u)(\Phi)^k$ for short if all the directions are the same, we have

$$\varphi(U^{n+1}) = \varphi(U^n) + \tau \left[D\varphi(U^n)\Phi^n + \frac{1}{2}\tau D^2\varphi(U^n)(\Phi^n)^2 \right] + \frac{1}{6}D^3\varphi(U^n)(\tau\Phi^n)^3 + R_n^{\Phi}$$

$$=: \varphi(U^n) + \tau \mathscr{L}^{\Phi}\varphi(U^n) + \frac{1}{6}D^3\varphi(U^n)(\tau\Phi^n)^3 + R_n^{\Phi}, \tag{6.14}$$

where $\Phi^n := \Phi(U^n, \tau, h, \delta_{n+1}\beta)$,

$$\mathscr{L}^{\Phi}\varphi(U^n) := D\varphi(U^n)\Phi^n + \frac{1}{2}\tau D^2\varphi(U^n)(\Phi^n)^2,$$

and

$$R_n^{\Phi} := \frac{1}{6}\int_0^1 s^3 D^4\varphi(sU^n + (1-s)U^{n+1})(\tau\Phi^n)^4 ds.$$

Here, $sU^n + (1-s)U^{n+1} \in B(0, 1)$. Adding (6.14) together from $n = 0$ to $n = N - 1$ for some fixed $N \in \mathbb{N}$, then dividing the result by $T = N\tau$, and noticing that $\mathscr{L}\varphi(U^n) = f(U^n) - \hat{f}$, we obtain

$$\frac{\varphi(U^N) - \varphi(U^0)}{N\tau} = \frac{1}{N}\left(\sum_{n=0}^{N-1}[\mathscr{L}^{\Phi}\varphi(U^n) - \mathscr{L}\varphi(U^n)] + \sum_{n=0}^{N-1}\mathscr{L}\varphi(U^n) \right.$$

$$\left. + \frac{1}{\tau}\sum_{n=0}^{N-1}\frac{1}{6}D^3\varphi(U^n)(\tau\Phi^n)^3 + \frac{1}{\tau}\sum_{n=0}^{N-1}R_n^{\Phi} \right)$$

$$= \frac{1}{N}\sum_{n=0}^{N-1}\left[\mathscr{L}^{\Phi}\varphi(U^n) - \mathscr{L}\varphi(U^n) + \frac{1}{6\tau}D^3\varphi(U^n)(\tau\Phi^n)^3\right]$$

$$+ \left(\frac{1}{N}\sum_{n=0}^{N-1}f(U^n) - \hat{f}\right) + \frac{1}{N\tau}\sum_{n=0}^{N-1}R_n^{\Phi},$$

which shows

$$\left| \mathbb{E}\left[\frac{1}{N}\sum_{n=0}^{N-1}f(U^n) - \hat{f}\right] \right| \leq \left|\frac{1}{N\tau}\mathbb{E}[\varphi(U^N) - \varphi(U^0)]\right| + \left|\frac{1}{N\tau}\sum_{n=0}^{N-1}\mathbb{E}R_n^{\Phi}\right|$$

$$+ \left|\frac{1}{N}\sum_{n=0}^{N-1}\mathbb{E}\left[\mathscr{L}^{\Phi}\varphi(U^n) - \mathscr{L}\varphi(U^n) + \frac{1}{6\tau}D^3\varphi(U^n)(\tau\Phi^n)^3\right]\right| =: I + II + III.$$

$$\tag{6.15}$$

The average $\frac{1}{N} \sum_{n=0}^{N-1} f(U^n)$ is regarded as an approximation of $\hat{f}$. We next begin to investigate the approximate error by estimating I, II and III respectively.

According to the fact that $\varphi \in \mathbf{W}^{4,\infty}(B(0,1))$ and Lemma 6.1, we have

$$I \le \frac{2\|\varphi\|_{0,\infty}}{N\tau} \le \frac{C}{T} \tag{6.16}$$

and

$$II \le \frac{1}{N\tau} \sum_{n=0}^{N-1} \mathbb{E}\left[\|\tau\Phi^n\|^4 \|D^4\varphi\|_{L^\infty}\right] \le \frac{C}{N\tau} \sum_{n=0}^{N-1} \mathbb{E}\left[\|U^{n+1} - U^n\|^4\right]$$

$$\le \frac{C}{N\tau} \sum_{n=0}^{N-1} \left(\tau^4 h^{-8} + \tau^2\right) \le C\left(\tau^3 h^{-8} + \tau\right), \tag{6.17}$$

where $\|\varphi\|_{\gamma,\infty} := \sup_{|\alpha|\le\gamma, u\in B(0,1)} |D^\alpha\varphi(u)|$, $\gamma \in \mathbb{N}$.

It then remains to estimate the term III. To this end, we need the estimate about the local weak convergence, which is stated in the following theorem.

Theorem 6.5 *Assume that $Q^{\frac{1}{2}} \in \mathscr{L}_2^{\frac{1}{2}+\varepsilon}$ for some $\varepsilon > 0$. On one hand, for a fixed $h \in (0,1)$, the local weak error satisfies*

$$\left|\mathbb{E}\left[\varphi(U(\tau)) - \varphi(U^1)\right]\right| \le C_h \tau^2$$

with some constant $C_h = C(\varphi, \eta, h)$. On the other hand, for any $h, \tau \in (0,1)$ satisfying $\rho := \tau^{\frac{1}{2}} h^{-2} < C$, it holds

$$\left|\mathbb{E}\left[\varphi(U(\tau)) - \varphi(U^1)\right]\right| \le C_\rho \tau^{\frac{3}{2}}$$

with some constant $C_\rho = C(\varphi, \eta, \rho)$.

Proof Based on the Taylor expansion, Lemmas 6.1 and 6.2, we obtain

$$\mathbb{E}\left[\varphi(U(\tau)) - \varphi(U^1)\right] = \mathbb{E}\left[D\varphi(U^1)(U(\tau) - U^1) + O\left(\|U(\tau) - U^1\|^2\right)\right]$$
$$= \mathbb{E}\left[D\varphi(U^0)(U(\tau) - U^1)\right] + \mathbb{E}\left[D^2\varphi(U^0)(U^1 - U^0, U(\tau) - U^1)\right]$$
$$+ O\left(\mathbb{E}\left[\|U^1 - U^0\|^2 \|U(\tau) - U^1\|\right] + \mathbb{E}\|U(\tau) - U^1\|^2\right)$$
$$=: \mathscr{A} + \mathscr{B} + \mathscr{C}.$$

We give the mild solution and the discrete mild solution of (6.2) and (6.7) respectively,

$$U(\tau) = e^{i\frac{1}{h^2}A\tau}U^0 + \int_0^\tau e^{i\frac{1}{h^2}A(\tau-s)}\left(i\lambda F(U(s))U(s) - \hat{E}U(s)\right)ds$$
$$+ \int_0^\tau e^{i\frac{1}{h^2}A(\tau-s)}iZ(U(s))d\beta(s),$$

$$U^1 = (Id - \frac{i\tau}{2h^2}A)^{-1}(Id + \frac{i\tau}{2h^2}A)U^0 + (Id - \frac{i\tau}{2h^2}A)^{-1}i\lambda\tau F\left(U^{\frac{1}{2}}\right)U^{\frac{1}{2}}$$
$$+ (Id - \frac{i\tau}{2h^2}A)^{-1}iZ\left(U^{\frac{1}{2}}\right)\delta_1\beta.$$

Estimation of $\mathscr{A}$. Considering the difference between above equations, we have

$$U(\tau) - U^1 = \left(e^{i\frac{1}{h^2}A\tau} - (Id - \frac{i\tau}{2h^2}A)^{-1}(Id + \frac{i\tau}{2h^2}A)\right)U^0$$
$$+ i\int_0^\tau \left[e^{i\frac{1}{h^2}A(\tau-s)} - (Id - \frac{i\tau}{2h^2}A)^{-1}\right]\lambda F(U(s))U(s)ds$$
$$+ i\int_0^\tau (Id - \frac{i\tau}{2h^2}A)^{-1}\lambda\left[F(U(s))U(s) - F\left(U^{\frac{1}{2}}\right)U^{\frac{1}{2}}\right]ds$$
$$+ i\int_0^\tau \left[e^{i\frac{1}{h^2}A(\tau-s)} - \left(Id - \frac{i\tau}{2h^2}A\right)^{-1}\right]Z(U(s))d\beta(s)$$
$$+ i\int_0^\tau \left(Id - \frac{i\tau}{2h^2}A\right)^{-1}Z(U(s) - U^0)d\beta(s)$$
$$- \left[\frac{i}{2}\left(Id - \frac{i\tau}{2h^2}A\right)^{-1}Z(U^1 - U^0)\delta_1\beta + \int_0^\tau e^{i\frac{1}{h^2}A(\tau-s)}\hat{E}U(s)ds\right]$$
$$=: \mathbf{a} + \mathbf{b} + \mathbf{c} + \mathbf{d} + \mathbf{e} + \mathbf{f},$$

which, together with the fact that $\mathbb{E}[D\varphi(U^0)\mathbf{d}] = \mathbb{E}[D\varphi(U^0)\mathbf{e}] = 0$, yields that

$$\mathscr{A} = \mathbb{E}\left[D\varphi(U^0)\mathbf{a}\right] + \mathbb{E}\left[D\varphi(U^0)\mathbf{b}\right] + \mathbb{E}\left[D\varphi(U^0)\mathbf{c}\right] + \mathbb{E}\left[D\varphi(U^0)\mathbf{f}\right]$$
$$=: A_1 + A_2 + A_3 + A_4.$$

Based on the estimates $e^x - (1 - \frac{x}{2})^{-1}(1 + \frac{x}{2}) = O(x^3)$ for $\|x\| < 1$, and

$$\left\|e^{i\frac{1}{h^2}A(\tau-s)} - (Id - \frac{i\tau}{2h^2}A)^{-1}\right\| \le C\left(\frac{\tau}{h^2}\|A\|\right) \le C\tau h^{-2}, \quad \forall s \in [0,\tau], \quad (6.18)$$

we have

$$|A_1| \le C\|\varphi\|_{1,\infty}\|\tau h^{-2}A\|^3 \mathbb{E}\|U^0\| \le C\tau^3 h^{-6} \tag{6.19}$$

and

$$|A_2| \le C\|\varphi\|_{1,\infty}\int_0^\tau \|\tau h^{-2}A\|\|F(U(s))U(s)\|ds \le C\tau^2 h^{-2}. \tag{6.20}$$

Term A_3 can be estimated based on Lemmas 6.1 and 6.2. In fact,

$$|A_3| = \left| \mathbb{E}\left[D\varphi(U^0) \int_0^\tau (Id - \frac{i\tau}{2h^2}A)^{-1}\left[(F(U(s))U(s) - F(U^0)U^0) \right. \right. \right.$$
$$\left. \left. \left. - \left(F\left(U^{\frac{1}{2}}\right)U^{\frac{1}{2}} - F(U^0)U^0\right) \right] ds \right] \right|.$$

Denote $g(V) := F(V)V$, $V \in B(0,1)$, which is a continuous differentiable function satisfying $|D^k g(V)| \le C$ for $\|V\| \le 1$ and $k \in \mathbb{N}$, such that

$$F(U(s))U(s) - F(U^0)U^0 = g(U(s)) - g(U^0)$$

$$= Dg(U^0)(U(s) - U^0) + \int_0^1 r D^2 g(rU^0 + (1-r)U(s))(U(s) - U^0)^2 dr$$

$$= Dg(U^0)\left(\int_0^s \frac{i}{h^2} AU(r) + i\lambda F(U(r))U(r) - \hat{E}U(r)dr + \int_0^s Z(U(r))d\beta(r) \right)$$

$$+ \int_0^1 r D^2 g(rU^0 + (1-r)U(s))(U(s) - U^0)^2 dr$$

for $s \in [0, \tau]$, and the same for the term $F\left(U^{\frac{1}{2}}\right)U^{\frac{1}{2}} - F(U^0)U^0$. Based on Lemma 6.2 with $\gamma = 1$ and the fact that $\mathbb{E}\left[Dg(U^0) \int_0^s Z(U(r))d\beta(r) \right] = 0$, we hence get

$$|A_3| \le C(\tau^2 h^{-2} + \tau^3 h^{-4} + \tau^2) \tag{6.21}$$

similar to the proof of Lemma 6.2. Rewrite

$$Z(U^1 - U^0)\delta_1\beta = \begin{pmatrix} u_1^1 - u_1^0 & & \\ & \ddots & \\ & & u_M^1 - u_M^0 \end{pmatrix} E_{MK}\Lambda\delta_1\beta$$

$$= \begin{pmatrix} \sum_{k=1}^K e_k(x_1)\sqrt{\eta_k}\delta_1\beta_k & & \\ & \ddots & \\ & & \sum_{k=1}^K e_k(x_M)\sqrt{\eta_k}\delta_1\beta_k \end{pmatrix} (U^1 - U^0)$$

$$=: G(U^1 - U^0),$$

where G satisfies that $\mathbb{E}[GU^0] = 0$. Utilizing that $\mathbb{E}[GF(U^0)U^0] = 0$ and $U^{\frac{1}{2}} = \frac{1}{2}(U^1 - U^0) + U^0$, we rewrite term A_4 as

$$A_4 = -\mathbb{E}\left[D\varphi(U^0)\left(\frac{i}{2}\left(Id - \frac{i\tau}{2h^2}A\right)^{-1} G(U^1 - U^0) + \int_0^\tau e^{i\frac{1}{h^2}A(\tau-s)}\hat{E}U(s)ds \right) \right]$$

$$= -\frac{i}{2}\mathbb{E}\left[D\varphi(U^0)\left(Id - \frac{i\tau}{2h^2}A\right)^{-1} G\left(i\frac{\tau}{h^2}AU^{\frac{1}{2}} + i\lambda\tau F(U^{\frac{1}{2}})U^{\frac{1}{2}} + iGU^{\frac{1}{2}}\right) \right]$$

$$- \mathbb{E}\left[D\varphi(U^0) \int_0^\tau e^{\mathbf{i}\frac{1}{h^2}A(\tau-s)} \hat{E}U(s)ds \right]$$

$$= \frac{\tau}{4h^2}\mathbb{E}\left[D\varphi(U^0)\left(Id - \frac{\mathbf{i}\tau}{2h^2}A \right)^{-1} GA(U^1 - U^0) \right]$$

$$+ \frac{1}{2}\lambda\tau\mathbb{E}\left[D\varphi(U^0)\left(Id - \frac{\mathbf{i}\tau}{2h^2}A \right)^{-1} G\left(F(U^{\frac{1}{2}})U^{\frac{1}{2}} - F(U^0)U^0 \right) \right]$$

$$+ \frac{1}{4}\mathbb{E}\left[D\varphi(U^0)\left(Id - \frac{\mathbf{i}\tau}{2h^2}A \right)^{-1} G^2(U^1 - U^0) \right]$$

$$+ \mathbb{E}\left[D\varphi(U^0)\left(\left(Id - \frac{\mathbf{i}\tau}{2h^2}A \right)^{-1} \frac{1}{2}G^2 U^0 - \int_0^\tau e^{\mathbf{i}\frac{1}{h^2}A(\tau-s)} \hat{E}U(s)ds \right) \right]$$

$$=: A_{4,1} + A_{4,2} + A_{4,3} + A_{4,4},$$

where

$$|A_{4,1}| \le C\tau h^{-2}\left(\mathbb{E}\|G\|^2\right)^{\frac{1}{2}}\left(\mathbb{E}\|U^1 - U^0\|^2\right)^{\frac{1}{2}} \le C\left(\tau^{\frac{5}{2}}h^{-4} + \tau^2 h^{-2}\right)$$

and

$$|A_{4,2}| \le C\tau\left(\mathbb{E}\|G\|^2\right)^{\frac{1}{2}}\left(\mathbb{E}\left\| F(U^{\frac{1}{2}})U^{\frac{1}{2}} - F(U^0)U^0 \right\|^2\right)^{\frac{1}{2}}$$

$$\le C\tau^{\frac{3}{2}}\left(\mathbb{E}\|U^1 - U^0\|^2\right)^{\frac{1}{2}} \le C\left(\tau^{\frac{5}{2}}h^{-2} + \tau^2\right).$$

According to $\mathbb{E}[G^3 U^0] = 0$, $A_{4,3}$ can be expressed as

$$\frac{1}{4}\mathbb{E}\left[D\varphi(U^0)\left(Id - \frac{\mathbf{i}\tau}{2h^2}A \right)^{-1} G^2\left(\mathbf{i}\frac{\tau}{h^2}AU^{\frac{1}{2}} + \mathbf{i}\tau\lambda F(U^{\frac{1}{2}})U^{\frac{1}{2}} + \frac{\mathbf{i}}{2}G(U^1 - U^0) \right) \right].$$

For any $U \in \mathbb{C}^M$, we have

$$\mathbb{E}\|GU\| = \mathbb{E}\|Z(U)\delta_1\beta\| \le C\mathbb{E}\left(\|U\|^2\left(\sum_{k=1}^K \sqrt{\eta_k}|\delta_1\beta_k| \right)^2 \right)^{\frac{1}{2}} \le C\tau^{\frac{1}{2}}\left(\mathbb{E}\|U\|^2\right)^{\frac{1}{2}}.$$

Hence, $\mathbb{E}\|G^3(U^1 - U^0)\| \le C\tau^{\frac{1}{2}}(\mathbb{E}\|G^2(U^1 - U^0)\|^2)^{\frac{1}{2}}$ can be further estimated based on Lemma 6.1 with $\gamma = 2$ and (6.11) with $\gamma = 4$, that is,

$$\mathbb{E}\|G^2(U^1 - U^0)\|^2 = \mathbb{E}\left[\sum_{m=1}^M \left| \left(\sum_{k=1}^K e_k(x_m)\sqrt{\eta_k}\delta_1\beta_k \right)^2 (u_m^1 - u_m^0) \right|^2 \right]$$

$$\leq \mathbb{E}\left[\|U^1 - U^0\|^2 \left(\sum_{k=1}^{K} \eta_k^{\frac{1}{2}} |\delta_1 \beta_k| \right)^4 \right]$$

$$\leq \left(\mathbb{E}\|U^1 - U^0\|^4 \right)^{\frac{1}{2}} \left(\mathbb{E} \left(\sum_{k=1}^{K} \eta_k^{\frac{1}{2}} |\delta_1 \beta_k| \right)^8 \right)^{\frac{1}{2}}$$

$$\leq C(\tau^4 h^{-4} + \tau^3).$$

Combining with Lemma 6.1 and $\|U^{\frac{1}{2}}\| \leq 1$, we get

$$|A_{4,1} + A_{4,2} + A_{4,3}| \leq C(\tau^{\frac{5}{2}} h^{-4} + \tau^2 h^{-2} + \tau^2).$$

For the term $A_{4,4}$, we have

$$\mathbb{E}\left[D\varphi(U^0) \left(Id - \frac{i\tau}{2h^2} A \right)^{-1} G^2 U^0 \right]$$

$$= \mathbb{E}\left[D\varphi(U^0) \left(Id - \frac{i\tau}{2h^2} A \right)^{-1} \begin{pmatrix} \sum_{k=1}^{K} e_k^2(x_1) \eta_k (\delta_1 \beta_k)^2 u_1^0 \\ \vdots \\ \sum_{k=1}^{K} e_k^2(x_M) \eta_k (\delta_1 \beta_k)^2 u_M^0 \end{pmatrix} \right]$$

and

$$\hat{E}U(s) = \frac{1}{2} \begin{pmatrix} \sum_{k=1}^{K} e_k^2(x_1) \eta_k u_1(s) \\ \vdots \\ \sum_{k=1}^{K} e_k^2(x_M) \eta_k u_M(s) \end{pmatrix}.$$

Thus, we obtain

$$A_{4,4} = \frac{1}{2} \mathbb{E}\left[D\varphi(U^0) \left(Id - \frac{i\tau}{2h^2} A \right)^{-1} \begin{pmatrix} \sum_{k=1}^{K} e_k^2(x_1) \eta_k (\delta_1 \beta_k)^2 u_1^0 \\ \vdots \\ \sum_{k=1}^{K} e_k^2(x_M) \eta_k (\delta_1 \beta_k)^2 u_M^0 \end{pmatrix} \right]$$

$$- \frac{1}{2} \mathbb{E}\left[D\varphi(U^0) \int_0^\tau e^{i\frac{1}{h^2} A(\tau - s)} \begin{pmatrix} \sum_{k=1}^{K} e_k^2(x_1) \eta_k u_1(s) \\ \vdots \\ \sum_{k=1}^{K} e_k^2(x_M) \eta_k u_M(s) \end{pmatrix} ds \right]$$

$$= \frac{1}{2} \mathbb{E}\left[D\varphi(U^0) \left(Id - \frac{i\tau}{2h^2} A \right)^{-1} \begin{pmatrix} \sum_{k=1}^{K} e_k^2(x_1) \eta_k ((\delta_1 \beta_k)^2 - \tau) u_1^0 \\ \vdots \\ \sum_{k=1}^{K} e_k^2(x_M) \eta_k ((\delta_1 \beta_k)^2 - \tau) u_M^0 \end{pmatrix} \right]$$

$$
+ \frac{1}{2} \mathbb{E} \left[D\varphi(U^0) \int_0^\tau \left(\left(Id - \frac{i\tau}{2h^2} A \right)^{-1} - e^{i\frac{1}{h^2} A(\tau - s)} \right) \begin{pmatrix} \sum_{k=1}^K e_k^2(x_1)\eta_k u_1^0 \\ \vdots \\ \sum_{k=1}^K e_k^2(x_M)\eta_k u_M^0 \end{pmatrix} ds \right]
$$

$$
- \frac{1}{2} \mathbb{E} \left[D\varphi(U^0) \int_0^\tau e^{i\frac{1}{h^2} A(\tau - s)} \begin{pmatrix} \sum_{k=1}^K e_k^2(x_1)\eta_k \left(u_1(s) - u_1^0 \right) \\ \vdots \\ \sum_{k=1}^K e_k^2(x_M)\eta_k \left(u_M(s) - u_M^0 \right) \end{pmatrix} ds \right]. \tag{6.22}
$$

Note that the first term in (6.22) vanishes as $\mathbb{E}(\delta_1 \beta_k)^2 = \tau$. Replacing $U(s) - U^0$ by the integral type of (6.2), we obtain

$$
|A_{4,4}| \leq C(\tau^2 h^{-2} + \tau^2)
$$

based on (6.18) and the technique used in (6.21). It then leads to

$$
|A_4| \leq C \left(\tau^{\frac{5}{2}} h^{-4} + \tau^2 h^{-2} + \tau^2 \right). \tag{6.23}
$$

We then conclude from (6.19)–(6.23) that

$$
|\mathscr{A}| \leq C \left(\tau^3 h^{-6} + \tau^{\frac{5}{2}} h^{-4} + \tau^2 h^{-2} + \tau^2 \right). \tag{6.24}
$$

Estimation of $\mathscr{C}$. Estimations of A_1 and A_2 show that

$$
\mathbb{E}\|\mathbf{a} + \mathbf{b}\|^2 \leq C(\tau^6 h^{-12} + \tau^4 h^{-4}). \tag{6.25}
$$

Based on Hölder's inequality, Itô isometry, Lemmas 6.1 and 6.2, together with the expression $Z(U^1 - U^0)\delta_1 \beta = G(U^1 - U^0)$, we have

$$
\mathbb{E}\|\mathbf{c} + \mathbf{d}\|^2 \leq C\tau \int_0^\tau \mathbb{E}\|U(s) - U^{\frac{1}{2}}\|^2 ds + \int_0^\tau C\tau^2 h^{-4} ds
$$
$$
\leq C(\tau^3 h^{-4} + \tau^3), \tag{6.26}
$$

$$
\mathbb{E}\|\mathbf{e}\|^2 \leq C\mathbb{E} \left[\int_0^\tau \left\| \left(Id - \frac{i\tau}{2h^2} A \right)^{-1} Z\left(U(s) - U^0 \right) \right\|_F^2 ds \right]
$$
$$
\leq C \int_0^\tau \mathbb{E} \left[\sum_{m=1}^M \sum_{k=1}^K \left| \left(u_m(s) - u_m^0 \right) e_k(x_m)\sqrt{\eta_k} \right|^2 \right] ds
$$
$$
\leq C \left(\tau^3 h^{-4} + \tau^2 \right)
$$

and

$$
\begin{aligned}
\mathbb{E}\|\mathbf{f}\|^2 &\leq C\mathbb{E}\|Z(U^1 - U^0)\delta_1\beta\|^2 + C\tau \int_0^\tau \mathbb{E}\|\hat{E}U(s)\|^2 ds \\
&\leq C\left(\mathbb{E}\|G\|^4\right)^{\frac{1}{2}} \left(\mathbb{E}\|U^1 - U^0\|^4\right)^{\frac{1}{2}} + C\tau^2 \\
&\leq C(\tau^3 h^{-4} + \tau^2).
\end{aligned}
$$

We then conclude that

$$
\mathbb{E}\|U(\tau) - U^1\|^2 \leq C\left(\tau^6 h^{-12} + \tau^3 h^{-4} + \tau^2\right), \tag{6.27}
$$

which yields

$$
\begin{aligned}
|\mathscr{C}| &= O\left(\left(\mathbb{E}\|U^1 - U^0\|^4\right)^{\frac{1}{2}} \left(\mathbb{E}\|U(\tau) - U^1\|^2\right)^{\frac{1}{2}} + \mathbb{E}\|U(\tau) - U^1\|^2\right) \\
&\leq C\left(\tau^6 h^{-12} + \tau^4 h^{-8} + \tau^3 h^{-4} + \tau^2\right). \tag{6.28}
\end{aligned}
$$

Estimation of $\mathscr{B}$. As for $\mathscr{B} = \mathbb{E}\left[D^2\varphi(U^0)\left(U^1 - U^0, \mathbf{a} + \mathbf{b} + \mathbf{c} + \mathbf{d} + \mathbf{e} + \mathbf{f}\right)\right]$, according to Hölder's inequality, (6.25) and (6.26), we have

$$
\begin{aligned}
&\left|\mathbb{E}\left[D^2\varphi(U^0)\left(U^1 - U^0, \mathbf{a} + \mathbf{b} + \mathbf{c} + \mathbf{d}\right)\right]\right| \\
&\leq C\left(\mathbb{E}\|U^1 - U^0\|^2\right)^{\frac{1}{2}} \left(\mathbb{E}\|\mathbf{a} + \mathbf{b} + \mathbf{c} + \mathbf{d}\|^2\right)^{\frac{1}{2}} \\
&\leq C\left(\tau^4 h^{-8} + \tau^{\frac{7}{2}} h^{-6} + \tau^{\frac{5}{2}} h^{-4} + \tau^2 h^{-2} + \tau^2\right).
\end{aligned}
$$

Note that

$$
\begin{aligned}
&\mathbb{E}\left[D^2\varphi(U^0)\left(U^1 - U^0, \mathbf{e} + \mathbf{f}\right)\right] \\
&= \mathbb{E}\left[D^2\varphi(U^0)\left(U^1 - U^0, \mathbf{i}\int_0^\tau \left(Id - \frac{\mathbf{i}\tau}{2h^2}A\right)^{-1} Z(U(s) - U^1)d\beta(s)\right)\right] \\
&\quad + \frac{1}{2}\mathbb{E}\left[D^2\varphi(U^0)\left(U^1 - U^0, \mathbf{i}\left(Id - \frac{\mathbf{i}\tau}{2h^2}A\right)^{-1} Z(U^1 - U^0)\delta_1\beta\right)\right] \\
&\quad - \mathbb{E}\left[D^2\varphi(U^0)\left(U^1 - U^0, \int_0^\tau e^{\mathbf{i}\frac{1}{h^2}A(\tau-s)}\hat{E}U(s)ds\right)\right] \\
&=: B_1 + B_2 + B_3,
\end{aligned}
$$

where

$$
|B_1| \leq C\left(\tau^6 h^{-12} + \tau^3 h^{-4} + \tau^2\right)
$$

according to (6.27) and Lemma 6.1. Furthermore,

$$B_2 = \frac{1}{2}\mathbb{E}\left[D^2\varphi(U^0)\left(\mathrm{i}\frac{\tau}{h^2}AU^{\frac{1}{2}} + \mathrm{i}\tau\lambda F(U^{\frac{1}{2}})U^{\frac{1}{2}}, \mathrm{i}\left(Id - \frac{\mathrm{i}\tau}{2h^2}A\right)^{-1}Z(U^1 - U^0)\delta_1\beta\right)\right]$$

$$+ \frac{1}{2}\mathbb{E}\left[D^2\varphi(U^0)\left(\mathrm{i}Z\left(\frac{U^1 - U^0}{2}\right)\delta_1\beta, \mathrm{i}\left(Id - \frac{\mathrm{i}\tau}{2h^2}A\right)^{-1}Z(U^1 - U^0)\delta_1\beta\right)\right]$$

$$+ \frac{1}{2}\mathbb{E}\left[D^2\varphi(U^0)\left(\mathrm{i}Z(U^0)\delta_1\beta, \mathrm{i}\left(Id - \frac{\mathrm{i}\tau}{2h^2}A\right)^{-1}Z(U^1 - U^0)\delta_1\beta\right)\right]$$

$$=: B_{2,1} + B_{2,2} + B_{2,3}$$

with

$$|B_{2,1} + B_{2,2}| \le C\left(\tau^{\frac{5}{2}}h^{-4} + \tau^2 h^{-2} + \tau^2\right)$$

similar to the estimate of $\mathbb{E}\|\mathbf{f}\|^2$. Replacing $U^1 - U^0$ again by (6.7) and using the notation $Z(U)\delta_1\beta = GU$, we obtain

$$|B_{2,3}| \le \left|\frac{1}{2}\mathbb{E}\left[D^2\varphi(U^0)\left(GU^0, \left(Id - \frac{\mathrm{i}\tau}{2h^2}A\right)^{-1}G^2U^{\frac{1}{2}}\right)\right]\right| + C(\tau^2 h^{-2} + \tau^2)$$

$$= \left|\frac{1}{4}\mathbb{E}\left[D^2\varphi(U^0)\left(GU^0, \left(Id - \frac{\mathrm{i}\tau}{2h^2}A\right)^{-1}G^2(U^1 - U^0)\right)\right]\right| + C(\tau^2 h^{-2} + \tau^2)$$

$$\le C(\tau^2 h^{-2} + \tau^2),$$

where we used the fact $\mathbb{E}[G^3 U^0] = 0$ since U^0 is $\mathscr{F}_0$-adapted. For term B_3,

$$|B_3| \le \left|\mathbb{E}\left[D^2\varphi(U^0)\left(\mathrm{i}\frac{\tau}{h^2}AU^{\frac{1}{2}} + \mathrm{i}\tau\lambda F(U^{\frac{1}{2}})U^{\frac{1}{2}}, \int_0^\tau e^{\mathrm{i}\frac{1}{h^2}A(\tau-s)}\hat{E}U(s)ds\right)\right]\right|$$

$$+ \left|\mathbb{E}\left[D^2\varphi(U^0)\left(\mathrm{i}Z(U^{\frac{1}{2}})\delta_1\beta, \int_0^\tau e^{\mathrm{i}\frac{1}{h^2}A(\tau-s)}\hat{E}\left(U(s) - U^0\right)ds\right)\right]\right|$$

$$+ \left|\mathbb{E}\left[D^2\varphi(U^0)\left(\mathrm{i}Z(U^{\frac{1}{2}})\delta_1\beta, \int_0^\tau e^{\mathrm{i}\frac{1}{h^2}A(\tau-s)}\hat{E}U^0 ds\right)\right]\right|$$

$$\le C(\tau^2 h^{-2} + \tau^2) + \frac{1}{2}\left|\mathbb{E}\left[D^2\varphi(U^0)\left(\mathrm{i}Z(U^1 - U^0)\delta_1\beta, \int_0^\tau e^{\mathrm{i}\frac{1}{h^2}A(\tau-s)}\hat{E}U^0 ds\right)\right]\right|$$

$$\le C(\tau^2 h^{-2} + \tau^2).$$

We finally obtain

$$|\mathscr{B}| \le C\left(\tau^6 h^{-12} + \tau^{\frac{5}{2}}h^{-4} + \tau^2 h^{-2} + \tau^2\right),$$

which, together with (6.24) and (6.28), leads to

$$\left|\mathbb{E}\left[\varphi(U(\tau)) - \varphi(U^1)\right]\right| \le C\left(\tau^6 h^{-12} + \tau^4 h^{-8} + \tau^3 h^{-6} + \tau^{\frac{5}{2}}h^{-4} + \tau^2 h^{-2} + \tau^2\right).$$

That is, for a fixed step-size $h \in (0, 1)$,

$$\left|\mathbb{E}\left[\varphi(U(\tau)) - \varphi(U^1)\right]\right| \leq C_h \tau^2.$$

However, for any $h, \tau \in (0, 1)$ satisfying $\rho := \tau^{\frac{1}{2}} h^{-2} \leq C$, we have

$$\left|\mathbb{E}\left[\varphi(U(\tau)) - \varphi(U^1)\right]\right| \leq C\left(\rho^6 \tau^3 + \rho^4 \tau^2 + (\rho^3 + \rho^2 + \rho)\tau^{\frac{3}{2}} + \tau^2\right) \leq C_\rho \tau^{\frac{3}{2}},$$

with some constant C_ρ depending on ρ instead of h. $\square$

We would like to mention that the dependence on h or ρ may be released if higher regularity of both $\{U(t)\}_{t \geq 0}$ and $\{U^n\}_{n \in \mathbb{N}}$ can be derived uniformly. More precisely, $\frac{1}{h^2} AU(t)$ and $\frac{1}{h^2} AU^n$ are uniformly bounded in strong sense.

Now we are in the position to show the approximate error between the time average of FDS and the ergodic limit of FDA.

Theorem 6.6 *Under the assumptions in Theorem 6.5 on any time interval $[0, T]$ and for any $f \in \mathbf{W}^{4,\infty}(\mathbb{S})$, for a fixed $h \in (0, 1)$, there exists a positive constant $C_h = C(f, \eta, h)$ such that*

$$\left|\mathbb{E}\left[\frac{1}{N}\sum_{n=0}^{N-1} f(U^n) - \hat{f}\right]\right| \leq C_h\left(\frac{1}{T} + \tau\right).$$

Proof Based on (6.15)–(6.17), it suffices to estimate term *III*. For any $f \in \mathbf{W}^{4,\infty}(\mathbb{S})$, we know from the statement above that the solution to the Poisson equation $\mathscr{L}\varphi = f - \hat{f}$ can be extended to $\varphi \in \mathbf{W}^{4,\infty}(B(0, 1))$. Based on (6.14), Lemma 6.1 and the condition $\tau = O(h^4)$, we have

$$\mathbb{E}\varphi(U^1) = \mathbb{E}\varphi(U^0) + \tau\mathbb{E}\left[\mathscr{L}^\Phi \varphi(U^0)\right] + \frac{1}{6}\mathbb{E}\left[D^3\varphi(U^0)(U^1 - U^0)^3\right] + O(\tau^2)$$
$$= \mathbb{E}\varphi(U^0) + \tau\mathbb{E}\left[\mathscr{L}^\Phi \varphi(U^0)\right] + O(\tau^2), \tag{6.29}$$

where in the last step we used the fact that

$$\mathbb{E}\left[D^3\varphi(U^0)(U^1 - U^0)^3\right] = \mathbb{E}\left[D^3\varphi(U^0)\left(i\frac{\tau}{h^2}AU^{\frac{1}{2}} + i\lambda\tau F(U^{\frac{1}{2}})U^{\frac{1}{2}} + iZ(U^{\frac{1}{2}})\delta_1\beta\right)^3\right]$$
$$= \mathbb{E}\left[D^3\varphi(U^0)\left(iZ(U^{\frac{1}{2}})\delta_1\beta\right)^3\right] + O(\tau^2 h^{-2} + \tau^2)$$
$$= \mathbb{E}\left[D^3\varphi(U^0)\left(\frac{i}{2}Z(U^1 - U^0)\delta_1\beta + iZ(U^0)\delta_1\beta\right)^3\right]$$
$$\quad + O(\tau^2 h^{-2} + \tau^2)$$
$$= O(\tau^2 h^{-2} + \tau^2) \tag{6.30}$$

based on the linearity of Z, Lemma 6.1 and the fact that $\mathbb{E}\left(iZ(U^0)\delta_1\beta\right)^3 = 0$. We can also get the following expression similar to (6.29) based on the Taylor expansion and Lemma 6.2

$$
\begin{aligned}
\mathbb{E}\varphi(U(\tau)) = & \mathbb{E}\varphi(U^0) + \int_0^\tau \mathbb{E}\left[D\varphi(U^0)b(U(t)) + \frac{1}{2}D^2\varphi(U^0)\left(\sigma(U(t))\right)^2 \right] dt \\
& + \mathbb{E}\left[\int_0^\tau D\varphi(U^0)\sigma(U(t))d\beta(t) \right] \\
& + \frac{1}{6}\mathbb{E}\left[D^3\varphi(U^0)(U(\tau) - U^0)^3 \right] + O(\tau^2) \\
= & \mathbb{E}\varphi(U^0) + \mathbb{E}\left[\int_0^\tau \tilde{\mathscr{L}}_t\varphi(U^0)dt \right] + O(\tau^2), \quad (6.31)
\end{aligned}
$$

where

$$
\tilde{\mathscr{L}}_t\varphi(U^0) := D\varphi(U^0)b(U(t)) + \frac{1}{2}D^2\varphi(U^0)\left(\sigma(U(t))\right)^2
$$

and $\mathbb{E}\left[\int_0^\tau D\varphi(U^0)\sigma(U(t))d\beta(t) \right] = 0$. Thus, subtracting (6.29) from (6.31), we derive

$$
\left| \mathbb{E}\left[\tau\mathscr{L}^\Phi\varphi(U^0) - \int_0^\tau \tilde{\mathscr{L}}_t\varphi(U^0)dt \right] \right| \le \left| \mathbb{E}\left[\varphi(U(\tau)) - \varphi(U^1) \right] \right| + C\tau^2. \quad (6.32)
$$

Note that

$$
\begin{aligned}
\left| \int_0^\tau \mathbb{E}\left[\tilde{\mathscr{L}}_t\varphi(U^0) - \mathscr{L}\varphi(U^0) \right] dt \right| \le & \left| \int_0^\tau \mathbb{E}\left[D\varphi(U^0)\left(b(U(t)) - b(U^0)\right) \right] dt \right| \\
& + \left| \frac{1}{2}\int_0^\tau \mathbb{E}\left[D^2\varphi(U^0)\left(\sigma(U(t)) - \sigma(U^0), \sigma(U(t)) + \sigma(U^0)\right) \right] dt \right|. \quad (6.33)
\end{aligned}
$$

For the first term in (6.33), we have

$$
\begin{aligned}
\left| \mathbb{E}\left[D\varphi(U^0)\left(b(U(t)) - b(U^0)\right) \right] \right| = & \left| \mathbb{E}\left[D\varphi(U^0)\left(i\frac{1}{h^2}A\left(U(t) - U^0\right)\right. \right.\right. \\
& \left.\left.\left. + i\lambda\left(F(U(t))U(t) - F(U^0)U^0\right) - \hat{E}(U(t) - U^0)\right) \right] \right| \le C(th^{-2} + t),
\end{aligned}
$$

where we used the fact that $g(V) = F(V)V$, $V \in \mathbb{S}$, is a continuous differentiable function and replaced $U(t) - U^0$ by the integral form of (6.2) similar to the estimate in Theorem 6.5. The second term in (6.33) can be estimated in the same way. Thus, we have

$$
\left| \int_0^\tau \mathbb{E}\left[\tilde{\mathscr{L}}_t\varphi(U^0) - \mathscr{L}\varphi(U^0) \right] dt \right| \le C(\tau^2 h^{-2} + \tau^2). \quad (6.34)
$$

We hence conclude based on (6.30), (6.32), (6.34) and Theorem 6.5 that

$$III = \left| \frac{1}{N} \sum_{n=0}^{N-1} \mathbb{E}\left[\mathscr{L}^\Phi \varphi(U^n) - \mathscr{L}\varphi(U^n) + \frac{1}{6\tau} D^3\varphi(U^n)(U^{n+1} - U^n)^3 \right] \right|$$

$$\leq \frac{1}{\tau} \sup_{U^0 \in S} \left\{ \left| \mathbb{E}\left[\tau\mathscr{L}^\Phi \varphi(U^0) - \int_0^\tau \tilde{\mathscr{L}}_t\varphi(U^0)dt \right] \right| + \left| \int_0^\tau \mathbb{E}\left[\tilde{\mathscr{L}}_t\varphi(U^0) - \mathscr{L}\varphi(U^0) \right] dt \right| \right\}$$

$$+ C(\tau h^{-2} + \tau) \leq C_h\tau. \tag{6.35}$$

From (6.16), (6.17) and (6.35), we finally obtain

$$\left| \mathbb{E}\left[\frac{1}{N} \sum_{n=0}^{N-1} f(U^n) - \hat{f} \right] \right| \leq C_h \left(\frac{1}{T} + \tau \right).$$

$\square$

Remark 6.3 Based on the theorem above and the ergodicity of (6.2), for a fixed h, we obtain

$$\left| \mathbb{E}\left[\frac{1}{N} \sum_{n=0}^{N-1} f(U^n) - \frac{1}{T} \int_0^T f(U(t))dt \right] \right| \leq C_h(B(T) + \tau),$$

which implies that the global weak error is of order one, i.e.,

$$\left| \mathbb{E}\left[f(U^n) - f(U(t)) \right] \right| \leq C_h(\tilde{B}(t) + \tau), \quad t \in [n\tau, (n+1)\tau],$$

where $B(T) \to 0$ and $\tilde{B}(T) \to 0$ as $T \to \infty$. On the other hand, a time independent weak error in turn leads to the result stated in Theorem 6.6.

6.4   Numerical Experiments

In this section, numerical experiments are given to test several properties of the fully discretization (6.7) of the stochastic NLSE (Eq. 3.8) in focusing case, i.e., $\lambda = 1$. Since these properties are considered in mean-square sense, we simulate noises $\delta_n\beta$ by identically distributed random variables $\sqrt{\tau}\xi_n$ with ξ_n being independent K-dimensional $N(0, 1)$-random variables, and approximate the expectation by taking averaged value over 1000 paths. In addition, the proposed scheme, which is implicit, is numerically solved utilizing the fixed point iteration.

According to the assumption $Q^{\frac{1}{2}} \in \mathscr{L}_2^{\frac{1}{2}+\varepsilon}$ in Theorem 6.5, the eigenvalues $\{\eta_k\}_{k\in\mathbb{N}}$ of Q should satisfy that $\sum_{k=1}^\infty k^{(1+\varepsilon)}\eta_k < \infty$. We choose $\eta_k = k^{-4}$ throughout the numerical experiments, and the first K terms will be used in the scheme. As we omit the boundary nodes in the simulation, we may choose the normalized initial value $U^0 = c_*(U^0(1), \cdots, U^0(M))^\top$ based on function $u_0(x)$ satisfying $U^0(m) = u_0(mh)$, $m = 1, \cdots, M$, in which $u_0(x)$ need not to satisfy the boundary condition

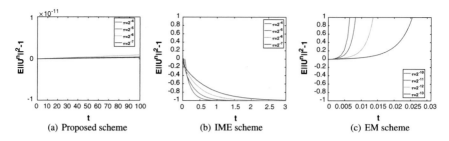

Fig. 6.1 Charge evolution $\mathbb{E}\|U^n\|^2 - 1$ for **a** the proposed scheme with $T = 100$ under step-sizes $\tau = 2^{-i}$ ($i = 4, 5, 6, 7$), **b** IME scheme with $T = 3$ under step-sizes $\tau = 2^{-i}$ ($i = 4, 5, 6, 7$), and **c** EM scheme with $T = 2^{-5}$ under step-sizes $\tau = 2^{-i}$ ($i = 10, 11, 12, 13$) ($h = 0.05$, $K = 30$)

in (6.1). Let $u_0(x) = 1$. Then the normalized initial value U^0 satisfies $\|U^0\| = 1$, which is used in Figs. 6.1, 6.4 and 6.3.

Charge evolution. To verify Proposition 6.2 and show the superiority of the proposed scheme, we first simulate the discrete charge for the proposed scheme compared with Euler–Maruyama (EM) scheme and implicit Euler (IE) scheme, respectively. Figure 6.1 shows that the proposed scheme possesses the discrete charge conservation law $\mathbb{E}\|U^n\|^2 = 1$, while both the EM scheme and the IE scheme deviate from the initial charge gradually. Since the EM scheme is not stable and will blow up in a short time, we choose the time step-size τ small enough for the EM scheme in the experiments.

Ergodic limit. We use the notation $\|U\|_\gamma^\gamma := \sum_{m=1}^{M} (|p_m|^\gamma + |q_m|^\gamma)$ for $U \in \mathbb{C}^M$ and $\gamma = 3, 4$ with $P = (p_1, \cdots, p_M)^\top$, $Q = (q_1, \cdots, p_M)^\top$ being the real and imaginary parts of U. To verify the ergodicity shown in Theorem 6.3, which states that the temporal averages will converge to the ergodic limit $\int_{\mathbb{S}} f \, d\mu_h$ for almost every initial value $U^0 \in \mathbb{S}$, we simulate the temporal averages $\frac{1}{N} \sum_{n=0}^{N-1} \mathbb{E}[f(U^n)]$ for the proposed scheme stating from different initial values. In Fig. 6.2, we choose test functions $f \in \mathbf{W}^{4,\infty}$ as (a) $f(U) = \|U\|_3^3$, (b) $f(U) = \sin(\|U\|_4^4)$ and (c) $f(U) = e^{-\|U\|_4^4}$, and choose five different initial values

$$U_l^0 = c_*(U_l^0(1), \cdots, U_l^0(M))^\top, \quad l = 1, \cdots, 5$$

based on the following five functions

$$u_{0,1}(x) = \frac{1}{\sqrt{2}} + \frac{\mathbf{i}}{\sqrt{2}}, \quad u_{0,2}(x) = 1, \quad u_{0,3}(x) = 2x,$$

$$u_{0,4}(x) = \left(1 - \sqrt{\frac{\pi}{2}(\exp\frac{1}{4} - 1)}\right)(1 - \exp(x(1 - x))),$$

$$u_{0,5}(x) = c_* \mathrm{sech}(\frac{x}{\sqrt{2}}) \exp(\mathbf{i}\frac{x}{2})$$

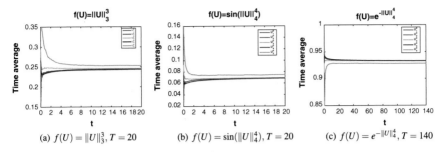

Fig. 6.2 The temporal averages $\frac{1}{N}\sum_{n=0}^{N-1}\mathbb{E}[f(U^n)]$ for the proposed scheme with **a** $f(U) = \|U\|_3^3$, **b** $f(U) = \sin(\|U\|_4^4)$ and **c** $f(U) = e^{-\|U\|_4^4}$ ($\tau = 2^{-6}$, $h = 0.05$, $K = 30$)

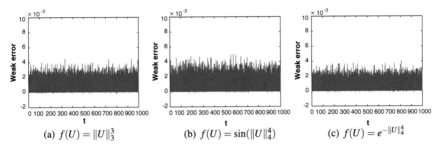

Fig. 6.3 The weak error $|\mathbb{E}[f(U^n)] - f(U(T))]|$ for **a** $f(U) = \|U\|_3^3$, **b** $f(U) = \sin(\|U\|_4^4)$ and **c** $f(U) = e^{-\|U\|_4^4}$ ($\tau = 2^{-12}$, $h = 0.05$, $T = 10^3$, $K = 30$)

with $U_l^0(m) = u_{0,l}(hm)$, $1 \leq m \leq M$ and c_* being normalized constants. Figure 6.2 shows that the proposed scheme starting from different initial values converges to the same value with error no more than $O(\tau)$ with $h = 0.05$ and $\tau = 2^{-6}$, which also coincides with the error analysis in Theorem 6.6.

Weak error and convergence order. For a fixed h, Figs. 6.3 and 6.4 show that the weak error over long time and weak convergence order in temporal direction, respectively, with test functions f being the same as those given above. Based on the ergodicity for both FDS and FDA, the weak error is supposed to be independent of time interval when time is large enough. To verify this property, we simulate the weak error over long time in Fig. 6.3. It shows that the weak error for the proposed scheme would not increase before $T = 1000$. Furthermore, Fig. 6.4 shows that the proposed scheme is of order one in the weak sense which coincides with the statement in Remark 6.3.

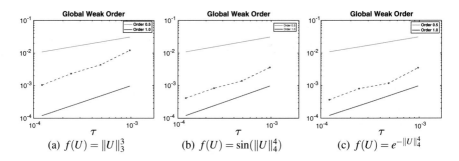

Fig. 6.4 The weak convergence order of $|\mathbb{E}[f(U^n) - f(U(T))]|$ with **a** $f(U) = \|U\|_3^3$, **b** $f(U) = \sin(\|U\|_4^4)$ and **c** $f(U) = e^{-\|U\|_4^4}$ ($\tau = 2^{-i}$, $10 \leq i \leq 13$, $h = 0.05$, $T = 2^{-1}$, $K = 30$)

Summary

In this chapter, stochastic NLSE with a linear multiplicative noise is studied. Similar to the deterministic case, this model possesses both the symplectic and multi-symplectic conservation laws and is charge-conserved almost surely. A finite dimensional approximation based on the central difference scheme is shown to be charge-conserved and ergodic on the unit sphere with charge one. A full discretization is then proposed possessing the discrete charge conservation law and the multi-symplectic conservation law. The numerical solution is also ergodic on the unit sphere, whose temporal approximation converges to the ergodic limit of the finite dimensional approximation. The ergodicity for the exact solution of stochastic systems without dissipative conditions is still unknown up to the knowledge of the authors.

It is introduced that the deterministic NLSE possesses an invariant Gibbs measure in Chap. 3. For stochastic cases, the behavior of the solution will be definitely influenced by the noise. There is a series of papers considering the blow-up phenomenon for stochastic NLSEs theoretically [62, 63, 66] and numerically [18, 68], and also the large deviation principle to give the asymptotics of the tails of the blow-up time [86–88]. We refer to [36, 45, 138] for large deviations for invariant measures, to [52, 72, 75] for more details about the theory and applications of large deviations. The large deviations related to numerical approximations for SDEs, as well as their invariant measures, are also worth considering as a way to investigate the behavior of the exact solution.

Appendix A
Basic Inequalities

This chapter is devoted to giving some basic inequalities (see e.g. [99]) which are frequently used in the study of the well-posedness and numerical approximations for differential equations.

Lemma A.1 (ε-Young inequality) *Let $\varepsilon > 0$, $p, q > 1$ such that $\frac{1}{p} + \frac{1}{q} = 1$. Then*

$$|a||b| \leq \frac{\varepsilon |a|^p}{p} + \frac{\varepsilon^{-\frac{q}{p}} |b|^q}{q} \leq \varepsilon |a|^p + \varepsilon^{-\frac{q}{p}} |b|^q$$

for any $a, b \in \mathbb{R}$.

For the case $\varepsilon = 1$, the inequality above is known as the Young inequality for products, which is also called Cauchy inequality if in particular $p = q = 2$. In what follows, we introduce the Young inequality for integral operators.

Lemma A.2 *Let X and Y be measurable spaces. Denote $Tf(x) := \int_Y K(x, y) f(y) dy$ with $K : X \times Y \to \mathbb{R}$ satisfying*

$$\sup_{x \in X} \| K(x, \cdot) \|_{\mathbf{L}^r(Y)} \leq C_0$$

and

$$\sup_{y \in Y} \| K(\cdot, y) \|_{\mathbf{L}^r(X)} \leq C_0$$

for $r \geq 1$. Then for $1 \leq p, q \leq \infty$ satisfying $1 + \frac{1}{q} = \frac{1}{r} + \frac{1}{p}$, it holds

$$\| Tf \|_{\mathbf{L}^q(X)} \leq C_0 \| f \|_{\mathbf{L}^p(Y)}.$$

Let $\mathcal{O} \subset \mathbb{R}^d$, $d \geq 1$, be a measurable set.

© Springer Nature Singapore Pte Ltd. 2019
J. Hong and X. Wang, *Invariant Measures for Stochastic Nonlinear
Schrödinger Equations*, Lecture Notes in Mathematics 2251,
https://doi.org/10.1007/978-981-32-9069-3

Lemma A.3 (Hölder's inequality) *Let* $1 \leq p, q \leq \infty$ *such that* $\frac{1}{p} + \frac{1}{q} = 1$. *For any* $f \in \mathbf{L}^p(\mathscr{O})$, $g \in \mathbf{L}^q(\mathscr{O})$, *it holds*

$$\|fg\|_{\mathbf{L}^1(\mathscr{O})} \leq \|f\|_{\mathbf{L}^p(\mathscr{O})} \|g\|_{\mathbf{L}^q(\mathscr{O})}.$$

For the case $p = q = 2$, *the inequality above is also called the Schwarz inequality.*

Lemma A.4 (Minkowski's inequality) *Let* $1 \leq p \leq \infty$. *For any* $f, g \in \mathbf{L}^p(\mathscr{O})$, *the following inequality holds*

$$\|f + g\|_{\mathbf{L}^p(\mathscr{O})} \leq \|f\|_{\mathbf{L}^p(\mathscr{O})} + \|g\|_{\mathbf{L}^p(\mathscr{O})}.$$

Sobolev Type Inequalities

Sobolev inequalities are frequently used to show the Sobolev embedding between different Sobolev spaces. We refer to [77] for more details.

Let k be a non-negative integer and $1 \leq p \leq \infty$. Denote by $\mathbf{W}^{k,p}(\mathscr{O})$ the space of equivalence classes of functions $u \in \mathbf{L}^p(\mathscr{O})$ such that $D^\alpha u \in \mathbf{L}^p(\mathscr{O})$ for all derivations of length $|\alpha| \leq k$, where

$$D^\alpha u = \frac{\partial^{\alpha_1}}{\partial x_1^{\alpha_1}} \cdots \frac{\partial^{\alpha_d}}{\partial x_d^{\alpha_d}} u$$

for each multi-index $\alpha = (\alpha_1, \cdots, \alpha_d)$.

The Gagliardo–Nirenberg interpolation inequality introduced below gives the estimates on weak derivatives of functions on Sobolev spaces.

Lemma A.5 (Gagliardo–Nirenberg inequality) *Let* $1 \leq p, q, r \leq \infty$ *and* j, m *be two integers such that* $0 \leq j < m$ *and*

$$\frac{1}{p} = \frac{j}{d} + a\left(\frac{1}{r} - \frac{m}{d}\right) + \frac{1-a}{q},$$

where $a \in [\frac{j}{m}, 1]$. *If* $r > 1$ *and* $m - j - \frac{d}{r} = 0$, *choose* $a < 1$. *Then there exists* $C = C(d, m, j, a, q, r)$ *such that for any* $u \in \mathbf{C}_0^\infty(\mathbb{R}^d)$, *one has*

$$\sum_{|\alpha|=j} \|\partial^\alpha u\|_{\mathbf{L}^p(\mathbb{R}^d)} \leq C\left(\sum_{|\alpha|=m} \|\partial^\alpha u\|_{\mathbf{L}^r(\mathbb{R}^d)}\right)^a \|u\|_{\mathbf{L}^q(\mathbb{R}^d)}^{1-a}.$$

Lemma A.6 (Gagliardo–Nirenberg–Sobolev inequality) *For any* $1 \leq p < d$, *there exists a constant* C *depending only on* d *and* p *such that for any* $u \in \mathbf{C}_0^\infty(\mathbb{R}^d)$, *one has*

$$\|u\|_{\mathbf{L}^{p^*}(\mathbb{R}^d)} \leq C\|Du\|_{\mathbf{L}^p(\mathbb{R}^d)}$$

with $\frac{1}{p^*} = \frac{1}{p} - \frac{1}{d}$.

The result above implies the Sobolev embedding $\mathbf{W}^{1,p}(\mathbb{R}^d) \hookrightarrow \mathbf{L}^{p^*}(\mathbb{R}^d)$. The general case is stated in the Sobolev embedding theorem.

Theorem A.1 (Sobolev embedding theorem) *Let* $\mathscr{O} = \mathbb{R}^d$, $d \geq 1$.

(i) *If* $k > l$ *and* $1 \leq p < q < \infty$ *satisfy* $(k-l)p < d$ *and*

$$\frac{1}{p} - \frac{k}{d} = \frac{1}{q} - \frac{l}{d},$$

then one has the embedding

$$\mathbf{W}^{k,p}(\mathbb{R}^d) \hookrightarrow \mathbf{W}^{l,q}(\mathbb{R}^d)$$

and the embedding is continuous.

(ii) *If* $d < p$ *and*

$$\frac{1}{p} - \frac{k}{d} = -\frac{r+\alpha}{d}$$

with $\alpha \in (0,1]$, *then one has the embedding*

$$\mathbf{W}^{k,p}(\mathbb{R}^d) \hookrightarrow \mathbf{C}^{r,\alpha}(\mathbb{R}^d).$$

The second part of the Sobolev embedding theorem is a direct consequence of the Morrey inequality.

Lemma A.7 (Morrey inequality) *Let* $d < p \leq \infty$ *and* $\gamma = 1 - \frac{d}{p}$. *Then there exists* $C = C(d,p)$ *such that*

$$\|u\|_{\mathbf{C}^{0,\gamma}(\mathbb{R}^d)} \leq C\|u\|_{\mathbf{W}^{1,p}(\mathbb{R}^d)}$$

for all $u \in \mathbf{C}^1(\mathbb{R}^d) \cap \mathbf{L}^p(\mathbb{R}^d)$.

The proof of the Sobolev embedding theorem relies on the following inequality, which is also known as the Hardy–Littlewood–Sobolev fractional integration theorem.

Lemma A.8 (Hardy–Littlewood–Sobolev inequality) *Assume that* $r > 1$, $1 < p < q < \infty$ *and*

$$1 + \frac{1}{q} = \frac{1}{r} + \frac{1}{p}.$$

Then there exists $C = C(p,q)$ *such that*

$$\|I_r f\|_{\mathbf{L}^q(\mathbb{R}^d)} \leq C\|f\|_{\mathbf{L}^p(\mathbb{R}^d)},$$

where $I_r f(x) = \int_{\mathbb{R}^d} |x-y|^{-\frac{d}{r}} f(y)dy$ *denotes the Riesz potential.*

Gronwall Type Inequalities

The classical Gronwall inequality is established by Gronwall [94]. It is then modified by Bellman [19], Bihari [21] and other researchers to deal with stability and uniqueness problems of differential equations.

Lemma A.9 (Gronwall–Bellman inequality) *Let $T > 0$ and u, $F : [0, T] \to \mathbb{R}$ be continuous and non-negative functions. If there exists $K > 0$ such that*

$$u(t) \le K + \int_0^t F(s)u(s)ds, \quad \forall\, t \in [0, T],$$

then

$$u(t) \le K \exp\left(\int_0^t F(s)ds\right), \quad \forall\, t \in [0, T].$$

Lemma A.9 has a discrete version, which is frequently used in the numerical analysis of differential equations.

Lemma A.10 *Let $\{u_n\}_{n\in\mathbb{N}}$ and $\{F_n\}_{n\in\mathbb{N}}$ be non-negative sequences. If there exists $K > 0$ such that*

$$u_N \le K + \sum_{n=1}^{N-1} F_n u_n, \quad N \in \mathbb{N},$$

then

$$u_N \le K \prod_{n=1}^{N-1}(1 + F_n) \le K \exp\left(\sum_{n=1}^{N-1} F_n\right), \quad N \in \mathbb{N}.$$

The Gronwall–Bellman inequality, together with its discrete version, is widely used to show the stability and uniqueness of the solutions of differential equations and their numerical discretizations. This inequality is then generalized by Bihari to establish a bound for the difference of solutions of differential equations with perturbations. The proof is given below for the readers' convenience.

Lemma A.11 (Bihari inequality) *Let u, F be non-negative and continuous functions in $[a, b] \subset \mathbb{R}$, and $K > 0$ be a constant. Assume that ϖ is a non-negative non-decreasing continuous function in $[0, \infty)$ such that*

$$u(t) \le K + \int_a^t F(s)\varpi(u(s))ds, \quad \forall\, t \in [a, b],$$

then one has

$$u(t) \le \Theta^{-1}\left(\Theta(K) + \int_a^t F(s)ds\right), \quad \forall\, t \in [a, T]$$

for $T \leq b$ such that $\Theta(K) + \int_a^t F(s)ds$ is within the domain of Θ^{-1} for any $t \in [a, T]$. Here,

$$\Theta(v) = \int_{v_0}^v \frac{1}{\varpi(x)} dx$$

for $v_0 > 0$, $v \geq 0$ and Θ^{-1} denotes the inverse function of Θ, which exists due to the monotony of ϖ.

Proof Denoting

$$y(t) := K + \int_a^t F(s)\varpi(u(s))ds > 0,$$

then we have $u(t) \leq y(t)$ for any $t \in [a, b]$. Furthermore, $\varpi(u(t)) \leq \varpi(y(t))$ and $\varpi(y(t)) > 0$ since ϖ is non-decreasing. It then yields that

$$\frac{y'(t)}{\varpi(y(t))} = \frac{F(t)\varpi(u(t))}{\varpi(y(t))} \leq F(t).$$

Integrating with respect to t, we have

$$\Theta(y(t)) - \Theta(y(a)) = \int_a^t \frac{1}{\varpi(y(s))} dy(s) \leq \int_a^t F(s)ds$$

with $\Theta(y(a)) = \Theta(K)$. As a result,

$$u(t) \leq y(t) \leq \Theta^{-1}\left(\Theta(K) + \int_a^t F(s)ds\right)$$

since Θ^{-1} is also non-decreasing.

Furthermore, we claim that the result above is independent of the choice of v_0 in the definition of function Θ. In fact, we define another function

$$\bar{\Theta}(v) := \int_{v_1}^v \frac{1}{\varpi(x)} dx.$$

By denoting

$$\delta := \int_{v_0}^{v_1} \frac{1}{\varpi(x)} dx,$$

we get

$$\bar{\Theta}(v) = \Theta(v) - \delta$$

and hence

$$\bar{\Theta}^{-1}(w) = \Theta^{-1}(w + \delta).$$

Finally, we derive

$$\bar{\Theta}^{-1}\left(\bar{\Theta}(K) + \int_a^t F(s)ds\right) = \Theta^{-1}\left(\bar{\Theta}(K) + \int_a^t F(s)ds + \delta\right)$$
$$= \Theta^{-1}\left(\Theta(K) + \int_a^t F(s)ds\right)$$

and complete the proof.   □

Example A.1 As an example, we choose $\varpi(x) := x$. Then

$$\Theta(v) = \ln(v) - \ln(v_0) \quad \text{for } v, v_0 > 0$$

and

$$\Theta^{-1}(w) = \exp(w + \ln(v_0)) \quad \text{for } w \in \mathbb{R}.$$

Assume that the condition in Lemma A.9 holds, that is,

$$u(t) \le K + \int_0^t F(s)u(s)ds.$$

Then according to the Bihari inequality, we derive

$$u(t) \le \Theta^{-1}\left(\Theta(K) + \int_0^t F(s)ds\right)$$
$$= \exp\left(\ln(K) - \ln(v_0) + \int_0^t F(s)ds + \ln(v_0)\right)$$
$$= K \exp\left(\int_0^t F(s)ds\right),$$

which verifies the result of the Gronwall–Bellman inequality.

Example A.2 If a continuous function $u : [a, b] \to \mathbb{R}$ satisfies

$$u^2(t) \le u_0^2 + \int_a^t F(s)|u(s)|ds$$

with $u_0 \in \mathbb{R}$ and F being non-negative and continuous in $[a, b]$, we claim that it holds

$$|u(t)| \le |u_0| + \frac{1}{2}\int_a^t F(s)ds.$$

In fact, by denoting $y_0 := u_0^2$, $y(t) := u^2(t)$ for $t \in [a, b]$ and choosing $\varpi(x) = x^{\frac{1}{2}}$ for $x \in \mathbb{R}_+$ such that

$$\Theta(v) = 2v^{\frac{1}{2}} - 2v_0^{\frac{1}{2}}$$

and

$$\Theta^{-1}(w) = \left(\frac{w}{2} + v_0^{\frac{1}{2}}\right)^2,$$

we have

$$y(t) \le y_0 + \int_a^t F(s)\varpi(y(s))ds.$$

Then according to the Bihari inequality, it leads to

$$y(t) \le \Theta^{-1}\left(\Theta(y_0) + \int_a^t F(s)ds\right)$$

$$= \left(y_0^{\frac{1}{2}} + \frac{1}{2}\int_a^t F(s)ds\right)^2,$$

which verifies the claim.

More generally, one can modify the Gronwall–Bellman inequality by replacing the constant K by a function depending on t, which is stated below.

Theorem A.2 *Assume that continuous functions u, Φ and F satisfy that F is non-negative in $[a, b]$ and*

$$u(t) \le \Phi(t) + \int_a^t F(s)u(s)ds, \quad \forall\, t \in [a, b].$$

Then

$$u(t) \le \Phi(t) + \int_a^t F(s)\Phi(s)\exp\left(\int_s^t F(r)dr\right)ds, \quad \forall\, t \in [a, b].$$

Proof Denoting

$$y(t) := \int_a^t F(s)u(s)ds$$

with $y(a) = 0$, we get

$$y'(t) = F(t)u(t) \le F(t)\left[\Phi(t) + \int_a^t F(s)u(s)ds\right]$$

$$= F(t)\Phi(t) + F(t)y(t).$$

Solving this ordinary differential equation by multiplying $\exp(-\int_a^t F(s)ds)$, we derive

$$\frac{d}{dt}\left(y(t)\exp\left(-\int_a^t F(s)ds\right)\right) \le F(t)\Phi(t)\exp\left(-\int_a^t F(s)ds\right).$$

It then leads to

$$y(t) \leq \int_a^t F(s)\Phi(s) \exp\left(\int_s^t F(r)dr\right) ds,$$

which completes the proof by using $u(t) \leq \Phi(t) + y(t)$.   □

The theorem above also has a discrete version.

Corollary A.1 *Let* $\{u_n\}_{n\in\mathbb{N}}$, $\{K_n\}_{n\in\mathbb{N}}$ *and* $\{F_n\}_{n\in\mathbb{N}}$ *be non-negative sequences which satisfy*

$$u_n \leq K_n + \sum_{0 \leq l < n} F_l u_l, \quad \forall n \geq 0.$$

Then

$$u_n \leq K_n + \sum_{0 \leq l < n} K_l F_l \exp\left(\sum_{l < j < n} F_j\right), \quad \forall n \geq 0.$$

However, even though the estimate above has a similar expression as the continuous case, it is not optimal. We get the following result, which also implies the result in Corollary A.1, under the same conditions.

Would like to mention that the following result is a sufficient condition of the above result under the same conditions.

Corollary A.2 *Under the conditions of Corollary A.1, it holds*

$$u_n \leq K_n + \sum_{0 \leq l < n} K_l F_l \prod_{l < j < n} (1 + F_j), \quad \forall n \geq 0.$$

Proof For any $n \geq 0$, we denote

$$G_n := \prod_{0 \leq j < n} (1 + F_j), \quad \forall n \geq 0.$$

Noticing that $G_0 = 1$, then for $n = 1, 2$, we get

$$G_1 = G_0 + F_0 G_0$$

and

$$G_2 = G_1 + F_1 G_1 = G_0 + F_0 G_0 + F_1 G_1.$$

By induction, we assume that it holds for some $n \in \mathbb{N}$ that

$$G_n = G_k + \sum_{k \leq j < n} F_j G_j, \quad \forall 0 \leq k \leq n.$$

Then

$$G_{n+1} = G_n(1 + F_n) = G_k + \sum_{k \le j < n} F_j G_j + G_n F_n$$

$$= G_k + \sum_{k \le j < n+1} F_j G_j, \quad \forall \, 0 \le k \le n + 1.$$

Based on these notations, we have

$$K_n + \sum_{0 \le l < n} K_l F_l \prod_{l < j < n} (1 + F_j) = K_n + \sum_{0 \le l < n} K_l F_l \frac{G_n}{G_{l+1}}.$$

We first show that if another sequence $\{x_n\}_{n \in \mathbb{N}}$ satisfies the equality

$$x_n = K_n + \sum_{0 \le k < n} F_k x_k, \quad \forall \, n \ge 0,$$

then it holds

$$x_n = K_n + \sum_{0 \le l < n} K_l F_l \frac{G_n}{G_{l+1}}.$$

Note that $x_0 = K_0$ satisfies the result above. Assume by induction that the above result holds for $\{x_k\}_{0 \le k \le n}$. Then we derive

$$x_{n+1} = K_{n+1} + \sum_{0 \le k < n+1} F_k x_k$$

$$= K_{n+1} + \sum_{0 \le k < n+1} F_k \left(K_k + \sum_{0 \le l < k} K_l F_l \frac{G_k}{G_{l+1}} \right)$$

$$= K_{n+1} + \sum_{0 \le k < n+1} F_k K_k + \sum_{0 \le l < n} \left(\sum_{l < k < n+1} F_k K_l F_l \frac{G_k}{G_{l+1}} \right)$$

$$= K_{n+1} + F_n K_n + \sum_{0 \le k < n} F_k K_k + \sum_{0 \le k < n} F_k K_k \left(\sum_{k < l < n+1} F_l \frac{G_l}{G_{k+1}} \right)$$

$$= K_{n+1} + F_n K_n + \sum_{0 \le k < n} F_k K_k \frac{G_{n+1}}{G_{k+1}}$$

$$= K_{n+1} + \sum_{0 \le k < n+1} F_k K_k \frac{G_{n+1}}{G_{k+1}}.$$

It then suffices to show that $u_n \le x_n$ for any $n \ge 0$. In fact, $u_0 \le K_0 = x_0$. Assume by induction that $u_k \le x_k$ for $0 \le k \le n$. Then according to the above assumptions and the assumption of Corollary A.1, we finally derive

$$u_{n+1} \leq K_{n+1} + \sum_{0 \leq l < n+1} F_l u_l \leq K_{n+1} + \sum_{0 \leq l < n+1} F_l x_l = x_{n+1}.$$

□

Following is another modification of the Gronwall–Bellman inequality which can be obtained based on Theorem A.2 and is used in Chaps. 5 and 6.

Theorem A.3 *If a non-negative and continuous function u satisfies*

$$u(t) \leq e^{-\alpha(t-t_0)} u(t_0) + \int_{t_0}^{t} e^{-\alpha(t-s)} (au(s) + b) ds \qquad (A.1)$$

with a, b and α being positive constants, then

$$u(t) \leq e^{a(t-t_0)} \left[u(t_0) e^{-\alpha(t-t_0)} + \frac{b}{\alpha} \left(1 - e^{-\alpha(t-t_0)} \right) \right].$$

Proof By multiplying $e^{\alpha t}$ to both sides of inequality (A.1), we have

$$e^{\alpha t} u(t) \leq e^{\alpha t_0} u(t_0) + \int_{t_0}^{t} e^{\alpha s} (au(s) + b) ds$$

$$= e^{\alpha t_0} \left(u(t_0) - \frac{b}{\alpha} \right) + \frac{b}{\alpha} e^{\alpha t} + a \int_{t_0}^{t} e^{\alpha s} u(s) ds.$$

Denoting $y(t) := e^{\alpha t} u(t)$ and

$$\Phi(t) := e^{\alpha t_0} \left(u(t_0) - \frac{b}{\alpha} \right) + \frac{b}{\alpha} e^{\alpha t},$$

then Theorem A.2 leads to

$$y(t) \leq \Phi(t) + a \int_{t_0}^{t} \Phi(s) e^{a(t-s)} ds$$

$$= e^{\alpha t_0} \left(u(t_0) - \frac{b}{\alpha} \right) + \frac{b}{\alpha} e^{\alpha t} + a e^{\alpha t_0} \left(u(t_0) - \frac{b}{\alpha} \right) \int_{t_0}^{t} e^{a(t-s)} ds$$

$$+ \frac{ab}{\alpha} e^{(\alpha+a)t} \int_{t_0}^{t} e^{-as} ds$$

$$= e^{\alpha t} e^{a(t-t_0)} \left[u(t_0) e^{-\alpha(t-t_0)} + \frac{b}{\alpha} \left(1 - e^{-\alpha(t-t_0)} \right) \right],$$

from which we finally get the result.

□

Appendix B
Proof of the Birkhoff–Khinchin Ergodic Theorem

The Endomorphism Case

For an endomorphism P and any $\varphi \in \mathbf{L}^1(\mathbb{M}, \mu)$, define

$$M_N\varphi(x) := \frac{1}{N} \sum_{n=0}^{N-1} \varphi(P^n x), \quad x \in \mathbb{M}, \ n \in \mathbb{N}_+$$

with $M_0\varphi(x) \equiv 0$ and

$$E(\varphi) := \left\{ x \in \mathbb{M}, \ \sup_{N \in \mathbb{N}} M_N\varphi(x) > 0 \right\}.$$

Then set E has the following property.

Lemma B.1 (Maximal ergodic theorem) *For any $\varphi \in \mathbf{L}^1(\mathbb{M}, \mu)$, it holds*

$$\int_{E(\varphi)} \varphi d\mu \geq 0.$$

Proof For any fixed $\varphi \in \mathbf{L}^1(\mathbb{M}, \mu)$, we use the notation

$$S_n(x) := nM_n\varphi(x) = \sum_{k=0}^{n-1} \varphi(P^k x)$$

with $S_0(x) \equiv 0$ for simplicity. Then the set $E(\varphi)$ can also be expressed as

$$E(\varphi) = \left\{ x \in \mathbb{M}, \ \sup_{N \in \mathbb{N}} S_N(x) > 0 \right\}.$$

© Springer Nature Singapore Pte Ltd. 2019
J. Hong and X. Wang, *Invariant Measures for Stochastic Nonlinear Schrödinger Equations*, Lecture Notes in Mathematics 2251,
https://doi.org/10.1007/978-981-32-9069-3

Noticing that for any $n \in \mathbb{N}$, it holds

$$S_n(Px) = \sum_{k=0}^{n-1} \varphi(P^{k+1}x) = \sum_{k=1}^{n} \varphi(P^k x)$$
$$= S_{n+1}(x) - \varphi(x)$$

for all $x \in \mathbb{M}$.

Taking the maximum over $n = 0, 1, \cdots, N-1$ of the above equality, we obtain

$$\Phi_N^+(Px) = \Phi_{N+1}(x) - \varphi(x),$$

where

$$\Phi_N^+(x) := \max_{0 \le n \le N-1} S_n(x) = \max_{0 \le n \le N-1} \{0, S_1(x), \cdots, S_{N-1}(x)\} \ge 0,$$
$$\Phi_{N+1}(x) := \max_{0 \le n \le N-1} S_{n+1}(x) = \max_{0 \le n \le N-1} \{S_1(x), \cdots, S_N(x)\}.$$

As a result, we have

$$\varphi(x) = \Phi_{N+1}(x) - \Phi_N^+(Px) \ge \Phi_N(x) - \Phi_N^+(Px). \tag{B.1}$$

Denote the set

$$A_N := \{x \in \mathbb{M}, \, \Phi_N^+(x) > 0\}.$$

For $x \in A_N$, it holds $\Phi_N(x) = \Phi_N^+(x)$, while for $x \notin A_N$ it holds $\Phi_N^+(x) = 0$. We then deduce that

$$\int_{A_N} \Phi_N d\mu = \int_{A_N} \Phi_N^+ d\mu = \int_{\mathbb{M}} \Phi_N^+ d\mu$$

and

$$\int_{A_N} \Phi_N^+(Px)\mu(dx) \le \int_{\mathbb{M}} \Phi_N^+(Px)\mu(dx)$$

due to the fact $\Phi_N^+ \ge 0$.

According to (B.1), we finally derive

$$\int_{A_N} \varphi d\mu \ge \int_{A_N} \Phi_N(x)\mu(dx) - \int_{A_N} \Phi_N^+(Px)\mu(dx)$$
$$\ge \int_{\mathbb{M}} \Phi_N^+ d\mu - \int_{\mathbb{M}} \Phi_N^+(Px)\mu(dx) = 0,$$

where we used the fact μ is an invariant measure for the endomorphism P. It then leads to the result by taking the limit for $N \to \infty$. $\qquad \square$

Now we give the proof of the Birkhoff–Khinchin ergodic theorem for the endomorphism case utilizing the above lemma.

For any rational number a and b satisfying $a < b$, denote

$$G_{a,b} := \left\{ x \in \mathbb{M}, \ \varliminf_{N \to \infty} M_N \varphi(x) < a < b < \varlimsup_{N \to \infty} M_N \varphi(x) \right\} \in \mathcal{G}.$$

Note that

$$M_N \varphi(Px) = \frac{1}{N} \sum_{n=0}^{N-1} \varphi(P^{n+1}x) = \frac{1}{N} \sum_{n=1}^{N} \varphi(P^n x)$$

$$= M_N \varphi(x) + \frac{\varphi(P^N x)}{N} - \frac{\varphi(x)}{N}.$$

As a result, the set $G_{a,b}$, as well as its complementary set $G_{a,b}^{\mathrm{c}}$, is invariant with respect to the endomorphism P for any $a < b$.

To show the existence of the limit

$$\lim_{N \to \infty} M_N \varphi(x),$$

it then suffices to show that $\mu(G_{a,b}) = 0$ for any $a < b$.

We define a function

$$f(x) := \begin{cases} \varphi(x) - b & \text{for } x \in G_{a,b}, \\ 0 & \text{for } x \in G_{a,b}^{\mathrm{c}}, \end{cases}$$

which satisfies

$$\int_{E(f)} f \, d\mu \geq 0$$

with

$$E(f) = \left\{ x \in \mathbb{M}, \ \sup_{N \in \mathbb{N}} M_N f(x) > 0 \right\} = \left\{ x \in \mathbb{M}, \ \sup_{N \in \mathbb{N}} M_N \varphi(x) > b \right\} \supset G_{a,b}.$$

On the other hand, for any $x \in G_{a,b}^{\mathrm{c}}$, we have $P^n x \in G_{a,b}^{\mathrm{c}}$ for any $n \in \mathbb{N}$ due to the invariance of the set $G_{a,b}^{\mathrm{c}}$. Hence, $M_N f(x) = 0$ for any $x \in G_{a,b}^{\mathrm{c}}$, that is, $x \in F(f)^{\mathrm{c}}$. We then conclude from above that $E(f) = G_{a,b}$ and thus

$$\int_{G_{a,b}} \varphi \, d\mu = \int_{E(f)} (f + b) \, d\mu \geq b \geq b\mu(G_{a,b}).$$

If we consider the function

$$g(x) := \begin{cases} a - \varphi(x) & \text{for } x \in G_{a,b}, \\ 0 & \text{for } x \in G_{a,b}^c \end{cases}$$

instead of f defined above, we will get

$$\int_{G_{a,b}} \varphi d\mu \leq a\mu(G_{a,b})$$

following the same procedure as above. It finally leads to

$$\int_{G_{a,b}} \varphi d\mu \leq a\mu(G_{a,b}) \leq b\mu(G_{a,b}) \leq \int_{G_{a,b}} \varphi d\mu,$$

in which the equality holds if and only if $\mu(G_{a,b}) = 0$.

The Semi-flow Case

The proof of the semi-flow case follows from the endomorphism case.

For any $T > 0$, we denote $n_T := [T]$ and $n_r \in [0, 1)$ such that

$$T = n_T + n_r.$$

For a semi-flow $\{P_t\}_{t \geq 0}$, we define

$$M(T)\varphi(x) := \frac{1}{T} \int_0^T \varphi(P_t x) dt,$$

which satisfies

$$M(T)\varphi(x) = \frac{1}{T} \sum_{k=0}^{n_T - 1} \int_k^{k+1} \varphi(P_t x) dt + \frac{1}{T} \int_{n_T}^T \varphi(P_t x) dt$$

$$= \frac{1}{T} \sum_{k=0}^{n_T - 1} \int_0^1 \varphi(P_{s+k} x) ds + \frac{1}{T} \int_0^{n_r} \varphi(P_{s+n_T} x) ds$$

$$= \frac{n_T}{T} \frac{1}{n_T} \sum_{k=0}^{n_T - 1} M(1)\varphi(P_1^k x) + \frac{r_T}{T} M(r_T)\varphi(P_1^{n_T} x).$$

Since $M(1)\varphi, M(r_T)\varphi \in L^1(\mathbb{M}, \mu)$ and

$$\lim_{T \to \infty} \frac{n_T}{T} = 1, \quad \lim_{T \to \infty} \frac{r_T}{T} = 0,$$

we get the result according to the proof of the endomorphism case.

Appendix C
Proofs of Propositions 5.1, 5.3 and 5.4

C.1 Proof of Proposition 5.1

(i) It was proved in Part 3 of Theorem 5.1 that $\mathbb{E}\|u_M(t)\|_0^2 < C$, so we assume further that $\mathbb{E}\|u_M(t)\|_0^{2n} < C$ for any $n = 1, \cdots, p - 1$. Denoting $dM_1 := 2\Re (u_M, \pi_M dW)$, then Itô's formula and (5.6) yield

$$d\|u_M(t)\|_0^{2p} = p\|u_M(t)\|_0^{2(p-1)} d\|u_M(t)\|_0^2 + \frac{1}{2}p(p-1)\|u_M(t)\|_0^{2(p-2)} d\langle M_1\rangle$$

$$\leq -2\alpha p\|u_M(t)\|_0^{2p} dt + p\|u_M(t)\|_0^{2(p-1)} dM_1(t)$$

$$+ 2p(2p-1)\sum_{k=1}^{M}\eta_k\|u_M(t)\|_0^{2(p-1)} dt,$$

where $\langle \cdot \rangle$ denotes the quadratic variation process and in the last step we used the fact

$$d\langle M_1\rangle = 4\left\langle \Re\sum_{k=1}^{M}\int_0^1 \bar{u}_M(s)\sqrt{\eta_k}e_k(x)dx(d\beta_k^1 + \mathbf{i}d\beta_k^2)\right\rangle$$

$$= 4\sum_{k=1}^{M}\left[\left(\Re\int_0^1 \bar{u}_M(t, x)\sqrt{\eta_k}e_k(x)dx\right)^2 + \left(\Im\int_0^1 \bar{u}_M(t, x)\sqrt{\eta_k}e_k(x)dx\right)^2\right] dt$$

$$\leq 8\sum_{k=1}^{M}\eta_k\|u_M(t)\|_0^2 dt.$$

Taking the expectation on both sides of the above equation, we obtain

© Springer Nature Singapore Pte Ltd. 2019
J. Hong and X. Wang, *Invariant Measures for Stochastic Nonlinear Schrödinger Equations*, Lecture Notes in Mathematics 2251,
https://doi.org/10.1007/978-981-32-9069-3

$$\frac{d}{dt}\mathbb{E}\|u_M(t)\|_0^{2p} \leq -2\alpha p\mathbb{E}\|u_M(t)\|_0^{2p} + 2p(2p-1)\sum_{k=1}^{M}\eta_k\mathbb{E}\|u_M(t)\|_0^{2(p-1)}$$

$$\leq -2\alpha p\mathbb{E}\|u_M(t)\|_0^{2p} + C$$

by induction. Then multiplying both sides of the above equation by $e^{2\alpha pt}$ yields the result.

(ii) The proof in this part is similar to that of Lemma 2.5 in [70]. According to the Gagliardo–Nirenberg interpolation inequality, there exists a positive constant c_0 such that

$$\frac{5}{8}\lambda\|u_M(t)\|_{\mathbf{L}^4}^4 \leq \|u_M(t)\|_{\mathbf{L}^4}^4 \leq \frac{1}{4}\|\nabla u_M(t)\|_0^2 + \frac{1}{2}c_0\|u_M(t)\|_0^6. \qquad \text{(C.1)}$$

Thus,

$$0 \leq H(u_M(t)) := \frac{1}{2}\|\nabla u_M(t)\|_0^2 - \frac{\lambda}{4}\|u_M(t)\|_{\mathbf{L}^4}^4 + c_0\|u_M(t)\|_0^6$$

$$\leq \frac{2}{3}\left(\|\nabla u_M(t)\|_0^2 - \lambda\|u_M(t)\|_{\mathbf{L}^4}^4 + 2c_0\|u_M(t)\|_0^6\right). \qquad \text{(C.2)}$$

Applying Itô's formula to $H(u_M(t))$, we get

$$dH(u_M(t)) = \left[-\alpha\|\nabla u_M(t)\|_0^2 + \alpha\lambda\|u_M(t)\|_{\mathbf{L}^4}^4 - 6\alpha c_0\|u_M(t)\|_0^6 \right.$$

$$- 2\lambda\int_0^1 |u_M|^2\sum_{k=1}^{M}\eta_k|e_k|^2 dx + 6c_0\|u_M(t)\|_0^4\sum_{k=1}^{M}\eta_k$$

$$\left. + \sum_{k=1}^{M}k^2\eta_k + 12c_0\|u_M(t)\|_0^2\|\pi_M Q^{\frac{1}{2}}u_M(t)\|_0^2\right]dt$$

$$+ 6c_0\|u_M(t)\|_0^4\Re\left(u_M, \pi_M dW\right) - \Re\left(\Delta u_M(t) + \lambda|u_M(t)|^2 u_M(t), \pi_M dW\right),$$

where we have used the fact $((Id - \pi_M)v, v_M) = 0$ for any $v \in \dot{\mathbf{H}}^0$, $v_M \in V_M$. By the following estimates based on Young's inequality

$$6c_0\|u_M(t)\|_0^4\sum_{k=1}^{M}\eta_k + 12c_0\|u_M(t)\|_0^2\|\pi_M Q^{\frac{1}{2}}u_M(t)\|_0^2 \leq 4\alpha c_0\|u_M(t)\|_0^6 + C$$

and (C.2), we have

$$dH(u_M(t)) \leq \left[-\alpha\|\nabla u_M(t)\|_0^2 + \alpha\lambda\|u_M(t)\|_{\mathbf{L}^4}^4 - 2\alpha c_0\|u_M(t)\|_0^6 + \sum_{k=1}^{M}k^2\eta_k + C\right]dt$$

$$+ 6c_0\|u_M(t)\|_0^4\Re\left(u_M(t), \pi_M dW(t)\right)$$

$$- \Re \left(\Delta u_M(t) + \lambda |u_M(t)|^2 u_M(t), \pi_M dW \right)$$

$$\leq - \frac{3}{2} \alpha H(u_M(t)) dt + C dt + dM_2, \tag{C.3}$$

where

$$dM_2 := 6c_0 \|u_M\|_0^4 \Re (u_M, \pi_M dW) - \Re \left(\Delta u_M + \lambda |u_M|^2 u_M, \pi_M dW \right).$$

Taking the expectation, we derive

$$d\mathbb{E}[H(u_M(t))] \leq - \frac{3}{2} \alpha \mathbb{E}[H(u_M(t))] dt + C dt.$$

Hence, by multiplying $e^{\frac{3}{2}\alpha t}$ to both sides of the equation above and then taking integral from 0 to t, we get the uniform boundedness for $p = 1$. By induction, we assume that the results hold for $p - 1$. Then, based on the following estimates (see [70])

$$\left\langle 6\|u_M\|_0^4 \Re (u_M, \pi_M dW) \right\rangle^2 \leq C \|Q^{\frac{1}{2}}\|_{HS(L^2,L^2)}^2 \|u_M\|_0^{10} dt,$$

$$\left\langle \Re \left(\Delta u_M + \lambda |u_M|^2 u_M, \pi_M dW \right) \right\rangle^2 \leq C \|Q^{\frac{1}{2}}\|_{HS(L^2,\dot{H}^1)}^2 \left(\|\nabla u_M\|_0^2 + \|u_M\|_0^{10} \right) dt$$

and (C.3), we have

$$dH(u_M(t))^p = pH(u_M(t))^{p-1} dH(u_M(t)) + \frac{1}{2} p(p-1) H(u_M(t))^{p-2} d\langle M_2 \rangle$$

$$\leq - \frac{3}{2} \alpha p H(u_M(t))^p dt + Cp H(u_M(t))^{p-1} dt + pH(u_M(t))^{p-1} dM_2$$

$$+ Cp(p-1) H(u_M(t))^{p-2} \left(\|\nabla u_M(t)\|_0^2 + \|u_M(t)\|_0^{10} \right) dt. \tag{C.4}$$

From (C.1), we deduce that

$$H(u_M(t)) \geq \begin{cases} \dfrac{1}{2} \|\nabla u_M(t)\|_0^2 + c_0 \|u_M(t)\|_0^6, & \lambda = 0 \text{ or } -1, \\[2mm] \dfrac{7}{16} \|\nabla u_M(t)\|_0^2 + \dfrac{7}{8} c_0 \|u_M(t)\|_0^6, & \lambda = 1. \end{cases}$$

As a result, the last term in (C.4) can be estimated as

$$Cp(p-1) H(u_M(t))^{p-2} \left(\|\nabla u_M(t)\|_0^2 + \|u_M(t)\|_0^{10} \right)$$

$$\leq \left(C H(u_M(t)) + C H(u_M(t))^{\frac{5}{3}} \right) H(u_M(t))^{p-2}$$

$$\leq C H(u_M(t))^{p-1} + \frac{1}{2} \alpha p H(u_M(t))^p, \tag{C.5}$$

where in the last step we used the inequality of arithmetic and geometric means

$$C(H(u_M(t))^2 \cdot H(u_M(t))^2 \cdot H(u_M(t)))^{\frac{1}{3}}$$
$$\leq \frac{\frac{3}{4}\alpha p H(u_M(t))^2 + \frac{3}{4}\alpha p H(u_M(t))^2 + CH(u_M(t))}{3}.$$

Gathering (C.4) and (C.5) and taking the expectation, we obtain

$$d\mathbb{E}[H(u_M(t))^p] \leq -\alpha p\mathbb{E}[H(u_M(t))^p]dt + Cdt$$

by induction. This completes the proof by multiplying $e^{\alpha pt}$ on both sides of above equation.

(iii) We define a functional

$$\Phi(u) = \int_0^1 |\Delta u|^2 dx + \lambda\Re \int_0^1 (\Delta\overline{u})|u|^2 u dx,$$

which satisfies

$$\|\Delta u\|_0^2 \leq 2\Phi(u) + C\|u\|_1^6 \tag{C.6}$$

based on the continuous embedding $\mathbf{H}^1 \hookrightarrow \mathbf{L}^6$ and the fact

$$\left| \lambda\Re \int_0^1 \Delta\overline{u}|u|^2 u dx \right| \leq \frac{1}{2}\|\Delta u\|_0^2 + \frac{1}{2}\|u\|_{\mathbf{L}^6}^6 \leq \frac{1}{2}\|\Delta u\|_0^2 + C\|u\|_1^6. \tag{C.7}$$

Itô's formula applied to $\Phi(u_M)$ yields

$$d\Phi(u_M) = D\Phi(u_M)\Big((i\Delta u_M + i\lambda|u_M|^2 u_M - \alpha u_M)\, dt \Big) + D\Phi(u_M)\Big(\pi_M dW\Big)$$
$$+ \frac{1}{2}D^2\Phi(u_M)(\pi_M dW, \pi_M dW)$$
$$=: \mathscr{A} + \mathscr{B} + \mathscr{C}, \tag{C.8}$$

where $\mathbb{E}[\mathscr{B}] = 0$. The first and second order derivatives satisfy

$$D\Phi(u)(\varphi) = \Re \int_0^1 \Big[2\Delta\overline{u}\Delta\varphi + 2\lambda(\Delta\overline{u})u\Re(\overline{u}\varphi) + \lambda(\Delta\overline{u})|u|^2\varphi + \lambda(\Delta(|u|^2 u))\overline{\varphi} \Big]dx$$

and

$$D^2\Phi(u)(\varphi, \psi) = \Re \int_0^1 \Big[2\Delta\overline{\varphi}\Delta\psi + 2\lambda(\Delta\overline{u})u\Re(\overline{\varphi}\psi) + 2\lambda(\Delta\overline{u})\varphi\Re(\overline{u}\psi)$$
$$+ 2\lambda(\Delta\overline{\varphi})u\Re(\overline{u}\psi) + 2\lambda(\Delta\overline{u})\psi\Re(\overline{\varphi}u) + 2\lambda(\Delta\overline{\psi})u\Re(\overline{u}\varphi)$$

$$+ \lambda(\Delta\overline{\varphi})|u|^2\psi + \lambda(\Delta\overline{\psi})|u|^2\varphi\Big]dx.$$

Now we estimate $\mathscr{A}$ and $\mathscr{C}$ respectively.

$$\mathbb{E}[\mathscr{A}] = -2\alpha\mathbb{E}[\Phi(u_M)]dt + \Re\mathbb{E}\int_0^1\Big[4\lambda\mathbf{i}(\Delta\overline{u}_M)u_M|\nabla u_M|^2 + 2\lambda\mathbf{i}(\Delta\overline{u}_M)\overline{u}_M(\nabla u_M)^2\Big]dxdt$$

$$+ \Re\mathbb{E}\int_0^1\Big[\lambda^2\mathbf{i}(\Delta\overline{u}_M)|u_M|^4 - 4\alpha\lambda(\Delta\overline{u}_M)u_M|u_M|^2\Big]dxdt$$

$$+ \Re\mathbb{E}\int_0^1\Big[-4\alpha\lambda|u_M|^2|\nabla u_M|^2 - 2\alpha\lambda(\nabla u_M)^2\overline{u}_M^2\Big]dxdt$$

$$=: -2\alpha\mathbb{E}[\Phi(u_M)]dt + \mathscr{A}_1dt + \mathscr{A}_2dt + \mathscr{A}_3dt,$$

where we have used the fact $\Delta(|u|^2u) = 2|u|^2\Delta u + 4u|\nabla u|^2 + 2\overline{u}(\nabla u)^2 + u^2\Delta\overline{u}$ and $\mathscr{A}_1$, $\mathscr{A}_2$ and $\mathscr{A}_3$ are estimated as follows.

$$|\mathscr{A}_1| := \left|\Re\mathbb{E}\int_0^1\Big[4\lambda\mathbf{i}(\Delta\overline{u}_M)u_M|\nabla u_M|^2 + 2\lambda\mathbf{i}(\Delta\overline{u}_M)\overline{u}_M(\nabla u_M)^2\Big]dx\right|$$

$$\leq \frac{\alpha}{16}\mathbb{E}\|\Delta u_M\|_0^2 + C\mathbb{E}\big[\|u_M\|_{\mathbf{L}^\infty}^2\|\nabla u_M\|_{\mathbf{L}^4}^2\big]$$

$$\leq \frac{\alpha}{16}\mathbb{E}\|\Delta u_M\|_0^2 + C\mathbb{E}\big[\|u_M\|_{\mathbf{L}^\infty}^4 + \|\Delta u_M\|_0\|\nabla u_M\|_0^3\big]$$

$$\leq \frac{\alpha}{8}\mathbb{E}\|\Delta u_M\|_0^2 + C\mathbb{E}\big[\|u_M\|_1^4 + \|u_M\|_1^6\big]$$

$$\leq \frac{\alpha}{8}\mathbb{E}\|\Delta u_M\|_0^2 + C,$$

where we have used the uniform boundedness of $\|u_M\|_1^{2p}$ for $p \geq 1$ in (ii), the continuous embedding $\mathbf{H}^1 \hookrightarrow \mathbf{L}^\infty$ and the interpolation of $\mathbf{L}^4$ between $\mathbf{L}^2$ and $\mathbf{H}^1$. Similarly, based on the continuous embedding $\mathbf{H}^1 \hookrightarrow \mathbf{L}^6$ and $\mathbf{H}^1 \hookrightarrow \mathbf{L}^8$, we have

$$|\mathscr{A}_2| := \left|\Re\mathbb{E}\int_0^1\Big[\lambda^2\mathbf{i}(\Delta\overline{u}_M)|u_M|^4 - 4\alpha\lambda(\Delta\overline{u}_M)u_M|u_M|^2\Big]dx\right|$$

$$\leq \frac{\alpha}{8}\mathbb{E}\|\Delta u_M\|_0^2 + C\mathbb{E}[\|u_M\|_{\mathbf{L}^8}^8 + \|u_M\|_{\mathbf{L}^6}^6] \leq \frac{\alpha}{8}\mathbb{E}\|\Delta u_M\|_0^2 + C$$

and

$$|\mathscr{A}_3| := \left|\Re\mathbb{E}\int_0^1\Big[-4\alpha\lambda|u_M|^2|\nabla u_M|^2 - 2\alpha\lambda(\nabla u_M)^2\overline{u}_M^2\Big]dx\right| \leq C\mathbb{E}\|u_M\|_1^4 \leq C.$$

Thus, we obtain

$$\mathbb{E}[\mathscr{A}] \leq -2\alpha\mathbb{E}[\Phi(u_M)]dt + \frac{\alpha}{4}\mathbb{E}\|\Delta u_M\|_0^2 + C.$$

The estimate of $\mathscr{C}$ is similar to that of $\mathscr{A}$, and we derive $\mathbb{E}[\mathscr{C}] \leq \frac{\alpha}{4}\mathbb{E}\|\Delta u_M\|_0^2 + C$. Taking the expectation on both sides of (C.8) yields

$$d\mathbb{E}\Phi(u_M) + 2\alpha\mathbb{E}\Phi(u_M)dt \leq \frac{\alpha}{2}\mathbb{E}\|\Delta u_M\|_0^2 dt + Cdt \leq \alpha\mathbb{E}\Phi(u_M)dt + Cdt.$$

Multiplying both sides of above equation by $e^{\alpha t}$ and integrating from 0 to t, we conclude the uniform boundedness of $\mathbb{E}\Phi(u_M(t))$

$$\mathbb{E}\Phi(u_M(t)) \leq e^{-\alpha t}\mathbb{E}\Phi(u_M(0)) + \frac{C}{\alpha}(1 - e^{-\alpha t}),$$

which yields the uniform boundedness of $\mathbb{E}\|\Delta u_M\|_0^2$ based on (C.6). As the norm $\|u_M\|_2$ is equivalent to $\|\Delta u_M\|_0$ under Dirichlet boundary condition, we complete the proof.

C.2   Proof of Proposition 5.3

The proof for $\lambda = 0$ is in the same procedure as that for $\lambda = -1$ and is much easier. Here we only give the proof for $\lambda = -1$

$$u_M^n - e^{-\alpha\tau}u_M^{n-1} = \left(\mathbf{i}\Delta u_M^n - \mathbf{i}\pi_M\left(\frac{|u_M^n|^2 + |e^{-\alpha\tau}u_M^{n-1}|^2}{2}u_M^n\right)\right)\tau + \pi_M\delta_n W.$$

(C.9)

(i) $p = 1$. Multiplying (C.9) by $\overline{u}_M^n - e^{-\alpha\tau}\overline{u}_M^{n-1}$, integrating with respect to x, taking the imaginary part and using the fact $((Id - \pi_M)v, v_M) = 0$ for any $v \in \dot{\mathbf{H}}^0$, $v_M \in V_M$, we have

$$\|\nabla u_M^n\|_0^2 + \|\nabla(u_M^n - e^{-\alpha\tau}u_M^{n-1})\|_0^2 - e^{-2\alpha\tau}\|\nabla u_M^{n-1}\|_0^2$$

$$= -\Re\int_0^1 \left(|u_M^n|^2 + |e^{-\alpha\tau}u_M^{n-1}|^2\right)u_M^n(\overline{u}_M^n - e^{-\alpha\tau}\overline{u}_M^{n-1})dx$$

$$+ \frac{2}{\tau}\Im\int_0^1 \pi_M\delta_n W(\overline{u}_M^n - e^{-\alpha\tau}\overline{u}_M^{n-1})dx$$

$$=: \mathscr{A} + \mathscr{B}.$$

(C.10)

Simple computations yield

$$\mathscr{A} = -\Re\left[\int_0^1 \left(|u_M^n|^2 + |e^{-\alpha\tau}u_M^{n-1}|^2\right)\left(\frac{u_M^n + e^{-\alpha\tau}u_M^{n-1}}{2} + \frac{u_M^n - e^{-\alpha\tau}u_M^{n-1}}{2}\right)\right.$$

$$\left.\cdot(\overline{u}_M^n - e^{-\alpha\tau}\overline{u}_M^{n-1})dx\right]$$

$$\leq -\frac{1}{2}\|u_M^n\|_{\mathbf{L}^4}^4 + \frac{1}{2}e^{-4\alpha\tau}\|u_M^{n-1}\|_{\mathbf{L}^4}^4 \leq -\frac{1}{2}\|u_M^n\|_{\mathbf{L}^4}^4 + \frac{1}{2}e^{-2\alpha\tau}\|u_M^{n-1}\|_{\mathbf{L}^4}^4$$

and

$$\mathscr{B} = \frac{2}{\tau}\Im\left[\int_0^1 \pi_M\delta_n W\left[-\mathbf{i}\tau\Delta\overline{u}_M^n + \mathbf{i}\tau\frac{|u_M^n|^2 + |e^{-\alpha\tau}u_M^{n-1}|^2}{2}\overline{u}_M^n + \pi_M\overline{\delta_n W}\right]dx\right]$$

$$= 2\Re\left[\int_0^1 \nabla(\pi_M\delta_n W)\cdot\nabla\left(\overline{u}_M^n - e^{-\alpha\tau}\overline{u}_M^{n-1}\right)dx\right]$$

$$+ 2\Re\left[\int_0^1 \nabla(\pi_M\delta_n W)\cdot\nabla\left(e^{-\alpha\tau}\overline{u}_M^{n-1}\right)dx\right]$$

$$+ \Re\left[\int_0^1 \left(|u_M^n|^2 + |e^{-\alpha\tau}u_M^{n-1}|^2\right)\overline{u}_M^n\cdot\pi_M\delta_n W dx\right]$$

$$\leq \frac{1}{4}\|\nabla(u_M^n - e^{-\alpha\tau}u_M^{n-1})\|_0^2 + C\|\nabla(\pi_M\delta_n W)\|_0^2$$

$$+ 2\Re\left[\int_0^1 \nabla(\pi_M\delta_n W)\cdot\nabla\left(e^{-\alpha\tau}\overline{u}_M^{n-1}\right)dx\right]$$

$$+ \Re\left[\int_0^1 \left(|u_M^n|^2 + |e^{-\alpha\tau}u_M^{n-1}|^2\right)\overline{u}_M^n\cdot\pi_M\delta_n W dx\right].$$

Denoting $H_n = \|\nabla u_M^n\|_0^2 + \frac{1}{2}\|u_M^n\|_{\mathbf{L}^4}^4$ and taking the expectation on both sides of (C.10), we deduce from the above three formulas that

$$\mathbb{E}H_n + \frac{3}{4}\mathbb{E}\|\nabla(u_M^n - e^{-\alpha\tau}u_M^{n-1})\|_0^2$$

$$\leq e^{-2\alpha\tau}\mathbb{E}H_{n-1} + C\tau + \Re\mathbb{E}\left[\int_0^1 \left(|u_M^n|^2 + |e^{-\alpha\tau}u_M^{n-1}|^2\right)\overline{u}_M^n\cdot\pi_M\delta_n W dx\right].$$
$$\tag{C.11}$$

Based on the formula

$$(|a|^2 + |b|^2)\overline{a} = \overline{a}|a-b|^2 + b(\overline{a}-\overline{b})^2 + 3|b|^2(\overline{a}-\overline{b}) + \overline{b}|a-b|^2 + \overline{b}^2(a-b) + 2|b|^2\overline{b},$$

the last term on the right hand side in (C.11) can be rewritten as

$$\Re\mathbb{E}\left[\int_0^1 \left(|u_M^n|^2 + |e^{-\alpha\tau}u_M^{n-1}|^2\right)\overline{u}_M^n\pi_M\delta_n W dx\right]$$

$$= \Re\mathbb{E}\int_0^1 \overline{u}_M^n\left|u_M^n - e^{-\alpha\tau}u_M^{n-1}\right|^2\pi_M\delta_n W dx$$

$$+ \Re\mathbb{E}\int_0^1 e^{-\alpha\tau}u_M^{n-1}\left(\overline{u}_M^n - e^{-\alpha\tau}\overline{u}_M^{n-1}\right)^2\pi_M\delta_n W dx$$

$$+ 3\Re\mathbb{E} \int_0^1 |e^{-\alpha\tau} u_M^{n-1}|^2 (\overline{u}_M^n - e^{-\alpha\tau} \overline{u}_M^{n-1}) \pi_M \delta_n W dx$$

$$+ \Re\mathbb{E} \int_0^1 e^{-\alpha\tau} \overline{u}_M^{n-1} \left| u_M^n - e^{-\alpha\tau} u_M^{n-1} \right|^2 \pi_M \delta_n W dx$$

$$+ \Re\mathbb{E} \int_0^1 (e^{-\alpha\tau} \overline{u}_M^{n-1})^2 (u_M^n - e^{-\alpha\tau} u_M^{n-1}) \pi_M \delta_n W dx$$

$$+ 2\Re\mathbb{E} \int_0^1 |e^{-\alpha\tau} u_M^{n-1}|^2 e^{-\alpha\tau} \overline{u}_M^{n-1} \pi_M \delta_n W dx$$

$$=: I + II + III + IV + V + VI.$$

Noting that $VI = 0$, it suffices to estimate the other five terms:

$$I + II + IV$$
$$\leq \mathbb{E} \left[\|u_M^n\|_0 \|u_M^n - e^{-\alpha\tau} u_M^{n-1}\|_{\mathbf{L}^4}^2 \|\pi_M \delta_n W\|_{\mathbf{L}^\infty} \right]$$
$$\quad + 2\mathbb{E} \left[\|e^{-\alpha\tau} u_M^{n-1}\|_0 \|u_M^n - e^{-\alpha\tau} u_M^{n-1}\|_{\mathbf{L}^4}^2 \|\pi_M \delta_n W\|_{\mathbf{L}^\infty} \right]$$
$$\leq \mathbb{E} \left[\left(\|u_M^n\|_0 + 2\|e^{-\alpha\tau} u_M^{n-1}\|_0 \right) \|\nabla(u_M^n - e^{-\alpha\tau} u_M^{n-1})\|_0^{\frac{1}{2}} \right.$$
$$\quad \left. \cdot \|u_M^n - e^{-\alpha\tau} u_M^{n-1}\|_0^{\frac{3}{2}} \|\pi_M \delta_n W\|_{\mathbf{L}^\infty} \right]$$
$$\leq \frac{1}{4} \mathbb{E} \left[\|\nabla(u_M^n - e^{-\alpha\tau} u_M^{n-1})\|_0 \|u_M^n - e^{-\alpha\tau} u_M^{n-1}\|_0 \right]$$
$$\quad + C\mathbb{E} \left[\left(\|u_M^n\|_0^2 + \|e^{-\alpha\tau} u_M^{n-1}\|_0^2 \right) \|u_M^n - e^{-\alpha\tau} u_M^{n-1}\|_0^2 \|\pi_M \delta_n W\|_{\mathbf{L}^\infty}^2 \right]$$
$$\leq \frac{1}{4} \mathbb{E} \|\nabla(u_M^n - e^{-\alpha\tau} u_M^{n-1})\|_0^2$$
$$\quad + C\mathbb{E} \left(\tau^{\frac{1}{2}} \left(\|u_M^n\|_0^2 + \|e^{-\alpha\tau} u_M^{n-1}\|_0^2 \right) \|u_M^n - e^{-\alpha\tau} u_M^{n-1}\|_0^2 \right)^2$$
$$\quad + C\mathbb{E} \left(\tau^{-\frac{1}{2}} \|\pi_M \delta_n W\|_{\mathbf{L}^\infty}^2 \right)^2$$
$$\leq \frac{1}{4} \mathbb{E} \|\nabla(u_M^n - e^{-\alpha\tau} u_M^{n-1})\|_0^2 + C\tau$$

according to Proposition 5.2, and

$$III + V \leq 4\mathbb{E} \left[\|e^{-\alpha\tau} u_M^{n-1}\|_{\mathbf{L}^4}^2 \|u_M^n - e^{-\alpha\tau} u_M^{n-1}\|_0 \|\pi_M \delta_n W\|_{\mathbf{L}^\infty} \right]$$
$$\leq \frac{1}{2} \mathbb{E} \|u_M^n - e^{-\alpha\tau} u_M^{n-1}\|_0^2 + 8\eta\tau e^{-4\alpha\tau} \mathbb{E} \|u_M^{n-1}\|_{\mathbf{L}^4}^4$$
$$\leq \frac{1}{2} \mathbb{E} \|u_M^n - e^{-\alpha\tau} u_M^{n-1}\|_0^2 + C\mathbb{E} \left[\eta\tau e^{-4\alpha\tau} \|\nabla u_M^{n-1}\|_0 \|u_M^{n-1}\|_0^3 \right]$$
$$\leq \frac{1}{2} \mathbb{E} \|u_M^n - e^{-\alpha\tau} u_M^{n-1}\|_0^2 + \alpha\tau e^{-2\alpha\tau} \mathbb{E} \|\nabla u_M^{n-1}\|_0^2 + C\tau.$$

Then (C.11) turns out to be

$$\mathbb{E}H_n \leq (1 + \alpha\tau)e^{-2\alpha\tau}\mathbb{E}H_{n-1} + C\tau \leq e^{-\alpha\tau}\mathbb{E}H_{n-1} + C\tau.$$

We finally obtain that

$$\mathbb{E}H_n \leq C.$$

(ii) $p = 2$. From the case $p = 1$, by $\|\cdot\|_{L^4}^4 \leq \|\nabla\cdot\|_0 \cdot \|_0^3$, we derive

$$H_n - e^{-2\alpha\tau}H_{n-1}$$
$$\leq C\|\nabla(\pi_M\delta_n W)\|_0^2 + C\Re\left[\int_0^1 \nabla(\pi_M\delta_n W) \cdot \nabla\left(e^{-\alpha\tau}\overline{u}_M^{n-1}\right)dx\right]$$
$$+ C\left(\tau^{\frac{1}{2}}\left(\|u_M^n\|_0^2 + \|e^{-\alpha\tau}u_M^{n-1}\|_0^2\right)\|u_M^n - e^{-\alpha\tau}u_M^{n-1}\|_0^2\right)^2$$
$$+ C\left(\tau^{-\frac{1}{2}}\|\pi_M\delta_n W\|_{L^\infty}^2\right)^2 + \alpha\tau e^{-2\alpha\tau}H_{n-1} + C\tau^{-1}\|u_M^{n-1}\|_0^6\|\pi_M\delta_n W\|_{L^\infty}^4.$$

Multiplying the above formula by H_n, we have

$$H_n^2 + (H_n - e^{-2\alpha\tau}H_{n-1})^2 - e^{-4\alpha\tau}H_{n-1}^2$$
$$\leq CH_n\|\nabla(\pi_M\delta_n W)\|_0^2 + CH_n\Re\left[\int_0^1 \nabla(\pi_M\delta_n W) \cdot \nabla\left(e^{-\alpha\tau}\overline{u}_M^{n-1}\right)dx\right]$$
$$+ C\tau H_n\left(\|u_M^n\|_0^2 + \|e^{-\alpha\tau}u_M^{n-1}\|_0^2\right)^2\|u_M^n - e^{-\alpha\tau}u_M^{n-1}\|_0^4$$
$$+ CH_n\left(\tau^{-\frac{1}{2}}\|\pi_M\delta_n W\|_{L^\infty}^2\right)^2 + \alpha\tau e^{-2\alpha\tau}H_nH_{n-1} + C\tau^{-1}H_n\|u_M^{n-1}\|_0^6\|\pi_M\delta_n W\|_{L^\infty}^4$$
$$=: I' + II' + III' + IV' + V' + VI',$$

where

$$\mathbb{E}[I' + II' + III' + IV'] \leq \frac{1}{4}\mathbb{E}(H_n - e^{-2\alpha\tau}H_{n-1})^2 + C\tau + C\tau e^{-2\alpha\tau}.$$

$$\mathbb{E}\left[H_{n-1}\left(\|u_M^n\|_0^2 + \|e^{-\alpha\tau}u_M^{n-1}\|_0^2\right)^2\|u_M^n - e^{-\alpha\tau}u_M^{n-1}\|_0^4\right]$$
$$\leq \frac{1}{4}\mathbb{E}(H_n - e^{-2\alpha\tau}H_{n-1})^2 + \frac{1}{2}\tau e^{-4\alpha\tau}\mathbb{E}H_{n-1}^2 + C\tau,$$

$$\mathbb{E}[V'] \leq \frac{1}{2}\mathbb{E}\left(H_n - e^{-2\alpha\tau}H_{n-1}\right)^2 + (\frac{1}{2}\alpha^2\tau^2 + \alpha\tau)e^{-4\alpha\tau}\mathbb{E}H_{n-1}^2$$
$$\leq \frac{1}{2}\mathbb{E}\left(H_n - e^{-2\alpha\tau}H_{n-1}\right)^2 + \frac{3}{2}\alpha\tau e^{-4\alpha\tau}\mathbb{E}H_{n-1}^2$$

and

$$
\begin{aligned}
\mathbb{E}[VI'] &\leq \frac{1}{4}\mathbb{E}\left(H_n - e^{-2\alpha\tau}H_{n-1}\right)^2 + C\tau^{-2}\mathbb{E}\left[\|u_M^{n-1}\|_0^{12}\|\pi_M\delta_n W\|_{L^\infty}^8\right] \\
&\quad + \alpha\tau e^{-4\alpha\tau}\mathbb{E}H_{n-1}^2 + C\tau^{-3}\mathbb{E}\left[\|u_M^{n-1}\|_0^{12}\|\pi_M\delta_n W\|_{L^\infty}^8\right] \\
&\leq \frac{1}{4}\mathbb{E}\left(H_n - e^{-2\alpha\tau}H_{n-1}\right)^2 + \alpha\tau e^{-4\alpha\tau}\mathbb{E}H_{n-1}^2 + C\tau.
\end{aligned}
$$

Then we conclude

$$
\mathbb{E}[H_n^2] \leq (1+3\alpha\tau)e^{-4\alpha\tau}\mathbb{E}[H_{n-1}^2] + C\tau \leq e^{-\alpha\tau}\mathbb{E}[H_{n-1}^2] + C\tau \leq C,
$$

where we have used $(1+3\alpha\tau)e^{-3\alpha\tau} \leq 1$ for $\alpha\tau < 1$.

(iii) For $p = 2^l$, $l \in \mathbb{N}$, the result can be proved by the above procedure, and hence holds for any $p \in \mathbb{N}$.

C.3 Proof of Proposition 5.4

We give the proof for $\lambda = -1$ only. Multiplying (C.9) by $\Delta(\overline{u}_M^n - e^{-\alpha\tau}\overline{u}_M^{n-1})$, integrating with respect to x, and then taking the imaginary part, we obtain

$$
\begin{aligned}
&\|\Delta u_M^n\|_0^2 + \|\Delta(u_M^n - e^{-\alpha\tau}u_M^{n-1})\|_0^2 - e^{-2\alpha\tau}\|\Delta u_M^{n-1}\|_0^2 \\
&= \Re\int_0^1 \left(|u_M^n|^2 + |e^{-\alpha\tau}u_M^{n-1}|^2\right)u_M^n\Delta(\overline{u}_M^n - e^{-\alpha\tau}\overline{u}_M^{n-1})dx \\
&\quad - \frac{2}{\tau}\Im\int_0^1 \pi_M\delta_n W\Delta(\overline{u}_M^n - e^{-\alpha\tau}\overline{u}_M^{n-1})dx \\
&=: \mathscr{A} + \mathscr{B}.
\end{aligned}
$$

According to the uniform boundedness of $\{u_M^n\}_{n\in\mathbb{N}}$ in the 0-norm and the 1-norm, we have the following estimates.

$$
\begin{aligned}
\mathbb{E}[\mathscr{A}] &= \Re\mathbb{E}\int_0^1 |u_M^n|^2 u_M^n\Delta(\overline{u}_M^n - e^{-\alpha\tau}\overline{u}_M^{n-1})dx \\
&\quad + e^{-3\alpha\tau}\Re\mathbb{E}\int_0^1 |u_M^{n-1}|^2 u_M^{n-1}\Delta(\overline{u}_M^n - e^{-\alpha\tau}\overline{u}_M^{n-1})dx \\
&\quad + e^{-2\alpha\tau}\Re\mathbb{E}\int_0^1 |u_M^{n-1}|^2(u_M^n - e^{-\alpha\tau}u_M^{n-1})\Delta(\overline{u}_M^n - e^{-\alpha\tau}\overline{u}_M^{n-1})dx \\
&= \Re\mathbb{E}\int_0^1 |u_M^n|^2 u_M^n\Delta\overline{u}_M^n dx \\
&\quad + e^{-2\alpha\tau}\Re\mathbb{E}\int_0^1 |u_M^{n-1}|^2(u_M^n - e^{-\alpha\tau}u_M^{n-1})\Delta(\overline{u}_M^n - e^{-\alpha\tau}\overline{u}_M^{n-1})dx
\end{aligned}
$$

$$-e^{-4\alpha\tau}\Re\mathbb{E}\int_0^1 |u_M^{n-1}|^2 u_M^{n-1}\Delta\overline{u}_M^{n-1}dx + \Re\mathbb{E}\int_0^1 u_M^n\Delta\overline{u}_M^n|u_M^n - e^{-\alpha\tau}u_M^{n-1}|^2 dx$$

$$+ 2\Re\mathbb{E}\int_0^1 \overline{u}_M^n(\nabla u_M^n)^2(\overline{u}_M^n - e^{-\alpha\tau}\overline{u}_M^{n-1})dx$$

$$+ 4\Re\mathbb{E}\int_0^1 u_M^n|\nabla u_M^n|^2(\overline{u}_M^n - e^{-\alpha\tau}\overline{u}_M^{n-1})dx$$

$$+ \Re\mathbb{E}\int_0^1 (u_M^n - e^{-\alpha\tau}u_M^{n-1})\Delta\overline{u}_M^n\left(|u_M^n|^2 - |e^{-\alpha\tau}u_M^{n-1}|^2\right)dx$$

$$=: A_a^n - e^{-4\alpha\tau}A_a^{n-1} + A_b + A_c + A_d + A_e + A_f.$$

We estimate the above terms respectively and obtain

$$-e^{-4\alpha\tau}A_a^{n-1} = -e^{-2\alpha\tau}A_a^{n-1} + e^{-2\alpha\tau}(1 - e^{-2\alpha\tau})A_a^{n-1}$$
$$\leq -e^{-2\alpha\tau}A_a^{n-1} + C\tau\mathbb{E}\|u_M^{n-1}\|_1^4 \leq -e^{-2\alpha\tau}A_a^{n-1} + C\tau,$$

$$A_b \leq e^{-2\alpha\tau}\mathbb{E}\left[\|u_M^{n-1}\|_{\mathbf{L}^\infty}^2\|u_M^n - e^{-\alpha\tau}u_M^{n-1}\|_0\|\Delta(u_M^n - e^{-\alpha\tau}u_M^{n-1})\|_0\right]$$
$$\leq \frac{1}{6}\mathbb{E}\|\Delta(u_M^n - e^{-\alpha\tau}u_M^{n-1})\|_0^2 + C\tau\mathbb{E}\|u_M^{n-1}\|_1^8 + C\tau^{-1}\mathbb{E}\|u_M^n - e^{-\alpha\tau}u_M^{n-1}\|_0^4$$
$$\leq \frac{1}{6}\mathbb{E}\|\Delta(u_M^n - e^{-\alpha\tau}u_M^{n-1})\|_0^2 + C\tau,$$

$$A_c \leq \mathbb{E}\left[\|u_M^n - e^{-\alpha\tau}u_M^{n-1}\|_{\mathbf{L}^4}^2\|u_M^n\|_{\mathbf{L}^\infty}\|\Delta u_M^n\|_0\right]$$
$$\leq C\tau^{-1}\mathbb{E}\left[\|\nabla(u_M^n - e^{-\alpha\tau}u_M^{n-1})\|_0\|u_M^n - e^{-\alpha\tau}u_M^{n-1}\|_0^3\|u_M^n\|_1^2\right] + \frac{1}{8}\alpha\tau\mathbb{E}\|\Delta u_M^n\|_0^2$$
$$\leq \frac{1}{6}\mathbb{E}\|\Delta(u_M^n - e^{-\alpha\tau}u_M^{n-1})\|_0^2 + C\tau^{-5}\mathbb{E}\|u_M^n - e^{-\alpha\tau}u_M^{n-1}\|_0^{12}$$
$$+ C\tau\mathbb{E}\|u_M^n\|_1^8 + \frac{1}{8}\alpha\tau\mathbb{E}\|\Delta u_M^n\|_0^2$$
$$\leq \frac{1}{6}\mathbb{E}\|\Delta(u_M^n - e^{-\alpha\tau}u_M^{n-1})\|_0^2 + \frac{1}{8}\alpha\tau\mathbb{E}\|\Delta u_M^n\|_0^2 + C\tau,$$

$$A_d = 2\Re\mathbb{E}\int_0^1 \overline{u}_M^n(\nabla u_M^n)^2\left[-\mathbf{i}\tau\Delta\overline{u}_M^n + \mathbf{i}\tau\pi_M\left(\frac{|u_M^n|^2 + |e^{-\alpha\tau}u_M^{n-1}|^2}{2}\overline{u}_M^n\right) + \pi_M\overline{\delta_n W}\right]dx$$

$$\leq \frac{1}{16}\alpha\tau\mathbb{E}\|\Delta u_M^n\|_0^2 + C\tau + 2\Re\mathbb{E}\int_0^1 \overline{u}_M^n(\nabla u_M^n)^2\pi_M\overline{\delta_n W}dx$$

$$\leq \frac{1}{16}\alpha\tau\mathbb{E}\|\Delta u_M^n\|_0^2 + C\tau + 2\Re\mathbb{E}\int_0^1 (\overline{u}_M^n - e^{-\alpha\tau}\overline{u}_M^{n-1})(\nabla u_M^n)^2\pi_M\overline{\delta_n W}dx$$

$$+ 2\Re\mathbb{E}\int_0^1 e^{-\alpha\tau}\overline{u}_M^{n-1}\left((\nabla u_M^n)^2 - (e^{-\alpha\tau}\nabla u_M^{n-1})^2\right)\pi_M\overline{\delta_n W}dx$$

$$\leq \frac{1}{16}\alpha\tau\mathbb{E}\|\Delta u_M^n\|_0^2 + C\tau + C\mathbb{E}\left[\|u_M^n - e^{-\alpha\tau}u_M^{n-1}\|_0\|\nabla u_M^n\|_{\mathbf{L}^4}^2\|\pi_M\delta_n W\|_{\mathbf{L}^\infty}\right]$$

$$+ C\mathbb{E}\left[\|\nabla(u_M^n - e^{-\alpha\tau}u_M^{n-1})\|_0\left(\|u_M^{n-1}\|_1\|u_M^n\|_1 + \|u_M^{n-1}\|_1^2\right)\|\pi_M\delta_n W\|_{\mathbf{L}^\infty}\right]$$

$$\leq \frac{1}{6}\mathbb{E}\|\Delta(u_M^n - e^{-\alpha\tau}u_M^{n-1})\|_0^2 + \frac{1}{8}\alpha\tau\mathbb{E}\|\Delta u_M^n\|_0^2 + C\tau,$$

and

$$
\begin{aligned}
A_f =& \Re\mathbb{E}\int_0^1 (u_M^n - e^{-\alpha\tau}u_M^{n-1})\Delta\overline{u}_M^n \Re\big[\,(u_M^n - e^{-\alpha\tau}u_M^{n-1})\,(\overline{u}_M^n + e^{-\alpha\tau}\overline{u}_M^{n-1})\big]dx \\
\leq& \mathbb{E}\big[\|u_M^n - e^{-\alpha\tau}u_M^{n-1}\|_{\mathbf{L}^4}^2(\|u_M^n\|_{\mathbf{L}^\infty} + \|u_M^{n-1}\|_{\mathbf{L}^\infty})\|\Delta u_M^n\|_0\big] \\
\leq& \frac{1}{6}\mathbb{E}\|\Delta(u_M^n - e^{-\alpha\tau}u_M^{n-1})\|_0^2 + \frac{1}{8}\alpha\tau\mathbb{E}\|\Delta u_M^n\|_0^2 + C\tau,
\end{aligned}
$$

where A_e has a similar estimation as A_d and we have used that $\|\nabla\cdot\|_0 \cong \|\cdot\|_1 \leq \|\cdot\|_2 \cong \|\Delta\cdot\|_0$. We then conclude

$$\mathbb{E}[\mathscr{A}] \leq \frac{5}{6}\mathbb{E}\|\Delta(u_M^n - e^{-\alpha\tau}u_M^{n-1})\|_0^2 + \frac{1}{2}\alpha\tau\mathbb{E}\|\Delta u_M^n\|_0^2 + C\tau.$$

For term $\mathscr{B}$, we have

$$
\begin{aligned}
\mathbb{E}[\mathscr{B}] =& \frac{2}{\tau}\Im\mathbb{E}\int_0^1 \Delta\left(\pi_M\delta_n W\right)\left(\mathbf{i}\tau\Delta\overline{u}_M^n - \mathbf{i}\tau\pi_M\left(\frac{|u_M^n|^2 + |e^{-\alpha\tau}u_M^{n-1}|^2}{2}\overline{u}_M^n\right) - \pi_M\overline{\delta_n W}\right)dx \\
=& 2\Re\mathbb{E}\int_0^1 \Delta\left(\pi_M\delta_n W\right)\Delta(\overline{u}_M^n - e^{-\alpha\tau}\overline{u}_M^{n-1})dx \\
& - \Re\mathbb{E}\int_0^1 \Delta\left(\pi_M\delta_n W\right)\left(|u_M^n|^2\overline{u}_M^n - |e^{-\alpha\tau}u_M^{n-1}|^2 e^{-\alpha\tau}u_M^{n-1}\right)dx \\
& - \Re\mathbb{E}\int_0^1 \Delta\left(\pi_M\delta_n W\right)|e^{-\alpha\tau}u_M^{n-1}|^2(\overline{u}_M^n - e^{-\alpha\tau}\overline{u}_M^{n-1})dx \\
\leq& \frac{1}{6}\mathbb{E}\|\Delta(u_M^n - e^{-\alpha\tau}u_M^{n-1})\|_0^2 + C\tau.
\end{aligned}
$$

Denoting $K_n := \|\Delta u_M^n\|_0^2 - \Re\int_0^1 |u_M^n|^2 u_M^n \Delta\overline{u}_M^n dx$, then we obtain

$$\mathbb{E}\|\Delta u_M^n\|_0^2 = \mathbb{E}[K_n] + \Re\mathbb{E}\int_0^1 |u_M^n|^2 u_M^n \Delta\overline{u}_M^n dx \leq \mathbb{E}[K_n] + \frac{1}{2}\mathbb{E}\|\Delta u_M^n\|_0^2 + C$$

$$(C.12)$$

from (C.7) and Proposition 5.3, where

$$\mathbb{E}[K_n] - e^{-2\alpha\tau}\mathbb{E}[K_{n-1}] \leq \frac{1}{2}\alpha\tau\mathbb{E}\|\Delta u_M^n\|_0^2 + C\tau \leq \frac{1}{2}\alpha\tau\mathbb{E}[K_n] + C\tau.$$

Finally, we deduce

$$\mathbb{E}[K_n] \leq (1 - \frac{1}{2}\alpha\tau)^{-1} e^{-2\alpha\tau} \mathbb{E}[K_{n-1}] + C\tau \leq C,$$

where we have used $(1 - \frac{1}{2}\alpha\tau)^{-1} e^{-2\alpha\tau} \leq e^{-\alpha\tau}$ for $\alpha\tau < 1$. It completes the proof together with (C.12).

References

1. A. Abdulle, D. Cohen, G. Vilmart, K.C. Zygalakis, High weak order methods for stochastic differential equations based on modified equations. SIAM J. Sci. Comput. **34**(3), A1800–A1823 (2012)
2. A. Abdulle, G. Vilmart, K.C. Zygalakis, High order numerical approximation of the invariant measure of ergodic SDEs. SIAM J. Numer. Anal. **52**(4), 1600–1622 (2014)
3. A. Abdulle, G. Vilmart, K.C. Zygalakis, Long time accuracy of Lie-Trotter splitting methods for Langevin dynamics. SIAM J. Numer. Anal. **53**(1), 1–16 (2015)
4. G.D. Akrivis, V.A. Dougalis, O.A. Karakashian, On fully discrete Galerkin methods of second-order temporal accuracy for the nonlinear Schrödinger equation. Numer. Math. **59**(1), 31–53 (1991)
5. S. Albeverio, B. Ferrario, Uniqueness of solutions of the stochastic Navier-Stokes equation with invariant measure given by the enstrophy. Ann. Probab. **32**(2), 1632–1649 (2004)
6. H.H. Andersen, M. Højbjerre, D. Sørensen, P.S. Eriksen, *Linear and Graphical Models*. Lecture Notes in Statistics, vol. 101 (Springer, New York, 1995). For the multivariate complex normal distribution
7. C. Anton, J. Deng, Y.S. Wong, Weak symplectic schemes for stochastic Hamiltonian equations. Electron. Trans. Numer. Anal. **43**, 1–20 (2014/15)
8. C.A. Anton, Y.S. Wong, J. Deng, Symplectic schemes for stochastic Hamiltonian systems preserving Hamiltonian functions. Int. J. Numer. Anal. Model. **11**(3), 427–451 (2014)
9. L. Arnold, W. Kliemann, On unique ergodicity for degenerate diffusions. Stochastics **21**(1), 41–61 (1987)
10. V.I. Arnol'd, *Mathematical Methods of Classical Mechanics*, Graduate Texts in Mathematics, vol. 60, 2nd edn. (Springer, New York, 1989). Translated from the Russian by K. Vogtmann and A. Weinstein
11. A. Aydın, B. Karasözen, Symplectic and multi-symplectic methods for coupled nonlinear Schrödinger equations with periodic solutions. Comput. Phys. Comm. **177**(7), 566–583 (2007)
12. V. Bally, E. Pardoux, Malliavin calculus for white noise driven parabolic SPDEs. Potential Anal. **9**(1), 27–64 (1998)
13. V. Bally, D. Talay, The Euler scheme for stochastic differential equations: error analysis with Malliavin calculus. Math. Comput. Simul. **38**(1–3), 35–41 (1995). Probabilités numériques (Paris, 1992)
14. V. Bally, D. Talay, The law of the Euler scheme for stochastic differential equations. II. Convergence rate of the density. Monte Carlo Methods Appl. **2**(2), 93–128 (1996)

© Springer Nature Singapore Pte Ltd. 2019
J. Hong and X. Wang, *Invariant Measures for Stochastic Nonlinear Schrödinger Equations*, Lecture Notes in Mathematics 2251,
https://doi.org/10.1007/978-981-32-9069-3

15. V. Barbu, M. Röckner, D. Zhang, Stochastic nonlinear Schrödinger equations with linear multiplicative noise: rescaling approach. J. Nonlinear Sci. **24**(3), 383–409 (2014)
16. V. Barbu, M. Röckner, D. Zhang, Stochastic nonlinear Schrödinger equations. Nonlinear Anal. **136**, 168–194 (2016)
17. M. Barton-Smith, Invariant measure for the stochastic Ginzburg Landau equation. NoDEA Nonlinear Differ. Equ. Appl. **11**(1), 29–52 (2004)
18. M. Barton-Smith, A. Debussche, L. Di Menza, Numerical study of two-dimensional stochastic NLS equations. Numer. Methods Partial Differ. Equ. **21**(4), 810–842 (2005)
19. R. Bellman, The stability of solutions of linear differential equations. Duke Math. J. **10**, 643–647 (1943)
20. B. Bidégaray, Invariant measures for some partial differential equations. Phys. D **82**(4), 340–364 (1995)
21. I. Bihari, A generalization of a lemma of Bellman and its application to uniqueness problems of differential equations. Acta Math. Acad. Sci. Hungar. **7**, 81–94 (1956)
22. G.D. Birkhoff, Proof of the ergodic theorem. Proc. Nat. Acad. Sci. **17**(12), 656–660 (1931)
23. L. Boltzmann, Ueber die Eigenschaften monocyclischer und anderer damit verwandter systeme. J. Reine Angew. Math. **98**, 68–94 (1885)
24. S. Bonaccorsi, M. Zanella, Existence and regularity of the density for solutions of stochastic differential equations with boundary noise. Infin. Dimens. Anal. Quantum Probab. Relat. Top. **19**(1), 1650007, 24 (2016)
25. N. Bou-Rabee, H. Owhadi, Long-run accuracy of variational integrators in the stochastic context. SIAM J. Numer. Anal. **48**(1), 278–297 (2010)
26. J. Bourgain, On the Cauchy and invariant measure problem for the periodic Zakharov system. Duke Math. J. **76**(1), 175–202 (1994)
27. J. Bourgain, Periodic nonlinear Schrödinger equation and invariant measures. Commun. Math. Phys. **166**(1), 1–26 (1994)
28. J. Bourgain, Invariant measures for the 2D-defocusing nonlinear Schrödinger equation. Commun. Math. Phys. **176**(2), 421–445 (1996)
29. J. Bourgain, *Global Colutions of Nonlinear Schrödinger Equations*. American Mathematical Society Colloquium Publications, vol. 46 (American Mathematical Society, Providence, RI, 1999)
30. J. Bourgain, Invariant measures for NLS in infinite volume. Commun. Math. Phys. **210**(3), 605–620 (2000)
31. C.-E. Bréhier, Approximation of the invariant measure with an Euler scheme for stochastic PDEs driven by space-time white noise. Potential Anal. **40**(1), 1–40 (2014)
32. C.-E. Bréhier, G. Vilmart, High order integrator for sampling the invariant distribution of a class of parabolic stochastic PDEs with additive space-time noise. SIAM J. Sci. Comput. **38**(4), A2283–A2306 (2016)
33. J. Bricmont, A. Kupiainen, R. Lefevere, Exponential mixing of the 2D stochastic Navier-Stokes dynamics. Commun. Math. Phys. **230**(1), 87–132 (2002)
34. T.J. Bridges, Multi-symplectic structures and wave propagation. Math. Proc. Camb. Philos. Soc. **121**(1), 147–190 (1997)
35. T.J. Bridges, S. Reich, Multi-symplectic integrators: numerical schemes for Hamiltonian PDEs that conserve symplecticity. Phys. Lett. A **284**(4–5), 184–193 (2001)
36. Z. Brzeźniak, S. Cerrai, Large deviations principle for the invariant measures of the 2D stochastic Navier-Stokes equations on a torus. J. Funct. Anal. **273**(6), 1891–1930 (2017)
37. Z. Brzeźniak, H. Long, I. Simão, Invariant measures for stochastic evolution equations in M-type 2 Banach spaces. J. Evol. Equ. **10**(4), 785–810 (2010)
38. Z. Brzeźniak, E. Motyl, M. Ondrejat, Invariant measure for the stochastic Navier-Stokes equations in unbounded 2D domains. Ann. Probab. **45**(5), 3145–3201 (2017)
39. L.A. Bunimovich, I.P. Cornfeld, R.L. Dobrushin, M.V. Jakobson, N.B. Maslova, Ya B. Pesin, Ya G. Sinaĭ, YuM Sukhov, A.M. Vershik, *Dynamical Systems, II*. Encyclopaedia of Mathematical Sciences, vol. 2 (Springer, Berlin, 1989). Ergodic theory with applications to dynamical systems and statistical mechanics

40. F. Cacciafesta, A.-S. de Suzzoni, Invariant measure for the Schrödinger equation on the real line. J. Funct. Anal. **269**(1), 271–324 (2015)
41. M.P. Calvo, J.M. Sanz-Serna, High-order symplectic Runge-Kutta-Nyström methods. SIAM J. Sci. Comput. **14**(5), 1237–1252 (1993)
42. T. Cass, M. Hairer, C. Litterer, S. Tindel, Smoothness of the density for solutions to Gaussian rough differential equations. Ann. Probab. **43**(1), 188–239 (2015)
43. P. Cattiaux, J.R. León, C. Prieur, Estimation for stochastic damping Hamiltonian systems under partial observation–I. Invariant Den. Stoch. Process. Appl. **124**(3), 1236–1260 (2014)
44. T. Cazenave, *Semilinear Schrödinger Equations*. Courant Lecture Notes in Mathematics. vol. 10 (New York University, Courant Institute of Mathematical Sciences, New York; American Mathematical Society, Providence, RI, 2003)
45. S. Cerrai, M. Röckner, Large deviations for invariant measures of stochastic reaction-diffusion systems with multiplicative noise and non-Lipschitz reaction term. Ann. Inst. H. Poincaré Probab. Statist. **41**(1), 69–105 (2005)
46. P. Chartier, E. Hairer, G. Vilmart, Numerical integrators based on modified differential equations. Math. Comput. **76**(260), 1941–1953 (2007)
47. C. Chen, J. Hong, Symplectic Runge-Kutta semidiscretization for stochastic Schrödinger equation. SIAM J. Numer. Anal. **54**(4), 2569–2593 (2016)
48. C. Chen, J. Hong, L. Ji, Mean-square convergence of a symplectic local discontinuous Galerkin method applied to stochastic linear Schrödinger equation. IMA J. Numer. Anal. **37**(2), 1041–1065 (2017)
49. C. Chen, J. Hong, A. Prohl, Convergence of a θ-scheme to solve the stochastic nonlinear Schrödinger equation with Stratonovich noise. Stoch. Partial Differ. Equ. Anal. Comput. **4**(2), 274–318 (2016)
50. C. Chen, J. Hong, X. Wang, Approximation of invariant measure for damped stochastic nonlinear Schrödinger equation via an ergodic numerical scheme. Potential Anal. **46**(2), 323–367 (2017)
51. C. Chen, J. Hong, L. Zhang, Preservation of physical properties of stochastic Maxwell equations with additive noise via stochastic multi-symplectic methods. J. Comput. Phys. **306**, 500–519 (2016)
52. X. Chen, *Random Walk Intersections. Mathematical Surveys and Monographs*, vol. 157 (American Mathematical Society, Providence, RI, 2010). Large deviations and related topics
53. I.P. Cornfeld, S.V. Fomin, Y.G. Sinaĭ, *Ergodic Theory*. Grundlehren der Mathematischen Wissenschaften [Fundamental Principles of Mathematical Sciences], vol. 245 (Springer, New York, 1982). Translated from the Russian by A. B. Sosinskiĭ
54. S. Cox, M. Hutzenthaler, A. Jentzen, Local Lipschitz continuity in the initial value and strong completeness for nonlinear stochastic differential equations. arXiv: 1309.5595v1
55. J. Cui, J. Hong, Z. Liu, Strong convergence rate of finite difference approximations for stochastic cubic Schrödinger equations. J. Differ. Equ. **263**(7), 3687–3713 (2017)
56. J. Cui, J. Hong, Z. Liu, W. Zhou, Stochastic symplectic and multi-symplectic methods for nonlinear Schrödinger equation with white noise dispersion. J. Comput. Phys. **342**, 267–285 (2017)
57. J. Cui, J. Hong, Z. Liu, W. Zhou, Strong convergence rate of splitting schemes for stochastic nonlinear Schrödinger equations. J. Differ. Equ. **266**(9), 5625–5663 (2019)
58. G. Da Prato, *An Introduction to Infinite-Dimensional Analysis*. Universitext (Springer, Berlin, 2006). Revised and extended from the 2001 original by Da Prato
59. G. Da Prato, A. Debussche, Ergodicity for the 3D stochastic Navier-Stokes equations. J. Math. Pures Appl. (9), **82**(8), 877–947 (2003)
60. G. Da Prato, J. Zabczyk, *Ergodicity for Infinite-Dimensional Systems*. London Mathematical Society Lecture Note Series, vol. 229 (Cambridge University Press, Cambridge, 1996)
61. A. de Bouard, A. Debussche, A stochastic nonlinear Schrödinger equation with multiplicative noise. Commun. Math. Phys. **205**(1), 161–181 (1999)
62. A. de Bouard, A. Debussche, Finite-time blow-up in the additive supercritical stochastic nonlinear Schrödinger equation: the real noise case, in *The Legacy of the Inverse Scattering*

Transform in Applied Mathematics, (South Hadley, MA, 2001). Contemporary Mathematics, vol. 301 (American Mathematical Society, Providence, RI, 2002), pp. 183–194

63. A. de Bouard, A. Debussche, On the effect of a noise on the solutions of the focusing supercritical nonlinear Schrödinger equation. Probab. Theory Relat. Fields **123**(1), 76–96 (2002)
64. A. de Bouard, A. Debussche, The stochastic nonlinear Schrödinger equation in H^1. Stoch. Anal. Appl. **21**(1), 97–126 (2003)
65. A. De Bouard, A. Debussche, A semi-discrete scheme for the stochastic nonlinear Schrödinger equation. Numer. Math. **96**(4), 733–770 (2004)
66. A. de Bouard, A. Debussche, Blow-up for the stochastic nonlinear Schrödinger equation with multiplicative noise. Ann. Probab. **33**(3), 1078–1110 (2005)
67. A. De Bouard, A. Debussche, Weak and strong order of convergence of a semidiscrete scheme for the stochastic nonlinear Schrödinger equation. Appl. Math. Optim. **54**(3), 369–399 (2006)
68. A. Debussche, L. Di Menza, Numerical simulation of focusing stochastic nonlinear Schrödinger equations. Phys. D **162**(3–4), 131–154 (2002)
69. A. Debussche, E. Faou, Weak backward error analysis for SDEs. SIAM J. Numer. Anal. **50**(3), 1735–1752 (2012)
70. A. Debussche, C. Odasso, Ergodicity for a weakly damped stochastic non-linear Schrödinger equation. J. Evol. Equ. **5**(3), 317–356 (2005)
71. A. Debussche, J. Printems, Weak order for the discretization of the stochastic heat equation. Math. Comput. **78**(266), 845–863 (2009)
72. A. Dembo, O. Zeitouni, *Large Deviations Techniques and Applications*. Applications of Mathematics (New York), vol. 38, 2nd edn. (Springer, New York, 1998)
73. Z. Dong, L. Xu, X. Zhang, Invariant measures of stochastic 2D Navier-Stokes equations driven by α-stable processes. Electron. Commun. Probab. **16**, 678–688 (2011)
74. D. Down, S.P. Meyn, R.L. Tweedie, Exponential and uniform ergodicity of Markov processes. Ann. Probab. **23**(4), 1671–1691 (1995)
75. P. Dupuis, R.S. Ellis. *A Weak Convergence Approach to the Theory of Large Deviations*. Wiley Series in Probability and Statistics: Probability and Statistics (Wiley, A Wiley-Interscience Publication, New York, 1997)
76. I. Ekren, I. Kukavica, M. Ziane, Existence of invariant measures for the stochastic damped Schrödinger equation. Stoch. PDE Anal. Comput. (2017)
77. L.C. Evans, *Partial Differential Equations*. Graduate Studies in Mathematics, vol. 19 (American Mathematical Society, Providence, RI, 1998)
78. L.C. Evans, *An Introduction to Stochastic Differential Equations* (American Mathematical Society, Providence, RI, 2013)
79. G. Falkovich, I. Kolokolov, V. Lebedev, V. Mezentsev, S. Turitsyn, Non-Gaussian error probability in optical soliton transmission. Phys. D **195**, 1–28 (2004)
80. G.E. Falkovich, I. Kolokolov, V. Lebedev, S.K. Turitsyn, Statistics of soliton-bearing systems with additive noise. Phys. Rev. E **63**, 025601 (2001)
81. E. Faou, *Geometric Numerical Integration and Schrödinger Equations*. Zurich Lectures in Advanced Mathematics (European Mathematical Society (EMS), Zürich, 2012)
82. K. Feng, M. Qin, *Symplectic Geometric Algorithms for Hamiltonian Systems* (Zhejiang Science and Technology Publishing House, Hangzhou, Springer, Heidelberg, 2010). Translated and revised from the Chinese original, With a foreword by Feng Duan
83. K. Feng, H. Wu, M. Qin, D. Wang, Construction of canonical difference schemes for Hamiltonian formalism via generating functions. J. Comput. Math. **7**(1), 71–96 (1989)
84. F. Flandoli, Dissipativity and invariant measures for stochastic Navier-Stokes equations. NoDEA Nonlinear Differ. Equ. Appl. **1**(4), 403–423 (1994)
85. G. Gallavotti, F. Bonetto, G. Gentile, *Aspects of Ergodic, Qualitative and Statistical Theory of Motion*, Texts and Monographs in Physics (Springer, Berlin, 2004)
86. E. Gautier, Large deviations and support results for nonlinear Schrödinger equations with additive noise and applications. ESAIM Probab. Stat. **9**, 74–97 (2005)
87. E. Gautier, Uniform large deviations for the nonlinear Schrödinger equation with multiplicative noise. Stoch. Process. Appl. **115**(12), 1904–1927 (2005)

88. E. Gautier, Stochastic nonlinear Schrödinger equations driven by a fractional noise well-posedness, large deviations and support. Electron. J. Probab. **12**(29), 848–861 (2007)
89. B. Gess, W. Liu, M. Röckner, Random attractors for a class of stochastic partial differential equations driven by general additive noise. J. Differ. Equ. **251**(4–5), 1225–1253 (2011)
90. Ĭ, Ī, Gĭhman, A.V. Skorohod, *Stochastic Differential Equations* (Springer, New York-Heidelberg, 1972). Translated from the Russian by Kenneth Wickwire, Ergebnisse der Mathematik und ihrer Grenzgebiete, Band 72
91. J. Ginibre, G. Velo, On a class of nonlinear Schrödinger equations. III. Special theories in dimensions 1, 2 and 3. Ann. Inst. H. Poincaré Sect. A (N.S.) **28**(3), 287–316 (1978)
92. H. Goldstein, *Classical Mechanics*. Addison-Wesley Series in Physics, 2nd edn. (Addison-Wesley Publishing Co., Reading, MA, 1980)
93. J.P. Gordon, H.A. Haus, Random walk of coherently amplified solitons in optical fiber transmission. Opt. Lett. **11**, 665–667 (1986)
94. T.H. Gronwall, Note on the derivatives with respect to a parameter of the solutions of a system of differential equations. Ann. Math. (2) **20**(4), 292–296, 1919
95. E. Hairer, C. Lubich, G. Wanner, *Geometric Numerical Integration*.Springer Series in Computational Mathematics, vol. 31 (Springer, Heidelberg, 2010). Structure-preserving algorithms for ordinary differential equations, Reprint of the second (2006) edition
96. M. Hairer, J.C. Mattingly, Ergodicity of the 2D Navier-Stokes equations with degenerate stochastic forcing. Ann. Math. (2) **164**(3), 993–1032 (2006)
97. M. Hairer, N.S. Pillai, Ergodicity of hypoelliptic SDEs driven by fractional Brownian motion. Ann. Inst. Henri Poincaré Probab. Stat. **47**(2), 601–628 (2011)
98. M. Hairer, N.S. Pillai, Regularity of laws and ergodicity of hypoelliptic SDEs driven by rough paths. Ann. Probab. **41**(4), 2544–2598 (2013)
99. G.H. Hardy, J.E. Littlewood, G. Pólya, *Inequalities*, 2d edn. (Cambridge, at the University Press, 1952)
100. J. Hong, C. Huang, X. Wang, Symplectic Runge-Kutta methods for Hamiltonian systems driven by Gaussian rough paths. Appl. Numer. Math. **129**, 120–136 (2018)
101. J. Hong, L. Ji, Energy evolution of multi-symplectic methods for Maxwell equations with perfectly matched layer boundary. J. Math. Anal. Appl. **439**(1), 256–270 (2016)
102. J. Hong, L. Ji, X. Wang, Convergence in probability of an ergodic and conformal multi-symplectic numerical scheme for a damped stochastic NLS equation. arXiv: 1611.08778
103. J. Hong, L. Ji, L. Zhang, A stochastic multi-symplectic scheme for stochastic Maxwell equations with additive noise. J. Comput. Phys. **268**, 255–268 (2014)
104. J. Hong, S. Jiang, C. Li, H. Liu, Explicit multi-symplectic methods for Hamiltonian wave equations. Commun. Comput. Phys. **2**(4), 662–683 (2007)
105. J. Hong, L. Kong, Novel multi-symplectic integrators for nonlinear fourth-order Schrödinger equation with trapped term. Commun. Comput. Phys. **7**(3), 613–630 (2010)
106. J. Hong, X. Liu, C. Li, Multi-symplectic Runge-Kutta-Nyström methods for nonlinear Schrödinger equations with variable coefficients. J. Comput. Phys. **226**(2), 1968–1984 (2007)
107. J. Hong, Y. Liu, H. Munthe-Kaas, A. Zanna, Globally conservative properties and error estimation of a multi-symplectic scheme for Schrödinger equations with variable coefficients. Appl. Numer. Math. **56**(6), 814–843 (2006)
108. J. Hong, L. Sun, X. Wang, High order conformal symplectic and ergodic schemes for the stochastic Langevin equation via generating functions. SIAM J. Numer. Anal. **55**(6), 3006–3029 (2017)
109. J. Hong, X. Wang, L. Zhang, Parareal exponential θ-scheme for longtime simulation of stochastic Schrödinger equations with weak damping. arXiv:1803.09188
110. J. Hong, X. Wang, L. Zhang, Numerical analysis on ergodic limit of approximations for stochastic NLS equation via multi-symplectic scheme. SIAM J. Numer. Anal. **55**(1), 305–327 (2017)
111. L. Hörmander, Hypoelliptic second order differential equations. Acta Math. **119**, 147–171 (1967)

112. M. Hutzenthaler, A. Jentzen, On a perturbation theory and on strong convergence rates for stochastic ordinary and partial differential equations with non-globally monotone coefficients. arXiv: 1401.0295v1

113. M. Hutzenthaler, A. Jentzen, P.E. Kloeden, Strong and weak divergence in finite time of Euler's method for stochastic differential equations with non-globally Lipschitz continuous coefficients. Proc. R. Soc. Lond. Ser. A Math. Phys. Eng. Sci. **467**(2130), 1563–1576 (2011)

114. M. Hutzenthaler, A. Jentzen, P.E. Kloeden, Strong convergence of an explicit numerical method for SDEs with nonglobally Lipschitz continuous coefficients. Ann. Appl. Probab. **22**(4), 1611–1641 (2012)

115. M. Hutzenthaler, A. Jentzen, X. Wang, Exponential integrability properties of numerical approximation processes for nonlinear stochastic differential equations. Math. Comput. **87**(311), 1353–1413 (2018)

116. N. Ikeda, S. Watanabe, *Stochastic Differential Equations and Diffusion Processes*. North-Holland Mathematical Library, vol. 24, 2nd edn. (North-Holland Publishing Co., Amsterdam, Kodansha, Ltd., Tokyo, 1989)

117. A.L. Islas, D.A. Karpeev, C.M. Schober, Geometric integrators for the nonlinear Schrödinger equation. J. Comput. Phys. **173**(1), 116–148 (2001)

118. A. Jentzen, P. Pušnik, Exponential moments for numerical approximations of stochastic partial differential equations. Stoch. Partial Differ. Equ. Anal. Comput. **6**(4), 565–617 (2018)

119. S. Jiang, L. Wang, J. Hong, Stochastic multi-symplectic integrator for stochastic nonlinear Schrödinger equation. Commun. Comput. Phys. **14**(2), 393–411 (2013)

120. R. Khasminskii, *Stochastic Stability of Differential Equations*. Stochastic Modelling and Applied Probability, vol. 66, 2nd edn. (Springer, Heidelberg, 2012). With contributions by G.N. Milstein and M. B, Nevelson

121. J.U. Kim, Invariant measures for a stochastic nonlinear Schrödinger equation. Indiana Univ. Math. J. **55**(2), 687–717 (2006)

122. P.E. Kloeden, E. Platen, *Numerical Solution of Stochastic Differential Equations*. Applications of Mathematics (New York), vol. 23 (Springer, Berlin, 1992)

123. S.A. Klokov, AYu. Veretennikov, On mixing and convergence rates for a family of Markov processes approximating SDEs. Random Oper. Stoch Equ. **14**(2), 103–126 (2006)

124. L. Kong, J. Hong, L. Wang, F. Fu, Symplectic integrator for nonlinear high order Schrödinger equation with a trapped term. J. Comput. Appl. Math. **231**(2), 664–679 (2009)

125. L. Kong, J. Hong, J. Zhang, Splitting multisymplectic integrators for Maxwell's equations. J. Comput. Phys. **229**(11), 4259–4278 (2010)

126. M. Kopec, Weak backward error analysis for Langevin process. BIT **55**(4), 1057–1103 (2015)

127. M. Kopec, Weak backward error analysis for overdamped Langevin processes. IMA J. Numer. Anal. **35**(2), 583–614 (2015)

128. S. Kuksin, A. Shirikyan, A coupling approach to randomly forced nonlinear PDE's. I. Commun. Math. Phys. **221**(2), 351–366 (2001)

129. K. Kuritz, D. Stöhr, N. Pollak, F. Allgöwer, On the relationship between cell cycle analysis with ergodic principles and age-structured cell population models. J. Theoret. Biol. **414**, 91–102 (2017)

130. D. Lamberton, G. Pagès, Recursive computation of the invariant distribution of a diffusion. Bernoulli **8**(3), 367–405 (2002)

131. D. Lamberton, G. Pagès, Recursive computation of the invariant distribution of a diffusion: the case of a weakly mean reverting drift. Stoch. Dyn. **3**(4), 435–451 (2003)

132. J.L. Lebowitz, H.A. Rose, E.R. Speer, Statistical mechanics of the nonlinear Schrödinger equation. J. Statist. Phys. **50**(3–4), 657–687 (1988)

133. B.J. LeMesurier, Dissipation at singularities of the nonlinear Schrödinger equation through limits of regularisations. Phys. D **138**(3–4), 334–343 (2000)

134. X. Li, An averaging principle for a completely integrable stochastic Hamiltonian system. Nonlinearity **21**(4), 803–822 (2008)

135. H. Liu, K. Zhang, Multi-symplectic Runge-Kutta-type methods for Hamiltonian wave equations. IMA J. Numer. Anal. **26**(2), 252–271 (2006)

136. J. Liu, Order of convergence of splitting schemes for both deterministic and stochastic non-linear Schrödinger equations. SIAM J. Numer. Anal. **51**(4), 1911–1932 (2013)
137. W. Liu, J.M. Tölle, Existence and uniqueness of invariant measures for stochastic evolution equations with weakly dissipative drifts. Electron. Commun. Probab. **16**, 447–457 (2011)
138. D. Martirosyan, Large deviations for invariant measures of the white-forced 2D Navier-Stokes equation. J. Evol. Equ. **18**(3), 1245–1265 (2018)
139. J.C. Mattingly, A.M. Stuart, D.J. Higham, Ergodicity for SDEs and approximations: locally Lipschitz vector fields and degenerate noise. Stoch. Process. Appl. **101**(2), 185–232 (2002)
140. J.C. Mattingly, A.M. Stuart, M.V. Tretyakov, Convergence of numerical time-averaging and stationary measures via Poisson equations. SIAM J. Numer. Anal. **48**(2), 552–577 (2010)
141. R.I. McLachlan, B.N. Ryland, Y. Sun, High order multisymplectic Runge-Kutta methods. SIAM J. Sci. Comput. **36**(5), A2199–A2226 (2014)
142. H. Mei, G. Yin, Convergence and convergence rates for approximating ergodic means of functions of solutions to stochastic differential equations with Markov switching. Stoch. Process. Appl. **125**(8), 3104–3125 (2015)
143. S.P. Meyn, R.L. Tweedie, *Markov Chains and Stochastic Stability*, 2nd edn. (Cambridge University Press, Cambridge, 2009). With a prologue by Peter W. Glynn
144. G.N. Milstein, Weak approximation of solutions of systems of stochastic differential equations. Teor. Veroyatnost. i Primenen. **30**(4), 706–721 (1985)
145. G.N. Milstein, YuM Repin, M.V. Tretyakov, Numerical methods for stochastic systems preserving symplectic structure. SIAM J. Numer. Anal. **40**(4), 1583–1604 (2002)
146. G.N. Milstein, YuM Repin, M.V. Tretyakov, Symplectic integration of Hamiltonian systems with additive noise. SIAM J. Numer. Anal. **39**(6), 2066–2088 (2002)
147. G.N. Milstein, M.V. Tretyakov, Quasi-symplectic methods for Langevin-type equations. IMA J. Numer. Anal. **23**(4), 593–626 (2003)
148. G.N. Milstein, M.V. Tretyakov, Computing ergodic limits for Langevin equations. Phys. D **229**(1), 81–95 (2007)
149. T. Misawa, Symplectic integrators to stochastic Hamiltonian dynamical systems derived from composition methods. Math. Probl. Eng. Art. ID 384937, 12 (2010)
150. C.M. Mora, R. Rebolledo, Basic properties of nonlinear stochastic Schrödinger equations driven by Brownian motions. Ann. Appl. Probab. **18**(2), 591–619 (2008)
151. P.-L. Morien, On the density for the solution of a Burgers-type SPDE. Ann. Inst. H. Poincaré Probab. Statist. **35**(4), 459–482 (1999)
152. C. Mueller, Coupling and invariant measures for the heat equation with noise. Ann. Probab. **21**(4), 2189–2199 (1993)
153. C. Mueller, D. Nualart, Regularity of the density for the stochastic heat equation. Electron. J. Probab. **13**(74), 2248–2258 (2008)
154. A. Mushtaq, A. Kvæ rnø, K. Olaussen. Higher-order geometric integrators for a class of Hamiltonian systems. Int. J. Geom. Methods Mod. Phys. **11**(1), 1450009, 20 (2014)
155. V. Nersesyan, Exponential mixing for finite-dimensional approximations of the Schrödinger equation with multiplicative noise. Dyn. Partial Differ. Equ. **6**(2), 167–183 (2009)
156. T. Oh, G. Richards, L. Thomann, On invariant Gibbs measures for the generalized KdV equations. Dyn. Partial Differ. Equ. **13**(2), 133–153 (2016)
157. T. Oh, L. Thomann, A pedestrian approach to the invariant Gibbs measures for the 2-*d* defocusing nonlinear Schrödinger equations. Stoch. Partial Differ. Equ. Anal. Comput. **6**(3), 397–445 (2018)
158. F. Otto, H. Weber, M.G. Westdickenberg, Invariant measure of the stochastic Allen-Cahn equation: the regime of small noise and large system size. Electron. J. Probab. **19**(23) 76 (2014)
159. G. Pagès, F. Panloup, Ergodic approximation of the distribution of a stationary diffusion: rate of convergence. Ann. Appl. Probab. **22**(3), 1059–1100 (2012)
160. E. Pardoux, A.Y. Veretennikov, On the Poisson equation and diffusion approximation. I. Ann. Probab. **29**(3), 1061–1085 (2001)

161. E. Pardoux, A.Y. Veretennikov, On Poisson equation and diffusion approximation. II. Ann. Probab. **31**(3), 1166–1192 (2003)

162. E. Pardoux, AYu. Veretennikov, On the Poisson equation and diffusion approximation. III. Ann. Probab. **33**(3), 1111–1133 (2005)

163. S. Reich, Backward error analysis for numerical integrators. SIAM J. Numer. Anal. **36**(5), 1549–1570 (1999)

164. S. Reich, Multi-symplectic Runge-Kutta collocation methods for Hamiltonian wave equations. J. Comput. Phys. **157**(2), 473–499 (2000)

165. M.G. Reznikoff, E. Vanden-Eijnden, Invariant measures of stochastic partial differential equations and conditioned diffusions. C. R. Math. Acad. Sci. Paris **340**(4), 305–308 (2005)

166. G. Richards, Invariance of the Gibbs measure for the periodic quartic gKdV. Ann. Inst. H. Poincaré Anal. Non Linéaire **33**(3), 699–766 (2016)

167. G.O. Roberts, R.L. Tweedie, Exponential convergence of Langevin distributions and their discrete approximations. Bernoulli **2**(4), 341–363 (1996)

168. L.C.G. Rogers, D. Williams, *Diffusions, Markov Processes, and Martingales*, vol. 2 (Cambridge Mathematical Library. Cambridge University Press, Cambridge, 2000). Itô calculus, Reprint of the second (1994) edition

169. J.M. Sanz-Serna, Symplectic integrators for Hamiltonian problems: an overview, in *Acta Numerica, 1992*, Acta Numer. (Cambridge University Press, Cambridge, 1992), pp. 243–286

170. J. Schenker, Diffusion in the mean for an ergodic Schrödinger equation perturbed by a fluctuating potential. Commun. Math. Phys. **339**(3), 859–901 (2015)

171. T. Shardlow, Modified equations for stochastic differential equations. BIT **46**(1), 111–125 (2006)

172. T. Shardlow, A.M. Stuart, A perturbation theory for ergodic Markov chains and application to numerical approximations. SIAM J. Numer. Anal. **37**(4), 1120–1137 (2000)

173. S.I. Sobolev, On an invariant measure for the nonlinear Schrödinger equation. Tr. Petrozavodsk. Gos. Univ. Ser. Mat. **2**, 113–124 (1995)

174. C. Soize, *The Fokker-Planck Equation for Stochastic Dynamical Systems and its Explicit Steady State Solutions*. Series on Advances in Mathematics for Applied Sciences, vol. 17 (World Scientific Publishing Co., Inc, River Edge, NJ, 1994)

175. D.W. Stroock, *Partial Differential Equations for Probabilists*. Cambridge Studies in Advanced Mathematics, vol. 112 (Cambridge University Press, Cambridge, 2012). Paperback edition of the 2008 original

176. G. Sun, Construction of high order symplectic Runge-Kutta methods. J. Comput. Math. **11**(3), 250–260 (1993)

177. D. Talay, Efficient numerical schemes for the approximation of expectations of functionals of the solution of a SDE and applications, in *Filtering and Control of Random Processes (Paris, 1983)*. Lecture Notes in Control and Information Sciences vol. 61 (Springer, Berlin, 1984), pp. 294–313

178. D. Talay, Second-order discretization schemes of stochastic differential systems for the computation of the invariant law. Stoch. Stoch. Rep. **29**(1), 13–36 (1990)

179. D. Talay, Stochastic Hamiltonian systems: exponential convergence to the invariant measure, and discretization by the implicit Euler scheme. Markov Process. Relat. Fields **8**(2), 163–198 (2002). Inhomogeneous random systems (Cergy-Pontoise, 2001)

180. D. Talay, L. Tubaro, Expansion of the global error for numerical schemes solving stochastic differential equations. Stoch. Anal. Appl. **8**(4), 483–509 (1991). 1990

181. L. Thomann, N. Tzvetkov, Gibbs measure for the periodic derivative nonlinear Schrödinger equation. Nonlinearity **23**(11), 2771–2791 (2010)

182. R.L. Tweedie, Sufficient conditions for ergodicity and recurrence of Markov chains on a general state space. Stoch. Process. Appl. **3**(4), 385–403 (1975)

183. N. Tzvetkov, Invariant measures for the nonlinear Schrödinger equation on the disc. Dyn. Partial Differ. Equ. **3**(2), 111–160 (2006)

184. N. Tzvetkov, Invariant measures for the defocusing nonlinear Schrödinger equation. Ann. Inst. Fourier (Grenoble) **58**(7), 2543–2604 (2008)

185. L. Wang, *Variational Integrators and Generating Functions for Stochastic Hamiltonian Systems*. Ph.D. thesis, Karlsruhe Institute of Technology (KIT Scientific Publishing, 2007)
186. L. Wang, J. Hong, Generating functions for stochastic symplectic methods. Discret. Contin. Dyn. Syst. **34**(3), 1211–1228 (2014)
187. L. Wang, J. Hong, R. Scherer, F. Bai, Dynamics and variational integrators of stochastic Hamiltonian systems. Int. J. Numer. Anal. Model. **6**(4), 586–602 (2009)
188. L. Wang, J. Hong, L. Sun, Modified equations for weakly convergent stochastic symplectic schemes via their generating functions. BIT **56**(3), 1131–1162 (2016)
189. Y. Yang, D. Jiang, Long-time behavior of a perturbed enzymatic reaction model under negative feedback process by white noise. J. Math. Chem. **54**(4), 854–865 (2016)
190. K. Yosida, *Functional Analysis*. Classics in Mathematics (Springer, Berlin, 1995). Reprint of the sixth (1980) edition
191. B.Z. Zangeneh, Semilinear stochastic evolution equations with monotone nonlinearities. Stoch. Stoch. Rep. **53**(1–2), 129–174 (1995)
192. P.E. Zhidkov, An invariant measure for the nonlinear Schrödinger equation. Dokl. Akad. Nauk SSSR **317**(3), 543–546 (1991)
193. P.E. Zhidkov, Invariant measures for infinite-dimensional dynamical systems with applications to a nonlinear Schrödinger equation, in *Algebraic and Geometric Methods in Mathematical Physics (Kaciveli, 1993)*. Mathematical Physics Studies, vol 19. (Kluwer Academic Publishers, Dordrecht, 1996), pp. 471–476
194. P.E. Zhidkov, On an infinite sequence of invariant measures for the cubic nonlinear Schrödinger equation. Int. J. Math. Math. Sci. **28**(7), 375–394 (2001)
195. W. Zhou, J. Zhang, J. Hong, S. Song, Stochastic symplectic Runge-Kutta methods for the strong approximation of Hamiltonian systems with additive noise. J. Comput. Appl. Math. **325**, 134–148 (2017)

Index

© Springer Nature Singapore Pte Ltd. 2019
J. Hong and X. Wang, *Invariant Measures for Stochastic Nonlinear
Schrödinger Equations*, Lecture Notes in Mathematics 2251,
https://doi.org/10.1007/978-981-32-9069-3

Printed in the United States
By Bookmasters